IRON WILL

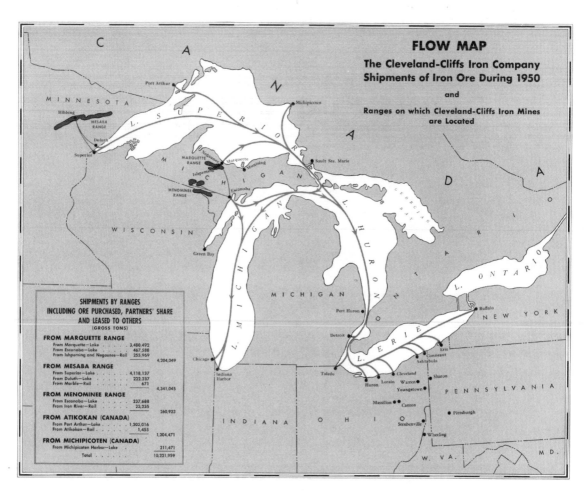

Flow map of Cleveland-Cliffs' iron ore shipments in 1950. *(From Cleveland-Cliffs, Annual Report, 1950. Courtesy of Cliffs Natural Resources, Inc.)*

IRON WILL

Cleveland-Cliffs and the Mining of Iron Ore, 1847–2006

Terry S. Reynolds and Virginia P. Dawson

Wayne State University Press
Detroit

© 2011 by Wayne State University Press, Detroit, Michigan 48201. All rights reserved. No part of this book may be reproduced without formal permission. Manufactured in the United States of America.

15 14 13 12 11   5 4 3 2 1

Library of Congress Cataloging-in-Publication Data

Reynolds, Terry S.
Iron will : Cleveland-Cliffs and the mining of iron ore, 1847–2006 / Terry S. Reynolds and Virginia P. Dawson.
    p. cm. — (Great Lakes books series)
Includes bibliographical references and index.
ISBN 978-0-8143-3511-6 (cloth : alk. paper)
1. Cleveland Cliffs, Inc.—History. 2. Iron industry and trade—United States—History.
3. Mineral industries—United States—History. 4. Iron ores—United States—History.
I. Dawson, Virginia P. (Virginia Parker) II. Title.

HD9519.C55R49 2011
338.7'622341097749—dc22

2010030421

∞

Designed and typeset by Anna Oler
Composed in Futura and Adobe Garamond

CONTENTS

PREFACE

I n early 2007 Cleveland-Cliffs contracted with us to write *Iron Will,* with the understanding that we would conduct research and write the narrative free from oversight. We are deeply grateful to the company for its support and forbearance. Terry S. Reynolds is a historian of technology who teaches at Michigan Tech in the Upper Peninsula of Michigan where the company's early mining history took place. He has published previously on the region's mining heritage. When Virginia P. Dawson, a Cleveland-based business historian, sought a collaborator on this project, she invited Terry to write the early chapters of the book.

Our book tells the story of the opening of the Lake Superior iron region, its rise, and its decline through the lens of the history of Cleveland-Cliffs, the region's last independent iron ore company. Terry wrote chapters 1–3 based on extensive research in a variety of repositories in Michigan and Cleveland, including the very large collection of internal reports and correspondence from the Cleveland Iron Mining Company, Iron Cliffs Company, and Cleveland-Cliffs stored at the regional repository of the Archives of Michigan at Northern Michigan University in Marquette. Virginia crafted chapters 5–7. Chapter 4, which provides the transition between the two parts of the book, was a joint effort, as was the introduction. Although Terry had extensive archival sources upon which to construct his narrative, Virginia had to combine investors' reports, board minutes, company documents, published accounts, and oral histories to extract the crucial story of the company's transition to pellets and its remarkable survival as the American integrated steel industry declined and liquidated. Although we did not see each other often, we regularly exchanged drafts, consulted on critical issues, and generally enjoyed working together on the history of one of Cleveland's oldest and most important companies and one of the mining industry's leading firms.

In the course of our research we received help from many people. We are particularly grateful to Marcus Robyns of the Northern Michigan University Archives, Rosemary Michelin of the Longyear Research Library, William Barrow and Lynn Duchez Bycko of the Department of Special Collections, Cleveland State University, Erik Nordberg and his staff at Michigan Tech, and Ann Sindelar and Margaret Burzynski-Bays of the Western Reserve Historical Society for their assistance. Cliffs' corporate records, including the minutes of the board of directors and the annual reports, are currently kept at the company's headquarters in downtown Cleveland. We are indebted to Kim Regan for making these available to us. We are likewise grateful to Dale Hemmila for giving us access to a small archive at the Empire mine. We were able to review the manuscript of a history of the company and miscellaneous files collected by Burton H. Boyum, who served as the company historian in the 1970s. This manuscript was made available to us by the staff of the Cliffs Shaft Mine Historical Museum in Ishpeming, Michigan. We would also like to thank current and retired employees of the company who agreed to be interviewed. Their names are listed in our bibliography. John Brinzo, Dana Byrne, William Calfee, Don Gallagher, Kim Regan, and Richard Tuthill read parts or all of the manuscript and offered helpful suggestions. Much of the scanning of the photographs was done by Kim Regan. Finally, we thank Charles K. Hyde and the anonymous referees used by Wayne State University Press for their extensive and very useful comments and suggestions.

INTRODUCTION

On New Year's Day, 2001, retired United Steelworkers' representative Ernie Ronn congratulated John Brinzo, then chief executive officer of Cleveland-Cliffs, Inc., for leading the company across the threshold of the new century. Ronn's letter was sent from Michigan's Upper Peninsula, the location of the company's Empire and Tilden mines. He proudly alluded to the five generations of the Ronn family who had worked in earlier mines owned by Cleveland-Cliffs and recalled the years when labor troubles had threatened the company's survival. "The yesteryear doom sayers and predictors of doom have been proven wrong," he wrote. "The Cleveland-Cliffs Iron Company and the United Steelworkers have not destroyed each other. The towns of Negaunee and Ishpeming are not the predicted ghost towns."[1]

The celebratory tone of Ronn's letter may have seemed somewhat incongruous to its recipient, for just then Brinzo was facing a challenge as daunting as any the company had encountered in more than 150 years as a mining enterprise. The company's problems had nothing to do with labor. At the end of the year 2000, Cliffs' steel company partners, LTV and Wheeling-Pittsburgh, had filed for Chapter 11 bankruptcy. Cliffs' other partners, Bethlehem, Acme, Stelco, and Algoma were also on the brink of failure.

North American steel companies had recovered from bankruptcies in the 1980s, but the situation in 2001 was entirely different. "Nobody anticipated that they would ever shut down and go out of business. Well, they did," Brinzo said in a 2006 interview. The most serious blow came with the sudden liquidation of LTV, which left Cleveland-Cliffs with a large inventory of unsold iron ore pellets. This was especially damaging because the steel company was a major partner and customer. LTV had just worked out an agreement with Cleveland-Cliffs to close down LTV's wholly owned mine in Hoyt Lakes, Minnesota, and take additional sales tonnage from Cliffs' Tilden mine. Without a customer for Tilden's pellets, Cleveland-Cliffs faced the prospect of shutting down the mine. "Now at that point, if we had stayed that way, we would have been dead," Brinzo recalled. "It would have been just a matter of time."[2] Although Cleveland-Cliffs' prospects looked grim, Brinzo had no intention of allowing the company to fail. As a stopgap solution to the Tilden ore problem, he proposed that Cliffs purchase LTV's Cleveland Works and restart its blast furnaces. By going into the steel business, the company could save the Tilden mine and keep its options open. Although this drastic step fortunately did not prove necessary, nevertheless the time was right for the implementation of Brinzo's plan to achieve majority ownership in the company's mines by assuming its struggling partners' liabilities—such as employee pensions and health care—in exchange for long-term sales contracts and the partners' shares of Cleveland-Cliffs' mines: a plan that, in essence, converted the former partners to customers.

With Cliffs freed from partnerships with North American steel companies, Brinzo envisioned a new chapter in the company's history. These and other actions, both calculated and serendipitous, taken between 2001 and 2005 positioned Cleveland-Cliffs to become the largest American producer of iron ore and a global merchant of iron pellets. Brinzo's plan enabled Cleveland-Cliffs to regain the independence it had enjoyed prior to the 1950s—before partnerships with steel companies became the industry norm.

If Brinzo had been a newcomer to Cliffs' management, it is unlikely that the company's board of directors would have approved such a risky proposition. A financial analyst by training, Brinzo had spent several years as controller of the Empire mine in the 1970s, where he embraced the company-centered lifestyle of Michigan's Upper Peninsula. After his return to the company's Cleveland headquarters, while serving as vice president and controller, he had weathered the first set of steel company bankruptcies and two corporate takeover attempts. Brinzo became senior vice president of finance and planning before being elected president and chief executive officer in 1997. Four years later, as the American steel industry collapsed, Cleveland-Cliffs began the systematic purchase of its partners' shares in its mines. The timing was impeccable. Cliffs' transition to iron ore merchant coincided with the increasing demand for iron ore by the People's Republic of China. The company was able to take advantage of this demand, and its revenues rebounded from a loss in 2001 to a record $1.9 billion in sales and an operating income of $366 million in 2006, the highest in its history.[3]

Brinzo's plan may have seemed unusual, but it fit a historical corporate pattern of dealing with adversity. Iron-mining companies, like their counterparts in other minerals resources industries, have continually been plagued by issues related to the cyclical nature of the metals business and the problem of ore depletion. Brinzo's predecessors faced challenges similar to his.

In 1847 a group of Clevelanders, undaunted by the prospect of establishing a mining enterprise in the wilds of Michigan's Upper Peninsula, created an association that led to the formation of the Cleveland Iron Mining Company, corporate predecessor of the Cleveland-Cliffs Iron Company. The Cleveland Iron Mining Company was incorporated in the State of Michigan in 1850. At that time Cleveland, located on Lake Erie, was a small port city. The so-called Cleveland Mountain, the mineral claim on which the company staked its future, was more than 600 miles from Cleveland; however, it was only fifteen miles from the south shore of Lake Superior. In 1855 Cleveland Iron shipped the first commercial load of ore from Marquette to Cleveland through the newly opened canal at Sault Ste. Marie. So began the shipping of iron ore down the "inland seas" to steel mills near the ports on the lower Great Lakes. This great inland system of transportation shielded Cleveland Iron's (and later Cleveland-Cliffs') market from competition from imported iron ore for a century.

Our history chronicles Cleveland-Cliffs' rise to prominence in the late nineteenth century, its struggle to survive the consolidation of the steel industry and the acquisition of ore properties by steel companies at the turn of the twentieth century, the shift from a labor-intensive to a capital-intensive business in the twentieth century, and the company's recent transition to a global iron merchant. From a midwestern company serving regional customers, Cleveland-Cliffs evolved into an international iron ore merchant, with mines not only in Michigan and Minnesota but also in Canada, Australia, and Brazil. One of the earliest mining companies in the Lake Superior region, Cleveland-Cliffs is the only independent iron-mining company still operating there in the twenty-first century.[4] We think our title, *Iron Will,* captures the company's resilience in the face of panics, depressions, strikes, technical bottlenecks, and bankrupt partners over the course of a long and distinguished history.

A popular history of Cleveland-Cliffs, *A Century of Iron and Men,* was published by Harlan Hatcher in 1950, but his informal style and lack of documentation limits the book's usefulness. The same can be said of Walter Havighurst's *Vein of Iron: The Pickands Mather Story,* published in 1958, which describes the history of one of the major iron ore merchants absorbed by Cliffs in the 1980s.[5] Only a few pages of Hatcher's book and only the last dozen pages of Havighurst's cover the development of pelletization, a technology that saved the Lake Superior iron ore industry. On a larger canvas, *Iron Will* is the story of the rise of the Lake Superior region as a major supplier of iron ore to the American steel industry, the rise of partnerships with steel producers to finance the development of new technology, and the redefining of these relationships at the beginning of the twenty-first century. Several publications treat the Lake Superior iron district, notably David A. Walker's *Iron Frontier,* Marvin G. Lamppa's *Minnesota's Iron Country,* and E. W. Davis's *Pioneering with Taconite,* but they focus on Minnesota's iron ranges, not the Michigan iron ranges that opened the Lake Superior region, established the paradigm of a steel industry dependent on ore supplies from hundreds of miles away, and were, until recently, the center of Cleveland-Cliffs' operations.[6]

Although there are numerous histories devoted to aspects of the American iron and steel industry, they have largely neglected the supply side of vertical integration, particularly the role of iron ore in the industry's rise to prominence. For instance, Kenneth Warren's history of the U.S. Steel Corporation, whose subsidiary the Oliver Iron Mining Company controlled over half of all American iron ore supplies for most of the twentieth century, devotes only a few pages to iron ore mining, as does his earlier overview of the American steel industry.[7] This neglect is also reflected in recent histories that focus on the decline of the American iron and steel industry at the end of the last century, such as *The Decline of American Steel* by Paul Tiffany, *An Economic History of the American Steel Industry* by Robert P. Rogers, and *Steel Phoenix: The Fall and Rise of the U.S. Steel Industry* by Christopher G. L. Hall.[8] This is unfortunate, for, in a sense, as the American steel industry floundered, the American iron ore industry triumphed, saving itself by shifting from increasingly more scarce direct-shipping ores to enriched and improved lean ores. No one has told this story from the point of view of a major supplier of iron ore—a survivor in a field of failed companies.

In *Scale and Scope: The Dynamics of Industrial Capitalism,* Alfred D. Chandler Jr. describes the evolution of modern industrial enterprise from small firms serving a few customers in local markets to large firms expanding rapidly to serve an ever-growing customer base on a national or even international level.[9] Natural resource companies may not completely fit into this paradigm of modern industrial enterprise. It is true that Cleveland-Cliffs grew, hiring more and more professional managers, but it remained small relative to its customers in the iron and steel industry. Even though Cleveland-Cliffs survived the backward integration of steel companies into iron ore mining at the turn of the twentieth century, becoming the largest *independent* iron ore–mining company, it did not service a growing list of customers. Instead, the company sold more ore to fewer and fewer customers as the iron and steel industry continued to consolidate into a few large enterprises.[10]

We touch on a number of interrelated themes in this book. Among these are management, the fear of mineral depletion, the cyclical nature of the iron and steel industry, labor, and technology. We also look at the particular responses, or survival strategies, that the company evolved to deal with these issues. They include a tradition of returning to its iron-mining core in times of economic stress, protecting mineral reserves at all costs, using acquisitions and joint ventures, or partnerships, to share development costs and combat ore depletion, and seeking amiable labor relations.

One of the first iron ore companies on Lake Superior, Cleveland Iron had to work out the difficulties of managing operations 600 miles from company headquarters in Cleveland in an era when transportation and communication were rudimentary. This was not accomplished easily, but the successful solution to this problem ensured the company's early survival and growth. The company benefited from the leadership provided by Samuel Livingston Mather and his son William Gwinn Mather. They served the company for over eighty years. A consummate nineteenth-century capitalist/entrepreneur, Samuel L. Mather took on the challenge of creating a market for the unfamiliar ores of Lake Superior while managing a mining company located in the remote wilds of Michigan. He was the company's salesman, marketing ore to hundreds of small companies manufacturing a range of products, from pots and hardware to rails and engines. In 1878, as the company commenced underground mining and struggled to adjust to competition from the newly opened Menominee iron range, he remarked that "only they that can mine cheaply & have light expenses will be able to keep alive."[11] Under Mather's leadership, first as secretary and later as president, the company solved its management, transportation, labor, and cash-flow problems. Mather guided the Cleveland Iron Mining Company as it made the transition from open-pit to more costly underground mining and adopted new mining techniques to remain viable.

His son William G. Mather would later develop the strategies necessary to ensure the company's survival when the world of iron ore mining changed around the turn of the century due to growing labor militancy, the emergence of cheap ore from the newly opened iron ranges of Minnesota, and the backward integration of giant steel firms into iron ore mining. William G. Mather countered by expanding the company's ore holdings, diversifying its operations, establishing joint mining ventures and other ties with small- and medium-sized steel firms, pacifying labor with a comprehensive paternalistic program, and, to some extent, integrating forward into transportation of ore and iron production. The Mathers were not the only Cleveland family to provide continuity to the company's leadership. Jeptha Homer Wade, better known for his association with Western Union, and his descendants served almost continuously on the company's board from the 1860s to the 1980s.

A second theme that pervades the company's history is concern over depletion. The company always marked two occasions in a mine's history: an evergreen tree planted on top of the ore in a railcar marked the first shipment of ore from a new mine; a broom heralded its last. This ritual, repeated many times during the company's first century, served as a periodic reminder that the company's—and indeed the industry's—future depended on finding new ore deposits. The need for additional mining properties drove Cleveland Iron Mining Company to absorb a rival—the Iron Cliffs Company—to create the Cleveland-Cliffs Iron Company in 1891. This acquisition, made shortly after William Gwinn Mather took over as president, provided the company with vast new iron reserves and extended its ability to own its mines "in fee" (that is, ownership of the land *and* the minerals in the ground). This often gave Cleveland-Cliffs leverage that rival iron ore–mining and merchant firms lacked, for most of them had to lease their mines from absentee property owners and pay royalties on ore extracted.

By the end of World War II, depletion had become the concern not just of Cleveland-Cliffs but of all the mining and steel companies of the Great Lakes region with the exception of U.S. Steel. Most of the high-grade iron ore on the Marquette, Menominee, Gogebic, and Mesabi ranges was gone. Even medium-grade ores, which required beneficiation (or enrichment) were growing scarce. In the late 1940s and 1950s Cleveland-Cliffs responded by employing a large cadre of geologists in a desperate search for new deposits of high-grade iron ore in North America and abroad. Ultimately, however, the salvation of the company, and indeed of the American iron ore industry as a whole, did not lie in the discovery of new, rich iron ore deposits. None were to be found in the United States. Fear of losing the economic viability of America's industrial heartland drove the development of a process for liberating iron from vast deposits of low-grade iron ore long known to exist near the old, now-depleted rich ores of Lake Superior. Convinced that the national security of the United States depended on a domestic supply of this vital raw material for making armaments, legislators modified federal and state tax structures to encourage the introduction of large-scale industrial processes for producing high-grade iron ore pellets from low-grade ore. Cleveland-Cliffs in the 1950s would embrace the new approach and play a leading role in the development of methods to exploit the large reserves of low-grade iron ore on its Michigan properties. In the 1980s and again in the early twenty-first century it would acquire new lean-ore properties in its continuing struggle to stay ahead of ore depletion.

A third theme addressed in this book is the cyclical nature of the iron and steel industry. Natural resource companies like Cleveland-Cliffs are subject to the same economic cycles that impact the industry they supply. As our narrative suggests, every major economic upheaval that struck the iron and steel industry rippled down the supply line and buffeted the iron ore industry. The company barely struggled through the economic recession that began in 1857, enjoyed the iron industry's boom years between 1862 and 1873, then suffered in the Panic of 1873 and its aftermath. When the iron industry transitioned to steel in the 1870s, ore suppliers faced new demands for ores of a different character than those previously delivered. Successfully navigating this transition, Cleveland-Cliffs and the iron ore industry as a whole enjoyed the steel industry's relatively steady growth in the 1880s, in the first three decades of the twentieth century, and after World War II, but they suffered through its downturns following the Panic of 1893, during the Great Depression of the 1930s, and during the steel industry's collapse in the 1980s and in the early years of the twenty-first century.

Several corporate traditions emerged in response to the boom-and-bust cycles of the industry and fear of depletion. Again and again the company returned to the core business—iron mining—in times of economic distress. In the depressions of the 1850s and 1870s, the company abandoned vertical integration after its early foray into transportation to refocus on iron mining. After diversifying into an assortment of industries between 1890 and 1930, including charcoal iron, wood chemicals, woodenwares, coal mines, and steel companies, the company sold off holdings and refocused on iron mining during the economic stress of the Great Depression of the 1930s. In the 1950s, when it desperately needed funds to expand into the processing of lean ores—something absolutely necessary to its long-term survival—the company completely withdrew from the coal business and sold some of its steel stocks, using them to fund the transition. J. S. Wilber, the company's vice president for sales, commented in 1955: "For our problems specialization, not diversification appears to be the solution."[12] After expanding into other mineral holdings such as oil shale, copper, oil drilling, and uranium in the 1960s and 1970s, the company again sold off these holdings to refocus on iron ore when the steel industry ran into problems in the 1980s.

Other key strategies the company regularly used to survive depletion concerns and the cyclical nature of its industry were acquisitions and partnerships (or joint ventures). In the early 1850s the company bought out a rival to expand operations on Lake Superior. Its acquisition of Iron Cliffs in 1891 significantly expanded its mineral reserves and alleviated growing concerns about ore exhaustion. And in the 1980s and the first decade of the present century acquisition of the ore holdings of rival Pickands Mather and the purchase of its bankrupt partners' shares in its mines made the company the largest iron ore producer in North America, exceeding the traditional giant U.S. Steel.

Partnerships or joint ventures were used frequently as a survival strategy, becoming the company's typical first response to ore depletion or developments in the iron and steel industry. The company's first partnership emerged at its very creation—an agreement with the Jackson Iron Company to share expenses in securing their claims and building a road from Lake Superior to their nearly adjacent holdings. Both Mathers used joint ventures to acquire new mining properties. William used them as well after 1900 to cement ties to small and medium steel companies when the backward integration of large steel firms into iron mining sharply reduced the number of firms to which the company could sell its ores. He also used joint ventures to diversify the company into coal mining and into nonmining ventures. Not all such strategic alliances were successful. When William agreed to join Cleveland-Cliffs to Cyrus Eaton's Republic Steel interests in a scheme to create a very large integrated steel company on the eve of the Great Depression, the effort failed. To stave off the bankruptcy that this partnership almost caused, the company invoked the time-honored Mather policy: sell any or all assets *except* the mines and properties potentially containing iron. The company returned to its core competency—its characteristic response in times of economic distress.

In the 1950s the company turned again to joint ventures to raise capital—even at the cost of becoming a minority partner in the ventures instead of the dominant or equal partner, as it had been in most of its earlier partnerships. Partnerships with steel companies became the preferred method of raising the capital necessary for processing low-grade ore. The new technology required construction of large, expensive grinding, separation, and pelletization facilities to achieve the economies of scale necessary for commercialization. These partnerships with steel companies shielded Cleveland-Cliffs from the cyclical ups and downs of the steel industry in the 1970s and 1980s, since its fees for managing the joint-venture mines rose with inflation and partners were obligated to take a fixed percentage of a mine's output, even when demand for steel fell. In addition, because Cleveland-Cliffs owned most of its mining properties in fee, it also received royalties from these partnerships. However, these joint ventures were advantageous only as long as the company's steel partners remained solvent. In the late 1980s the unthinkable occurred: Cleveland-Cliffs' steel partners began to go bankrupt as high-quality, low-cost imported steel and new domestic competition broke the integrated steel industry's lock on the domestic steel market.

Another theme we tackle is the history of the company's labor relations. For the first seventy years of the company's history the bulk of the company's workforce was new immigrants, first overwhelmingly Irish, then heavily Swedish and Cornish. By the 1890s around 90 percent of Cleveland-Cliffs' personnel was immigrant, coming from a diversity of backgrounds, with Finns, Italians, French Canadians, Germans, Slovenes, and half a dozen other nationalities added to the early mix. Generally, the company's labor relations were no better or worse than other companies' in the region through most of the nineteenth century. In 1895, in the midst of the depression following the Panic of 1893, miners on the Marquette Range, where the company's mines were all located, organized and struck, precipitating young William Mather's decision to bring in strikebreakers and petition the governor to use the National Guard to quell the strike. Mather subsequently atoned for this act by becoming a model employer, creating a comprehensive paternalistic system that remedied some earlier evils, split labor, and achieved a labor peace that lasted more than forty years. The company's labor relations were marred by a violent strike in 1946 and long strikes in 1959 and 1977, but as Ernie Ronn pointed out in his letter to John Brinzo, the company, the union, and the towns of Negaunee and Ishpeming, whose economies depended on iron mining, all survived labor contention.

A final theme that runs throughout our story is a tradition of openness to new technology. First Cleveland Iron and later Cleveland-Cliffs practiced a careful, deliberative approach to innovation. Almost as soon as the company began shipping ore, Cleveland Iron installed pocket docks for more efficient vessel loading, taking a technology first applied to iron ore transportation a year earlier and improving on the concept. In the late 1860s it was among the pioneers in the region to experiment with the diamond drill for exploration for new ore deposits. After some early skepticism, around 1870 Cleveland Iron was the first or among the first companies in the region to adopt high explosives to reduce mining costs. When it had to go underground and abandon its early open pits, it did so carefully. The company's agent in Marquette commented in 1876: "We were perfectly well aware that we knew nothing of that method of mining [underground], and consequently went to work with great caution and care."[13] Cleveland Iron was among the first companies to embrace electric lighting in 1880 and electric haulage in the early 1890s, both after systematic tests. Between 1900 and 1920 Cleveland-Cliffs was among the earliest mining companies to experiment with the use of concrete in mine shafts and with concrete shaft houses. Likewise, after World War II, the company made the transition from underground, direct-shipping ores to open-pit, enriched, and improved lean ores. Cleveland-Cliffs made this transition incrementally, waiting to see whether the technology had a reasonable chance for success before investing in the development of Michigan jasper. Once the commitment was made, the company tenaciously continued in the face of daunting technical obstacles.

The narrative that develops these themes is arranged chronologically. Chapter 1, "Foundations and Traditions, 1846–1865," covers the opening of the Lake Superior iron ore district and the history of the Cleveland Iron Mining Company up to the end of the Civil War in 1865. It reviews the problems involved in creating and managing a profitable enterprise in a very isolated locale. The chapter discusses the emergence of several of the long-standing tensions that characterize the company's history, notably that between pressures to operate primarily as an ore supplier for other entities and pressures to integrate forward into transportation and iron production as well as mining. In addition, this period saw the company establish several traditions, notably that of seeking partners for capital-intensive projects and that of buying out business rivals.

Chapter 2, "Technology, Transport, and Transformation, 1865–1891," discusses the company's expansion in conjunction with the American iron and steel industry's increased dependence on Lake Superior iron ores. It covers the company's success in reinventing itself as it shifted from an open-pit mining company to an underground mining company in the late 1870s, and how it dealt with declining iron prices, labor problems, and the competition of new iron ore ranges (Menominee, Gogebic, Vermilion). It also examines the acquisition of Iron Cliffs by key figures in the Cleveland Iron Mining Company in 1890 and the consolidation of the companies the following year under the umbrella of the Cleveland-Cliffs Iron Company.

Chapter 3, "Crises, Diversification, and Integration, 1891–1930," covers the first four decades of the consolidated company, down to the Great Depression. The period began with the transfer of leadership in the company from Samuel L. Mather to his son William G. Mather. Within a few years William G. Mather and Cleveland-Cliffs faced a series of overlapping crises: a massive economic depression; growing labor militancy; competition from cheap, open-pit mines on the Mesabi; and the consolidation of the steel industry and the backward integration of steel companies into iron ore mining. To solve this myriad of problems, the company reinvented itself as a broad-based, diversified corporation. It emerged by 1910 as virtually the only remaining major independent iron ore producer in the Great Lakes region. In the decades after 1891 Cliffs not only survived but inaugurated a series of new initiatives: an expanded system of corporate welfare, expansion outside of the Marquette iron range, diversification out of ore mining, and attempts to integrate forward into iron production. By the 1920s Cliffs was, in many ways, a vertically integrated company. While still primarily an ore producer, it drew substantial income from a host of other ventures, including railroads, coal, shipping, lumber, electric power, wood chemicals, woodenware, paper, and both charcoal and coke iron. The drive to integrate forward from mining into ownership of blast furnaces and steel production culminated in 1929 and 1930 with acquisition of controlling interest in a major steel firm.

Chapter 4, "Depression, War, and Depletion, 1930–1950," covers the company's response to the Great Depression and its operations in World War II as well as its post-war reactions to the depletion of high-grade ore reserves and the emergence of a militant labor force. Cleveland-Cliffs survived the depression of the 1930s—but only barely. On the other hand, its decades-long efforts to acquire iron- and steel-making capacity and become a fully integrated iron and steel company did *not* survive. By World War II Cleveland-Cliffs had begun the process, once more, of refocusing on iron ore mining and divesting itself of outside interests. During the Depression and World War II, the problems it had "solved" between 1891 and 1910 reemerged. Since its absorption of Iron Cliffs in 1891 the company had not had to worry seriously about future ore supplies. The rapid rate at which direct-shipping ores were used in World War II, however, made depletion of reserves a critical postwar problem. Labor militancy in the company's iron mines, in abeyance since 1900, remerged with a vengeance in 1946 with a long and bitter rangewide strike. Even after settlement of this strike, labor problems continued to plague the company for the next several decades, culminating in another long strike in 1959. The solution to both the company's labor and ore-quality problems, however, would be found through technology, discussed in the next chapter.

Chapter 5, "Pellets and Partnerships, 1950–1974," traces the early history of pelletization in Michigan. Cleveland-Cliffs introduced this technology on a limited, semi-experimental basis in the 1950s, gradually phasing out underground mining in favor of more cost-effective open-pit operations. Cliffs' transition to pellets culminated in the development of the large-scale Empire and Tilden mines in the 1960s and 1970s. Tilden's technology was especially challenging because it required a unique chemical process that involved producing high-grade iron ore pellets from nonmagnetic hematite. The chapter examines the role of Chief Executive Officer Walter Sterling and his successor, Stuart Harrison, in finding partners for these capital-intensive technology ventures. During this period the board of directors, still dominated by the Wade and Mather families, foiled Cleveland financier Cyrus Eaton's attempt to take over the company.

Chapter 6, "Great Expectations and Unexpected Challenges, 1974–2000," covers the expansion of the Empire and Tilden mines in anticipation of an increased demand for steel in the 1980s. This demand failed to materialize. The chapter examines the reasons for the decline of the American steel industry and describes Cliffs' attempt to evolve from an iron ore company into a natural resources company through diversification into gas and oil exploration, uranium, and oil shale. The company acquired a longtime rival, Pickands Mather, at a time when Cliffs' own survival was uncertain. In the 1990s, as the steel industry's recovery helped to lift Cliffs' revenues, the company again succeeded in fending off corporate takeover bids. However, it was forced to abandon its plan to become a natural resources company at that time.

Chapter 7, "Reinventing Cleveland-Cliffs, 2000–2006," focuses on the transformation of Cleveland-Cliffs from dependency on steel industry partners to self-sufficiency under Chief Executive Officer John Brinzo. Pushed by steel bankruptcies, Cleveland-Cliffs adopted a new paradigm for doing business. Cliffs acquired the mining interests of bankrupt partners, returning to its old role as merchant seller of iron ore. After the consolidation of the mining industry, the company emerged as the largest iron ore producer in the United States—surpassing even the former behemoth, U.S. Steel. This chapter chronicles the events and decisions that propelled the company's turnaround as well as missteps, such as the failure of a new venture for producing hot briquetted iron in Trinidad and Tobago. An agreement with the Chinese company Laiwu to jointly acquire the Eveleth mine (renamed United Taconite) in Minnesota and Cliffs' pursuit of Portman, Ltd., an Australian mining company, positioned the company to play a growing role in the global market for iron ore in the twenty-first century.

1

FOUNDATIONS AND TRADITIONS, 1846–1865

The founders of the company that would eventually become the Cleveland-Cliffs Iron Company did not intend to mine iron. They hoped for copper or, perhaps, silver. How, then, did they end up mining a base metal like iron?

In 1841 Douglass Houghton, Michigan's first state geologist, issued a report suggesting that the copper deposits long known to exist in the state's Upper Peninsula were sufficient to be commercially exploited. Houghton's report precipitated America's first major metal mining rush, as several thousand prospectors set out in the early 1840s on the long trek to the south shore of Lake Superior, a region previously considered "barren and worthless" and so distant from civilization that Kentucky senator Henry Clay called being there the equivalent to being on the moon.[1]

Against this backdrop, a group of Clevelanders meeting in 1845 and 1846 in the home of physician Morgan L. Hewitt began to discuss getting in on the metal rush. Hewitt, born in Hartford, New York, in 1807, had studied medicine at Castleton, Vermont, graduating in 1832. In 1833 he moved to Cleveland, population around 2,500, to practice medicine, eventually becoming the head of the city's marine hospital.[2] Probably at Hewitt's recommendation, the group pooled funds to hire J. Lang Cassels to lead an expedition to Michigan's Upper Peninsula to locate mineral claims. Born in Scotland in 1827, Cassels, like Hewitt, was a Cleveland physician. He was well known in the area for his public lectures on chemistry, and he had cofounded the Cleveland Medical College. More important to Hewitt's group, Cassels had earlier worked on the New York geological survey.[3] He knew rocks and minerals.

Morgan L. Hewitt (1807–89), first president of the Cleveland Iron Mining Company (CIMC) and one of a group of Clevelanders whose interest in mineral discoveries on Lake Superior led to the formation of the company. Hewitt moved to Michigan's Upper Peninsula in 1857, remained on CIMC's board of directors until 1889, and provided some local oversight of operations. *(Courtesy of the Marquette County Historical Museum, Marquette.)*

Reflecting the group's initial mineral interests, the Clevelanders called their informal association the Dead River Silver and Copper Company.[4] The preference indicated by this name was rational. Douglass Houghton's geological explorations of the Upper Peninsula had identified the area of copper deposits, but Houghton had not penetrated much inland and had little awareness of the region's immense iron ore deposits. Moreover, silver sells by the ounce and copper by the pound. Although the most utilitarian of the earth's metals, used to make everything from pots and pans to steam boilers and ships, iron ore sells by the ton. Per unit of weight, iron ore was not worth much. Moreover, in the 1840s the United States had no pressing need for new iron ore mines. Existing ones seemed quite sufficient.[5]

Setting out for Lake Superior in the late spring or early summer of 1846, Cassels's exploration party was a bit late in getting in on the copper rush. Cassels probably followed the usual route traveled by those hoping to stake claims: by water to Detroit and then up Lake Huron and the twenty-mile-long St. Mary's River, which linked Lake Huron to Lake Superior, to Sault Ste. Marie. At Sault Ste. Marie, a small trading village near the mouth of the St. Mary's on the far eastern end of Lake Superior, a mile of rapids interrupted the all-water route and forced a short portage. Expeditions normally paused here for a few days to buy provisions and hire a vessel and guide. Only then did they set out along the wild and unsettled south shore of Lake Superior for the Keweenaw Peninsula some 250 miles to the west, the location of the copper deposits identified by Houghton. It is not clear whether Cassels and his party were on the way up to the copper region or on their way back down when a chance encounter at Sault Ste. Marie with a representative of a similar party from Jackson, Michigan, provided

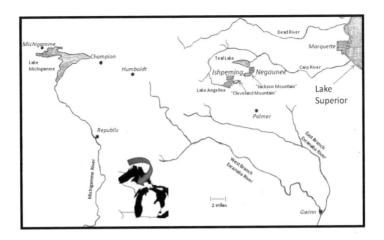

Map of the Marquette iron range. *(By the authors.)*

the impetus that lured some of Hewitt's Cleveland group from copper and silver to iron.

In June 1845, roughly a year before Cassels's group reached the Sault, when the copper fever was at its peak, a group of Jackson, Michigan, businessmen—much like the Cleveland group that met at Hewitt's house—had formed an association and sent an exploratory party north to place mineral claims. This group had paused, like Cassels's, at Sault Ste. Marie. While seeking a boat and a guide, the leader of the group, Philo Everett, encountered Louis Nolan, of French Canadian and Indian ancestry. Nolan mentioned shiny, metal-like outcrops in an area where he had traveled as a boy only about 150 miles west along the shore of Lake Superior and some 10–15 miles inland. This was closer than the copper district by about 100 miles. Hoping that Nolan's shiny outcrops were copper, silver, or perhaps lead, the Jackson group hired Nolan to guide them to the deposits. With the assistance of a local Ojibwa clan chief, Marjijesick, the Jackson group found near Teal Lake, where the town of Negaunee would eventually grow, "a mountain of solid iron ore, 150 feet high" that looked "as bright as a bar of iron just broken."[6] It was not copper, silver, or lead, but it seemed potentially valuable. The Jackson group went on to the copper region, where the government's land agency was then located, and in September 1845 they tried to file a claim on a square mile around the ore outcrop, although unable precisely to locate it on the rough government maps available.[7] With cold weather threatening, the Jackson party went back south.

In the winter of 1845–46, Everett's group had ore from their claim tested. The results were encouraging. Members of the group also learned that the area in which they had located their claim had been surveyed the year before. In September 1844 a government survey crew, headed by William Austin Burt, had laid out range and township lines in the Teal Lake area. If they could locate Burt's markers, they could precisely delineate their claim. Unknown to the Jackson group, Burt and his survey crew had also stumbled across evidence of iron ore in the area. While running survey lines near Teal Lake, Burt's crew members had noticed wild variations in their magnetic compasses. Suspecting iron deposits were the cause, Burt had called out, "Boys look around and see what you can find." His crew quickly returned with specimens of iron ore. Burt, however, was first and foremost a surveyor and inventor. He was more impressed with the ability of his patented solar compass to operate in regions where iron deposits rendered conventional compasses useless than by the ore specimens. Neither he nor his crew did anything about the discovery.[8]

Representatives of the Jackson group returned north in the spring of 1846 to locate Burt's markers. They had secured additional mineral claim permits by this time and intended to explore the region around their iron ore discovery and collect more samples. In the Teal Lake area, the expedition members found Burt's markers near their claim and discovered several additional large ore outcrops, including one two miles further inland. They built a log cabin on their initial claim, a step then necessary to reserve the claim, and returned to the shore of Lake Superior with 300 pounds of ore, leaving one member of the group to hold the location. At the lakeshore, the group split again. Some headed westward to Copper Harbor and the copper district to clarify their claim at Teal Lake and use the group's remaining permits in the copper region (the copper dream died hard). Abram V. Berry, president of the Jackson association, meanwhile, headed for Sault Ste. Marie with the ore samples. Berry intended to use the last permit in his possession to file on the other major iron ore outcrop his group had identified two miles further inland.[9]

Cleveland Interests Stake a Claim

It was while pausing at the Sault that Berry, by chance, met J. Lang Cassels and the Cleveland exploratory party. Berry and his Jackson associates were not wealthy. They knew that exploiting their iron ore claim would be difficult. Iron was a bulk commodity. Without good transportation the deposits—rich as they were—were useless. Building the roads and docks needed to get the ore efficiently to the lakeshore was likely to stretch their financial means. The association had thus instructed Berry that if he encountered a party in whom he had confidence, a party who would promise to work with the Jackson group in securing claims and building a road inland from the shore of Lake Superior, he could reveal the other large iron outcrop his group had discovered.[10]

Berry apparently regarded Cassels's group as just such a party. He showed Cassels the ore samples and told him of the other "mountain" of iron, roughly two miles west of the one the Jackson group had claimed. In return, Cassels agreed that his group would share expenses for holding possession of the claims and constructing a road. Cassels borrowed Berry's canoe, secured guidance from an Ojibwa, and visited the location, perhaps with several members of his group, such as W. A. Adair, who filed the Cleveland group's claim on the location, and Lorenzo Dow Burnell, who returned to the location and made arrangements in September 1846 with a local figure, Edmund C. Rogers, to build a cabin and hold the property.[11]

Thus, a combination of two serendipitous encounters in Sault Ste. Marie planted the seeds of the Cleveland-Cliffs Iron Company: Everett with Nolan, leading the Jackson group to the iron "mountains," and then Berry with Cassels, leading the Jackson group to seek a loose partnership with the Cleveland group. Cassels's agreement to work jointly with the Jackson group on the development of claims and transportation could be considered the beginning of a tradition that was to become part of the corporate culture of Cleveland-Cliffs: seeking partners for risky ventures in which capital costs for a single company were high.

Based on Cassels's reports, at least some of Hewitt's Cleveland associates elected to shift the focus of their enterprise from silver and copper to iron. On April 29, 1847, twelve men signed an agreement relating to the iron location that Cassels, Adair, and Burnell had visited. Around six months later nine of these twelve, plus George E. Freeman, prepared formal articles of association for what they now dubbed the Cleveland Iron Company.[12] No surviving document reveals why the members of this group shifted their interest to iron. Perhaps they saw Cleveland as the logical place to combine ore from Upper Michigan with coal from western Pennsylvania to produce iron rails for the railroad networks then expanding in Ohio.[13] Perhaps they had been impressed with Cassels's report on the location. The ore deposits on which most American forges and blast furnaces depended in the 1840s were relatively modest in scope, most sufficient only to feed one or a small number of local enterprises. The Michigan deposits were extraordinary. Early visitors to the region referred to them as "inexhaustible" and "the richest and best" iron deposits yet discovered anywhere.[14] The quality of the Michigan ores matched their quantity. Most North American iron mines then produced ores that were, at best, around 40 to 50 percent iron, and many only 20 to 30 percent.[15] The ores from the newly named Cleveland Mountain (really better described as a large hill) contained over 60 percent iron.[16]

As the Clevelanders began to transform their loose mineral exploration syndicate into a mining company, the man left to secure their claim, Edmund Rogers, constructed a small cabin on the site, planted potatoes, and began cutting a crude road to the Jackson company's location two miles to the east. In October 1847 Rogers left the site to travel to Cleveland to meet with the trustees of the new company but made arrangements for Charles Johnson, brother-in-law of Philo Everett of the Jackson group, to stay at the cabin for $90.[17] In mid-May 1848 Johnson left the location to meet and guide several directors of the new Cleveland Iron Company who were coming up to view their "iron mountain." On May 21, while he was absent, three mineral prospectors—Robert Graveraet, Samuel Moody, and John Mann—entered the seemingly vacant claim. When Johnson returned on May 28, accompanied by Cleveland Iron Company trustees George Freeman and John Outhwaite, he found the house burned and a new house occupied by Moody and Mann nearby. Moody and Mann refused to relinquish the claim, vowing to hold it by force. Two weeks later Johnson, acting as agent for the Cleveland Iron Company, with

James Peters built a new house on the site of the burned cabin. As it neared completion, Mann and Moody approached the two with firearms and ordered them to leave the property within ten minutes. Johnson and Peters refused, but a short time later, with Johnson absent from the location and Peters away from the cabin to gather materials to fill cracks between the shanty's logs, Moody burned the cabin down again.[18] From mid-1848 to mid-1850 Moody and Mann occupied the Cleveland Iron Company's claim.

During this period many of the group that in 1846 and 1847 had been enthusiastic about the iron ore claim fell away. In 1850 when the Cleveland Iron Company applied to the Michigan legislature for a charter to operate as the Cleveland Iron Mining Company of Michigan (CIMC), only two of the original twelve who had signed the articles of agreement in April 1847 remained: Morgan L. Hewitt and John Outhwaite.[19] Hewitt, as previously noted, was a physician; Outhwaite was a young chemist recently emigrated from England who operated a grocery.

Surviving documentation does not permit an exact assessment of the reasons so many dropped away between 1847 and 1850, but one may surmise that they included concern over the validity of their preempted claim, growing recognition of the magnitude of capital investment required to exploit an iron deposit in a completely unsettled area with a forbidding climate, and the desire to invest in copper instead.[20] The concern over the preempted claim could only have been exacerbated by the involvement of a new, seemingly well-funded company in the title dispute: the Marquette Iron Company.

The Marquette Iron Company

In the summer of 1848, two years after Cassels had staked his claim on the Cleveland Mountain and at the same time Graveraet, Moody, and Mann preempted the claim, Boston investors sent Edward Clark of Worcester, Massachusetts, to Lake Superior to look for copper. He, too, was lured to the iron deposits by a chance encounter at Sault Ste. Marie. There he met Graveraet, who persuaded him to stop and see the Jackson company's iron ore outcrops and newly erected forge and the Cleveland claim occupied by Mann and Moody before going on to the copper district. Clark did so and, like Everett, Cassels, and the Clevelanders, was impressed. On returning east, he took an iron bloom and some ore from the Jackson operations. After the bloom produced high-quality wire, Clark and some of his backers abandoned their plans for copper. Instead, on March 4, 1849, Clark, Graveraet (who had traveled east in the winter of 1848–49), A. R. Harlow (a mechanic who owned a machine shop), and Waterman Fisher, a Worcester, Massachusetts, textile mill owner who put up most of the capital, formed the Marquette Iron Company. They immediately launched plans to build a forge on the shores of Lake Superior to smelt ore from the rich deposits Clark had observed, some of which the company believed it owned, thanks to agreements with Graveraet and the group that had preempted the Cleveland Mountain.[21]

While the Clevelanders hesitated, the new Marquette Iron Company acted. In May

Sketch of the village of Marquette, c. 1850. The Cleveland Mountain was located roughly fifteen miles inland and almost 1,000 feet higher in altitude and was linked initially to Marquette by a rough road through dense forests, swampy areas, and steep rocky outcrops. *(From J. W. Foster and J. D. Whitney,* Report on the Geology of the Lake Superior Land District, part 2, The Iron Region *[Washington, DC: A. Boyd Hamilton, 1851], frontispiece.)*

1849 a party dispatched by the Marquette Iron Company took possession of the Cleveland Mountain from Moody and Mann and began to prepare it for mining. In July A. R. Harlow, who had purchased equipment for building a forge, including a steam engine, landed with a small party in the bay nearest the ore deposits. Harlow named the resulting lakeshore settlement Worcester, after Clark and Fisher's hometown, but it would soon be renamed Marquette. The bay on which the village was situated was protected on three sides from the often tempestuous weather of Lake Superior, providing a good location for shipping out raw ore or finished iron products, since the Jackson and Cleveland ore "mountains" were located only around twelve to fifteen miles inland.[22]

Just as the Jackson Iron Company had done two years earlier, the Marquette Iron Company erected a bloomery forge. Bloomeries used a hearth, power blowers, and power hammers to convert iron ore directly to wrought-iron blooms. The alternative technology—the blast furnace—was larger, more complicated, and more expensive, and produced pig iron that had to be further refined to manufacture wrought iron.[23] In the winter of 1849–50 twenty double teams hauled ore to the bloomery under construction on the shore.[24] By July 1850 Harlow had Marquette Iron's forges in production, drawing on Cleveland and Jackson location ores.[25] The fires of its ten forges quickly became a landmark to vessels traveling on Lake Superior at night.[26]

The Marquette Iron Company's forge, c. 1852. In perhaps the earliest photograph of Marquette, the forge is the large white structure on the left. In operation in July 1850, it burned in December 1853. *(Courtesy of the Marquette County Historical Museum, Marquette.)*

Following approval of its charter in April 1850 by the Michigan legislature, the Cleveland Iron Mining Company belatedly moved to reassert title to its claim. Probably in the spring or summer of 1850 it gathered depositions from those involved in claiming and attempting to hold the Cleveland Mountain. When the government held hearings at Sault Ste. Marie in November 1850, CIMC officials had documents and witnesses present to attest to the priority of their claim. Those contesting the claim never arrived. Mann drowned in Lake Superior on the way; Graveraet and Moody, delayed by weather conditions on Lake Superior, did not arrive in time. The court awarded the right to purchase the property from the government to Lorenzo Dow Burnell, an original member of the Cleveland group, who, in turn, assigned his rights to the new Cleveland Iron Mining Company of Michigan.[27]

Despite this legal victory, CIMC's small group of investors still hesitated. Major investors, especially John Outhwaite and Morgan Hewitt, paid visits to the region, but little happened.[28] Between 1846, when they claimed the Cleveland Mountain, and 1850 the only mining operations carried out were those of the Marquette Iron Company. As late as December 1852, two years after securing title to the Cleveland Mountain and more than six years after first staking claim, CIMC investors had put up only around $15,000 and mined no more than fifty tons of ore.[29]

In September 1852, however, the company finally began to move. Its directors sent a resident agent—Tower Jackson—to begin serious work. His first report was not encouraging. He found few people to hire, for there had been no reason for anyone, other than the aboriginal inhabitants of the region, to live there prior to the discovery of the iron ore. The few people available for hire were, according to Jackson, "uncertain, independent, dreadfully poor and constitutionally opposed to work."[30] Jackson and a small crew with considerable difficulty moved around 1,200 tons of ore to the shores of Lake Superior in the winter of 1852–53.[31]

The Marquette Iron Company, meanwhile, had problems of its own. Loss of its claim to the Cleveland Mountain was not fatal, because both the Jackson and Cleveland companies were quite eager to sell it ore, and Marquette Iron had secured a lease on other ore deposits a short distance west of the Cleveland company's location, later to become the Lake Superior mine. Although its forges began to produce high-quality wrought-iron bars in July 1850, the primitive road between the forge and the ore deposits meant ore could be hauled only during winter, when the region's abundant snowfall made sleighing possible.[32] Amos Harlow, the Marquette Iron Company's local agent, described the road as "the worst . . . I ever saw . . . all stumps and stone and short pitches as steep as a roof."[33] Even after the company's lakeshore forge processed the ore into wrought-iron blooms, the product had to be shipped by water to Sault Ste. Marie, unloaded, portaged around the rapids, reloaded on vessels for shipment to ports on Lake Erie, unloaded, and then reloaded on rail for transportation to Pittsburgh. By the time the wrought-iron blooms reached Pittsburgh, transportation had driven costs to $200 per ton. They could be sold in Pittsburgh for only $35 to $80.[34] To make matters worse, the company faced shortages of the key fuel for the bloomery: charcoal. Charcoal production, a labor-intensive task in a labor-short region, proved troublesome, and key figures in the company feuded.[35] In 1851 Marquette Iron had trouble paying its employees. By 1852 the company's leading investor, Waterman Fisher, who had put $60,000 into the venture with no sign of profitable return, faced bankruptcy and was ready to sell out.[36]

Absorbing a Rival

Cleveland Iron directors began talks with Fisher. In April 1853 the negotiations culminated in the purchase of the Marquette Iron Company for $100,000.[37] In preparation for this acquisition, the Cleveland Iron Mining Company in February 1853 reincorporated under Michigan's new general mining laws. It retained the same name, but increased authorized capitalization from $100,000 to $500,000 to fund the purchase.

By this time Cleveland Iron had attracted several new and more aggressive investors. Among them was Samuel L. Mather, born in 1817, descendant of famous Puritan

Samuel L. Mather (1817–90). Although not among the earliest investors in the Cleveland Mountain, Mather joined them in 1850. He became secretary of the Cleveland Iron Mining Company in 1853, serving in that position until 1869, when he became president. Mather was a central figure in the company from the early 1850s until his death. *(Courtesy of the Western Reserve Historical Society, Cleveland.)*

preachers Increase and Cotton Mather. In the 1790s Mather's grandfather (also named Samuel) had helped organize the Connecticut Land Company, which had purchased 3 million acres in northern Ohio that it hoped to sell at a profit. In 1835, after his graduation from Wesleyan University, Samuel L. Mather was taken into his father's commission and importing business. After a few years he set up on his own. In 1843, however, at age twenty-six, Mather moved at his father's request to the small but growing town of Cleveland (population around 8,000) to dispose of the lands his grandfather had left the family and to serve as local agent or property manager for other Connecticut families with landholdings in the area.

In Cleveland Mather took up the study of law and was admitted to the bar.[38] He struggled financially early and by 1845 was depressed over his future prospects.[39] But his legal and real estate businesses put him in contact with several who had become involved in the iron claim. By 1850 he had invested in the nascent Cleveland Iron Mining Company and entered on the course that was occupy him the remainder of his life: serving on that company's board beginning in 1850, becoming its secretary and treasurer when the company reorganized in 1853, and then serving as its president from 1869 until his death in 1890.[40]

Another addition to the investors was William J. Gordon, who had moved to Cleveland in 1839 at age twenty-one, becoming a commission merchant and wholesale grocer. Considered serious minded, intelligent, and stubborn, Gordon was in his mid-thirties when he invested in the newly reorganized Cleveland Iron Mining Company. In 1854 he traveled to the Cleveland Mountain and the following year he became a director of the firm. He would serve as its president from 1856 to 1866.[41]

Although Waterman Fisher briefly became CIMC's largest shareholder (holding

William J. Gordon (1818–92). A wholesale grocer in Cleveland, Gordon may have initially become interested in the mineral district of Lake Superior as a possible market for his staples. He was the Cleveland Iron Mining Company's second president (1856–66) and served on the company's board from 1855 until his death in 1892. *(Courtesy of the Western Reserve Historical Society, Cleveland.)*

5,000 of the company's 20,000 shares) as a result of the financial settlement that brought Marquette Iron into the hands of the Cleveland Iron Mining Company, he did not retain that distinction for long. He sold out as quickly as he could. By mid-1854 Morgan Hewitt, Samuel L. Mather, William J. Gordon, and other Clevelanders had purchased most of Fisher's shares, ensuring that control of the company remained in Cleveland.[42]

The acquisition of Marquette Iron Company seems to have galvanized the Clevelanders into action. The purchase had given CIMC substantial land along Marquette's bay and an established forge, machine shops, and houses.[43] They could, moreover, take advantage of the small amount of work Marquette Iron had done at the Cleveland Mountain prior to ownership clarification in 1850. Acquisition of the forge, moreover, seemed to settle the issue of whether CIMC would simply mine ore or be a vertically integrated company, owning both mines and iron production facilities. In June 1853, two months after the acquisition of Marquette Iron, company president Morgan Hewitt traveled to Marquette. He took with him James J. St. Clair as the new agent, replacing Tower Jackson. Under St. Clair's direction, company employees quickly built a warehouse in Marquette and expanded the Marquette Iron Company's dock. The newly acquired forge operated using ore purchased from the Jackson company's location while St. Clair worked to put the company's own mine in order. He also completed a road through the very heavy woods separating the Cleveland Mountain from the Jackson mine and improved the crude road already built inland from the tiny port settlement of Marquette on Lake Superior.[44]

What prompted the previously conservative directors of the Cleveland Iron Company to purchase the Marquette Iron Company and make a serious investment in their long-neglected claim in 1853 is not clear, but several converging factors likely influenced their decisions. One was probably growing recognition, based on multiple visits to the site by members of the Cleveland group, of how cheaply they could mine ore. At the Cleveland Mountain freeze-thaw cycles over the years had loosened thousands of tons of ore, breaking it off from the bluffs on which it was exposed. Thus little preparatory work—and hence little capital investment—was needed to begin mining ore. All one had to do was break the fallen ore into small enough chunks for loading onto a conveyance. Sledgehammers wielded by unskilled laborers were sufficient.[45]

Probably even more central to the decision to begin putting money into the Cleveland Mountain, however, was the appearance of impending solutions to the region's transportation woes. As a contemporary observer noted: "The immense wealth of Lake Superior is not waiting for a market but for swift and uninterrupted transport."[46] The mine-to-harbor transport problem was the first to have a potential solution. In 1851 Heman Ely, associated with Rochester, New York's leading flour milling family, working with John Burt, surveyor William A. Burt's brother, proposed a steam railroad from Marquette to the Jackson and Cleveland mining locations, using his own funds since Michigan had not yet passed a general railroad law that would have allowed formation of a limited-liability stock company. This was a capital-intensive proposition, since the rail line would have to climb nearly 1,000 feet in the twelve to fifteen miles from Lake Superior at Marquette to the iron ore deposits, cutting through dense forests and large rock outcrops and passing over numerous swamps. Ely, however, would not begin the work without commitments from the Jackson and Cleveland mines. He got them. In a contract signed on November 11, 1851, CIMC promised Ely's railroad a monopoly on its ore shipments, rights of way over company lands, construction materials, and land for depots and sidings. Ely and CIMC also settled on a declining rate scale for ore shipments, dependent on magnitude.[47] This contract may have been what initiated negotiations between CIMC and Waterman Fisher for the purchase of the Marquette Iron Company's forge. The following spring Ely began work, assembling an engineering staff that included William Ferguson, formerly of the Cleveland and Toledo Railroad, later to work for CIMC as its local agent.[48]

Almost coinciding with Ely's initiative, heavy lobbying by the state of Michigan, copper-mining investors, and iron interests laid the groundwork for a solution to the other major obstacle: the rapids at Sault Ste. Marie that prevented all-water shipments from Lake Superior to lower Great Lakes ports. On August 21, 1852, a congressional act promised 750,000 acres of public land to Michigan to construct a canal around the rapids. This was critical. The *New York Tribune*'s Horace Greeley had noted in 1847 that lack of a canal at the Sault had "practically shut the Superior region against emigration and settlement" and "greatly embarrasse[d] the development of the Mineral wealth of the region."[49] The state accepted the land grant in February 1853 and awarded it to a company headed by Charles T. Harvey to build the canal. Harvey broke ground in June 1853, with completion scheduled in two years.[50]

Just as all the pieces of the puzzle seemed to be falling into place, disaster struck. In

December 1853 fire destroyed Cleveland Iron's Marquette forge complex, leaving "nothing but a heap of blackened ruins and some chimneys," from which the company's local agent was able to salvage only around seventy tons of iron blooms.[51] The heavy financial losses reopened the issue of whether the Cleveland Iron Mining Company should process ores as well as simply mine and ship them.

The choice was not easy. Both the Jackson and Marquette forges had demonstrated that quality iron could be produced in forges from Lake Superior iron ores using locally produced charcoal as a fuel. Several dozen other companies in the next several decades, seeing the promise, would attempt to produce iron locally using either forge or blast furnace.[52] The greatest potential demand for Lake Superior ores, however, was from already established blast furnaces in Ohio and western Pennsylvania. However, in early 1854, when the company faced the quandary of what to do, it was not clear that Lake Superior ores could be used in blast furnaces. The first samples sent to Pittsburgh for testing were pronounced worthless—too hard for furnace operations. Further tests, including some involving CIMC ores, failed in 1853, 1854, and 1855.[53] Eventually, additional tests, some involving slightly redesigned furnaces and mixing Lake Superior ores with much leaner and softer local ores, proved successful and convinced Ohio and western Pennsylvania iron masters of the quality and tractability of iron from Lake Superior ores, but in early 1854 the picture was still murky.[54]

Further complicating matters, the backers of Cleveland Iron had to envision a new paradigm for the relationship between iron ore mines and blast furnaces. The dominant paradigm had small-scale iron mines linked in common ownership with blast furnaces located nearby.[55] The concept of an independent iron-mining company shipping ore hundreds of miles to supply blast furnaces owned by others was relatively novel. The concept of a blast furnace depending on an iron ore supply hundreds of miles away—the other side of the coin—was equally novel. John Fritz, later one of the most important figures in the American steel industry, visited the Cleveland and Jackson locations in 1852 and tried to persuade eastern ironmen to invest in the region. He was told that he "might as well talk about bringing iron ore from Kamchatka [in Siberia]."[56] In the mid-1850s most ironmen did not comprehend that the combination of the relative richness of the Lake Superior ores with the economies of an efficient waterborne transportation system would make this arrangement possible. The Cleveland Iron Mining Company's early investors were among the first to envision the possibility.

Focusing on Mining

In the spring or early summer of 1854, fresh from the loss of their forge, taking a chance that could have proven disastrous, the company's directors abandoned iron making to focus on mining. Company president Morgan Hewitt undoubtedly played a key role in the decision. In the summer of 1853 he visited the company's holdings before fire had destroyed its forge. On return he argued that the isolation of the region, the high cost of provisions, the lack of a regional mineral coal supply, the difficulties of securing a reliable supply of the alternate fuel (charcoal), and the heavy outlay of capital required for iron production all suggested focusing on producing ore and shipping it out to furnace operators elsewhere.[57] The decision that Cleveland Iron's investors made in early 1854 to go this route began another corporate tradition: in times of economic distress, such as that caused by the loss of its forge, the company would refocus on mining and dispose of other properties.

Reflecting this decision, CIMC shifted its attention to developing its mining location and the road leading to it; it abandoned the burned-out forge site on the lakefront. In February and March 1854, Samuel L. Mather, company secretary and already a central figure, instructed St. Clair, the company's Upper Peninsula agent, to erect a two-story structure at the iron mountain, build a store in Marquette, and begin, with the Jackson Iron Company, to construct a "Plank and Rail Road" to the Cleveland Mountain from the Jackson Mountain to replace the old road the company had built earlier.[58] In June and July Mather urged St. Clair to "rush our Road through this season if possible" and to "turn all energy" to enlarging the dock and constructing the road to the mine since "there is no difficulty in contracting for all the ore that we can get out."[59] To attract labor for the expanding mining operations, St. Clair built twenty houses and a "new, large and commodious" company store, tearing down the old log dwellings erected in 1848 and 1849.[60] CIMC also began planning a new and improved dock in partnership with the Jackson Iron Company, but the companies later decided to build separate docks instead.[61]

The purchase of Marquette Iron and the improvements needed to begin large-scale mining required the company's small band of stockholders to increase investment. Cleveland Iron's capitalization rose from $15,000 in December of 1852 to $100,000 in July 1853 to $240,000 by July 1855.[62] These funds were raised by assessing stockholders, something that hit most of them hard, for none were terribly wealthy at the time. Mather commented to St. Clair in October 1854, "We all feel the assessment to be heavy" and indicated that he hoped for no further calls until the spring.[63] The assessment, however, was not enough. W. J. Gordon, a large investor and future president of the company, wrote St. Clair in October 1855 that he was not to draw any more drafts, since "there is not a cent in the Treasury."[64]

The canal at Sault Ste. Marie opened in June 1855. That year Cleveland Iron shipped 1,447 tons of ore through the canal to Cleveland; it was the only company to ship ore during the canal's first season.[65] But the canal did its work. Within three years the rate for transporting a ton of ore from Marquette to Cleveland dropped from $5 to $2.09.[66]

Even as the new canal resolved the biggest hurdle in waterborne transportation of ore and the company began to receive some income from its investments—though not nearly as much as its expenses—new problems emerged in three areas: management, land transportation, and finances. In all three the company at first struggled, but by 1860 had come up with workable solutions.

Management Issues

Management was a crucial problem. How would the company effectively manage operations some 600 miles from Cleveland, especially when the area was effectively cut off from communications for five to six months annually when Lake Superior froze? The Cleveland directors at first wanted to closely manage their Upper Peninsula agents. In February 1854 Mather instructed St. Clair that he was to "communicate by *every mail* & other opportunity *direct* to the office here, through *me* as their Secretary," that he was to write "at length & give full particulars of everything going on, that can, in any way, interest our company." Mather followed with very specific instructions, including the exact grade for the plank road and the number of rooms in a house to be erected at the mine location.[67] Little was left to St. Clair's discretion. Several indicators suggest that management of properties did not, at first, go well. Four of the first five agents lasted only about a year each.[68] In August 1855 William J. Gordon wrote to company president Morgan L. Hewitt: "It makes me sick, to think how awfully our interest has been neglected, & how we have been robbed on all sides."[69] This may have been one of the reasons that Hewitt, one of the company's founders and its first president, moved permanently to Marquette in 1857.[70] Hewitt lived only around a dozen miles from the company's mining operations, and was soon deeply involved in the growing town's social and commercial life. He traveled to the company's mine with some regularity, writing back to company secretary Samuel L. Mather to report briefly on what he saw. Hewitt remained on the Cleveland Iron Company's board until his death in 1889, providing the board with a local perspective on its operations to supplement the views of Peter White, another Marquette resident sometimes consulted by board members.

Hewitt's move, however, did not completely solve the oversight problem. In 1858 the company sent one of its directors, H. B. Tuttle, to Marquette to serve as local agent. Tuttle, too, observed: "We have lost large amounts . . . by the remissness of previous agents."[71] The following year he commented on the "waste" and "carelessness evinced with our property," and Mather commented in response: "You see *now* where our money has all gone to, & I don't wonder that you grumble & complain."[72]

Adding to problems with its agents, in 1856 the company had its first significant labor dispute. As noted previously, Tower Jackson had difficulty securing labor when he moved to the Upper Peninsula in 1852. This problem continued.[73] In July 1856, just as the demand for the region's ores began to grow, the largely Irish immigrant employees of both the Jackson and Cleveland mines struck.[74] Cleveland Iron's directors seem to have learned from previous attempts at overly close management from Cleveland. They let their local agent act with some discretion. Within a week or two that agent, William Ferguson, had settled the strike, making, apparently on his own authority, partial concessions to the strikers while discharging those who would not agree to the compromise. New CIMC president William J. Gordon later wrote to Ferguson that he "fully indorse[d] the course you adopted let the consequence be what it may."[75]

The instructions sent to H. B. Tuttle on his appointment as the company's new agent at Marquette in early 1857 further illustrate the increasing autonomy being conceded to Upper Peninsula operations by Cleveland headquarters. The instructions given Tuttle were much less specific than those given St. Clair three years earlier, leaving him with considerable autonomy, perhaps because Tuttle was from Cleveland, had been made a director of the company, and was well known to the other directors. Thus, when the company wished to dispose of its smith shop in Marquette, a remnant of its purchase of the Marquette Iron Company, it left the final decision to Tuttle, letting him find a "good party" to purchase it.[76] By 1859 the expression of "you . . . can judge best" had become steadily more frequent in correspondence between Cleveland and company agents in Marquette and mining superintendents at the Cleveland Mountain.[77]

Land Transport and the Plank Road

Besides management problems, Cleveland Iron also struggled with land transportation problems between 1853 and 1857. The solution to the overland transportation problem that seemed so imminent in 1852 and 1853 had turned sour by 1854. Heman Ely, who had promised to construct a railroad from the port of Marquette to the mines, had begun work quickly in 1852, but problems soon cropped up. On December 5, 1852, Tower Jackson, then CIMC's resident manager, perhaps aware of negotiations with Waterman Fisher for the purchase of the Marquette Iron Company's forge, urged his company to abandon Ely and build a plank road on its own, arguing that it would help settle the area and bring in people who would produce the charcoal needed for iron production at the forge. Jackson, however, also knew that Ely was so short of funds that he had been forced to sell some of his company's provisions simply to get enough money for boat passage to search for investment capital. He was thus skeptical that the railroad would be built at all and argued that the company could construct a plank road much more quickly than Ely could complete his railroad.[78]

Tower Jackson's analysis of Ely's financial situation proved accurate. By 1854 Cleveland Iron's forge was gone—burned to the ground—but construction of Ely's railroad still floundered from lack of capital. At this point Ely informed the Jackson and Cleveland companies that unless they provided $50,000 in cash, he could not continue the project. They were willing to put up the capital, but Cleveland Iron's managers would do this only if they and their Jackson partners secured controlling interest in the railroad, something Ely was not willing to concede.[79]

What made the Cleveland and Jackson companies skeptical of Ely's request was Ely's success in finding other partners, particularly John Burt, to assist him in creating a rival iron-mining company—the Lake Superior Iron Company—located on land a mile to the west of the Cleveland location.[80] Believing that the creation of a rival mining company under the same ownership as the railroad invalidated the contract they had signed with Ely, both the Jackson and Cleveland companies declared their prior agreements with Ely null and void. Anxious to have a transportation system in place in time for the opening of the canal at the Sault, and recognizing that the Ely railroad would not be completed in time, the two companies now turned to Tower Jackson's suggestion of a plank road.[81]

The companies began building a plank road in the summer of 1853, following closely the line of the old wagon road leading to the Jackson location. They soon modified the original design by laying cross ties capped by straps of iron longitudinally on the plank roadbed, creating a strap railway that relied on horses and mules for power.[82] Crews completed the railway to the Jackson mine in 1854 and reached the Cleveland in 1855, in time for the opening of the canal at Sault Ste. Marie. The Cleveland and Jackson companies formalized their joint venture, creating the Iron Mountain Railway Company in March 1855 to operate the new line.[83]

Firms, especially those operating in high-risk environments, have long sought ways to leverage capital for investment and share risks. Strategies have included merger with or acquisition of rivals, joint ventures or partnerships, minority investments, and joint activities.[84] In purchasing the Marquette Iron Company, CIMC had taken one of these routes: acquisition of a rival. In undertaking the creation of a horse railroad line with the Jackson Iron Company it pursued a second option: the joint venture or partnership. The partnership or joint venture as a tool for survival in risky environments has a very long history, dating back to the ancient world when merchants pooled resources and shared risks on long-distance trading ventures by ship or by caravan. It continued into premodern Europe, used by Italian city-states in trade in the Mediterranean basin and by merchants in early European oceanic expansion.[85] The rise of the limited-liability corporation as a means of raising capital eclipsed the partnership or joint venture in the late eighteenth century. It would reemerge in America as a significant force only about 1890.[86]

Almost from its infancy, however, the Cleveland Iron Mining Company made use of the technique for the very same reasons Venetian merchants had used it in the fourteenth and fifteenth centuries: to pool its limited resources with others to undertake capital-intensive ventures and simultaneously to share risks and potential losses with other firms. The joint venture, typified by CIMC's early working partnership with the Jackson Iron Company, would eventually become almost a tradition when the company was faced with capital-intensive projects.

With the canal at Sault Ste. Marie open, growing acceptance of Lake Superior ores by blast furnaces in Ohio and Pennsylvania, and a strap railway operative, the Cleveland Iron Mining Company made "large additions" to its forces for the 1856 shipping season. In May 1856 Gordon took one of the earliest vessels to the region, bringing with

Iron ore shipments on the Marquette, Houghton, and Ontonagon Railroad, c. 1870s. The MH&O RR (initially called the Iron Mountain Railroad) linked the iron mines of the Cleveland, Jackson, Lake Superior, and other Marquette Range companies with the ore docks of Marquette from the mid-1850s onward. *(Courtesy of Superior View, Marquette.)*

him twenty-six mules and four horses, purchased out of Kentucky at up to $1,400 a pair, to be employed hauling ore on the new railway.[87] By 1857 quarrying operations at the mine employed thirty men who produced forty to fifty tons daily.[88] Prospects seemed bright.

Profits, however, remained elusive. While better than nothing, the strap railway was *not* a resounding success. A span of horses or mules could make only one round trip per day, drawing only four to five tons of ore.[89] The grades were so steep and the braking system so crude that accidents killed several of the expensive imported horses and mules.[90] Making matters worse, the Cleveland-Jackson road paralleled and frequently crossed the line being prepared by Ely for his railroad, prompting a host of lawsuits and injunctions.[91]

In 1855 Michigan's legislature finally passed a general railroad law, making it easier for Heman Ely to raise capital. He now incorporated his venture, adopting the name Iron Mountain Rail*road,* to distinguish it from the Cleveland-Jackson companies' joint venture, the Iron Mountain Rail*way.* Heman Ely died in 1856, but his brother Samuel Ely pushed the railroad to completion. In August 1855 the road's first locomotive, the Sebastopol, arrived by water at Marquette; in September 1857 the railroad finally reached the Cleveland and Jackson mines. Quickly recognizing the comparative inefficiency of their two-year-old horse-powered railroad, the Cleveland and Jackson companies soon abandoned it. Rates for moving a ton of ore from Negaunee to the docks at Marquette

Bird's-eye view of Ishpeming, 1871, looking south. The Cleveland Iron Mining Company's open-pit operations are at the left (identified by the number 6), with the Cleveland Mountain just above. The New York mine operated the lower pit on the left (7). Also visible is company housing near the pits and the rival Lake Superior (4) and Barnum (5) mines on the opposite side of Ishpeming. The company's mining superintendent lived in the dwelling on the hill to the southwest of the company's pits just above the railroad roundhouse. *(Courtesy of the Marquette County Historical Museum, Marquette.)*

dropped from around $3 a ton in 1855 to around $1.27 with the plank road to 87¢ with the coming of Ely's railroad.[92]

By mid-1857, with the company's management problems seemingly solved by granting autonomy to more carefully selected local mine agents and its transportation problems alleviated by the opening of the canal at Sault Ste. Marie and the completion of Ely's railroad, prospects seemed bright for the Cleveland Iron Mining Company. At Cleveland Mountain, a local observer reported, an "immense amount of ore" lay exposed above the surface and work had uncovered a rich new opening.[93] Moreover, the ore now had plentiful markets. *Mining Magazine* reported in February 1856 that Lake Superior ores could now compete with ores from New York and that CIMC intended to ship 700 tons a day in the coming season, even though only 2,000 tons total had been shipped down all of the previous year.[94] The company's blotter book entry for August 15, 1857, noted that the superintendent of the Cleveland and Mahoning Railroad in Ohio was ready to take all the ore the company could deliver.[95]

View of one of the Cleveland Iron Mining Company's open-pit mines, 1870, possibly after a rainstorm. Several of the horse carts used to haul ore out of the mine stand
near the rock face. Note the absence of trees in this once heavily forested region. The
timber had been used to build and heat local homes and to fire the steam engines
that had begun to be employed for pumping water and hoisting ore from these pits.
(Courtesy of the Marquette County Historical Museum, Marquette.)

The Panic of 1857

Just as the company seemed on the verge of returning profits to its investors, the long
period of prosperity that had followed the Mexican War and had fueled the growth of
the American iron industry ended. In August 1857, just as Ely's railroad was approach-

ing the Cleveland mine, the New York branch of the Ohio Life Insurance and Trust Company failed. Investors in other American banks and in railroads panicked, pulling out their funds. Land speculation based on the growing number of railroad land grants abruptly halted, undermining railroad investment. Railroad stocks and bonds plummeted, provoking further runs on banks. Demand for iron and iron ore dropped dramatically.[96]

When the panic shut down the iron mills of Pittsburgh and surrounding areas, the ripple effect soon spread the disaster to Cleveland, in the form of defaulted payments for ore already delivered and declining ore orders, and then to the Marquette iron range. By the late fall of 1857 Gordon (the company's new president) and Mather needed $50,000 in cash to keep the company solvent and continue operations. Mather visited New York and Boston bankers seeking these funds without success. By December, the company was closing down everything it could.[97]

Thus, on the verge of turning a profit, the Cleveland Iron Mining Company was abruptly pushed into crisis mode. The biggest problem now was neither management nor transportation, but finances. Due to the isolation of Lake Superior, hard currency had never been readily available in the Upper Peninsula. Even in tranquil economic times the company had problems collecting and transmitting sufficient currency to pay its workforce.[98] When the panic hit, money became scarce everywhere, and in Michigan's Upper Peninsula, where currency was always scarce, matters became critical.

As a temporary expedient some iron-mining and furnace companies in the Upper Peninsula, including the Cleveland, had, as early as 1853 and 1854, printed promissory notes, locally called "iron money," in varying amounts, payable at banks in Cleveland or New York in thirty or sixty days, sometimes with interest. Local mine agents signed the drafts, and merchants accepted these notes from workers. In cash-strapped Upper Michigan, the notes circulated like real currency, allowing iron companies to meet payrolls, giving them a breathing space until they could secure real currency to pay off the promissory notes. Local merchants and bankers collected the notes, sorted them by company, bundled them, and presented them to the agents of the companies, who then forwarded them to home offices for payment.[99] The use of such notes, usually referred to in historical literature as scrip, was relatively common in isolated mining communities in the nineteenth century.[100]

Unable to get currency and struggling to keep the company afloat, Mather and Gordon initiated an expanded system of "iron money" as a temporary fix for their inability to raise funds to pay wages. On high-quality paper stock they printed small drafts, down to $1 notes, against the treasury of the Cleveland Iron Mining Company. They then sent these notes up to the Cleveland mine to pay employees. Local confidence in the company was sufficient that the notes circulated among local merchants and banks as regular currency, only slowly drifting back to Cleveland to be turned in for cash. This delay provided Mather and company officers, desperately struggling to keep Cleveland Iron financially

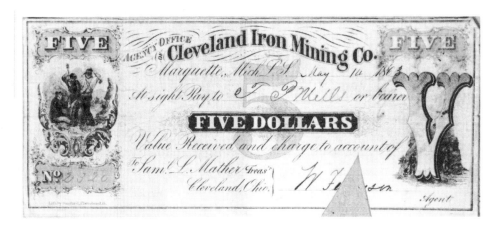

Cleveland Iron Mining Company's "iron money." The company used drafts like this one, signed by the company's local agent or mine superintendent, to pay employees during economic difficulties or when sufficient legal currency was not available in the isolated Lake Superior iron district. From around 1853 to the 1870s "iron money" circulated in the communities of Michigan's Upper Peninsula much like regular currency. *(Courtesy of Cliffs Natural Resources, Inc.)*

afloat, time to get affairs in order. Problems securing sufficient currency to pay its miners plagued CIMC for years to come, but "iron money" acted as a bandage that allowed the company's financial wounds slowly to heal.[101]

The use of the scrip seems to have sparked no worker resistance, perhaps because the company's employees recognized that the alternatives were worse: closed mines, no employment, or, at best, delayed paydays. In addition, CIMC's iron money differed in some respects from the scrip issued in other areas, particularly the coalfields of Appalachia later in the century where it was sometimes abused. The Cleveland company's scrip sometimes paid interest and did not have to be used at a company store; merchants in the region accepted it like regular currency.

In the summer of 1858, as the depression continued and with its treasury low, Cleveland Iron delayed payment of bills to the railroad that delivered its ores to Marquette.[102] To get funds to meet these and other financial obligations, the company's directors assessed shareholders $1 a share and issued additional stock.[103] Assessments, additional investment, and "iron money" enabled CIMC to survive the Panic of 1857, and sometime between 1858 and 1860, as railroad construction nationwide again picked up steam, the company eased again toward profitability. Tuttle, the company's agent in Marquette, wrote in the summer of 1858 that he had been able to sell ore to everyone to whom the company "desired" to sell.[104] In February 1859 Mather commented that the demand for ores in the coming year would be beyond the abilities of all Lake Superior companies to supply, and in May he noted that the company could "readily sell all the ore we can get down."[105]

Problems did not end with the resumption of economic prosperity in the late 1850s, for now the company unexpectedly faced increased mining costs. Cleveland officials had assumed that their "mountain" was made entirely of relatively pure iron oxide. By 1859, however, the company had discovered this was not the case. The desirable ore, specular hematite (over 60 percent iron content), ran in thick veins sandwiched between veins of much lower-quality jasper rock (or jaspilite, which was only around 30 to 35 percent iron).[106] This caused dual problems. First, the useless jasper frequently got mixed in with the specular ore, leading to complaints from users and threats to turn to the company's rivals (the Jackson and Lake Superior mining companies) for ore. Second, quarrying the ore became much more difficult and expensive since the "mountain" was not uniform. Complaints and concerns about both of these issues began to accumulate in 1859.[107]

Improvements in the company's transportation nexus, always critical in dealing with a bulk commodity like iron ore, however, helped offset the unexpected increase in the cost of mining and the need for ore sorting. Loading ore at the docks in Marquette was initially a major bottleneck. Loading the 200- to 300-ton cargoes typical of the period took twenty to thirty men using shovels and wheelbarrows from three to six days. The completion of Ely's railroad in 1857 had only made this bottleneck worse by accelerating the pace of ore deliveries. In 1857 the Lake Superior Iron Company, CIMC's neighbor to the west, introduced a new dock design, drawing on practice from the coalfields of Pennsylvania. In Pennsylvania operators used overhead storage compartments equipped with chutes to quickly load railroad cars parked below them. The Lake Superior Iron Company built its new dock well above water level and installed similar storage bins and chutes to dump iron ore directly into or onto vessels moored below, eliminating the need for wheelbarrows and large loading crews. In 1858 CIMC modified its dock by raising its height and equipping it with bins and chutes, improving on the Lake Superior Company's innovation with higher-capacity storage pockets and an improved chute system with better control over angle and rate of ore flow. The new docking system reduced loading time from days to hours.[108]

With this bottleneck eliminated, the decade of the 1860s opened with the Cleveland Iron Mining Company prospering as never before. Already by 1860 a number of furnaces in Ohio and western Pennsylvania had begun using Lake Superior ores, some exclusively.[109] In June 1860 Tuttle noted that the company had in "actual view" two years of ore, far more than ever before, and had nearly 100 "good" men "under good discipline," directed at the mine by Frank Mills, an experienced and qualified mine superintendent.[110] In 1855 the company had shipped only around 1,500 tons of ore; in 1860 it shipped over 40,000.[111] By mid-1860 the company was entirely free from debt and had a "fair surplus." The Cleveland Iron Mining Company could have begun paying dividends for the first time. But its directors, always cautious, refrained, wanting a surplus large enough to guarantee a continuation of dividends once begun.[112]

The Cleveland Iron Mining Company's dock at Marquette, 1863. Note the ore chutes on the dock (vaguely visible projecting into the air toward the end of the dock) that speeded up loading of vessels. *(Courtesy of Superior View, Marquette.)*

The Civil War Years

The caution proved justified. In April 1861 the South Carolina militia fired on Fort Sumter in Charleston harbor, igniting the Civil War. Uncertainty over the effect of the hostilities on the iron trade led furnace men to reduce orders for ore. "Our country seems to be in a distracted state yet no one can tell how it is going to end," Hewitt wrote to Mather. "No business calculations can be made with any certainty until it is settled." One of the company's customers commented similarly about the "fog" surrounding business, noting: "We really do not know how to move."[113] Most of the company's customers responded by reducing orders. The firm's ore shipments dropped precipitously from 40,000 tons in 1860 to 12,000 tons in 1861.[114]

During the war years (1861–65) Cleveland Iron resumed its practice of seeking partnerships for capital-intensive ventures. In 1863–64, desiring to maintain its focus on iron mining but recognizing the profits to be made in processing ore into pig iron, then at record high prices, it entered an agreement with Middlesex Furnace in Pennsylvania. The furnace gave the company a ton of pig iron for every three tons of raw ore provided by the company. In 1863–64 nearly a third of CIMC's production went to Middlesex.[115] Another, related venture was the creation of a new company under an old name: the Marquette Iron Company. In the midst of the Civil War, with charcoal iron selling at double its prewar price and demand exceeding supply by a factor of three, linking mines to furnaces seemed logical. Marquette-area entrepreneur Peter White had long urged the company to build a blast furnace at the Cleveland Mountain to use less desirable grades of iron ore, offering to invest in the project himself.[116] The company did not bite, at least immediately. But the new Marquette Iron Company was a bow in that direction. It combined the local charcoal blast furnaces, forges, and mineral lands owned by the Collins Iron Company with unused Cleveland Iron ore deposits. In early 1864 CIMC directors sold 250 acres of mineral land and 100 feet of bay front in Marquette to the new company for $100,000. The Cleveland Iron Mining Company's local agent and mine superintendent jointly managed the new Marquette Iron Company's mining operations, and Samuel Mather, CIMC's secretary and treasurer, served as the venture's president.[117]

To raise money for expanding operations, Cleveland Iron also made a move that it would come to regret. In 1864 it sold for $30,000 the lease it held for mining forty acres on the Lake Superior Iron Company's property, a legacy of its acquisition of the Marquette Iron Company in 1853. The company justified the sale by declaring that the few hundred acres of mineral land it already owned around its present openings would, "beyond reasonable doubt, supply all the Ore the Company will ever want."[118] This error in judgment soon became abundantly clear and may have contributed to the emergence of another element of the company's future culture: sell everything else but do not sell iron reserves. Although it proved a false alarm, as early as 1865 corporate agents and executives became concerned about the company's supposedly inexhaustible mines being used up.[119] Moreover, the Lake Superior Iron Company, to which the property was sold, regularly outproduced the Cleveland most years for the next several decades.

Renewed Labor Problems

Ultimately, wartime high prices provided sufficient financial resources to the company. But high wartime employment in the industrial areas to the south and the need for men to fill Northern armies meant wartime labor problems. Labor had been a problem from the area's opening. Other than a small band of local Ojibwa, no one lived in the region of the iron deposits prior to 1845. All labor had to be imported. Edward Clark, in order to get the Marquette Iron Company's forge and mining operations going in 1850, went to Milwaukee to recruit French and German immigrants.[120] Tower Jackson, CIMC's first agent, complained in 1852 about the quality and quantity of labor, as previously noted. The Civil War made a bad problem worse.

Attracting workers to Upper Peninsula mines and keeping them were difficult tasks even in calm times. The region was remote and isolated, far from the amenities of civilization, especially once Lake Superior froze over and before railroads from the south reached the region in 1872. "This mineral world," an 1853 visitor commented, "is a region by itself."[121] If the isolation was not sufficient to frighten off workers, the climate was. Snowfall could reach over 200 inches annually. In April 1857 one early pioneer commented that snow had been abundant since the middle of November, varying from three to five feet deep and was still, in early April, three feet deep in the forest.[122] The cold that accompanied the snow could make life miserable and dangerous in the company's mining pits. Mining superintendent Frank Mills wrote in January 1872 that the ladders miners used to climb down into the mining pits were "loaded with ice & it is all a mans [*sic*] life is worth to be careless in going down."[123] In January 1873 outside temperatures of -32 idled the company's entire workforce.[124] Hours, moreover, were long. Miners in the period 1855–63 typically worked eleven or twelve hours per day, six days a week.

In these circumstances, native-born laborers were not interested in moving to the settlement, soon to be named Ishpeming, growing up near the Cleveland company's mine. Not surprisingly, the laborers lured to the mines were mostly immigrants. No figures exist for the ethnic composition of CIMC's labor force in the 1850s or 1860s, but it probably paralleled the region as a whole. In the late 1850s and 1860s the workforce at Michigan iron mines was largely Irish. By the early 1870s, however, it was a bit more diverse. In 1870 employees at three large mines of the region, probably including the Cleveland, were Irish (31 percent), Cornish or other English (27 percent), and Swedish (18 percent). French Canadian, German, and a variety of other nationalities formed the remainder. Only around 5 percent were American born. In Ishpeming in 1870 over 70 percent of the population was foreign born.[125]

Serious labor shortages began to appear in 1859, as the company's output approached 40,000 tons and as some operations below ground level commenced. Tuttle, then the company's agent, warned Cleveland officials that men were not plentiful in the region and even sent handbills to be posted in Cleveland to try to attract labor to company mines.[126] But this was simply a foretaste of what was to come. Although labor was relatively plentiful in 1860 and early 1861, the situation quickly deteriorated as wartime demand for iron ramped up and as men enlisted in the federal army. Aggravating matters, in the fall of 1861 the region's copper mines, 100 miles to the west, suffering labor shortages of their own, sent agents to the iron region offering higher wages. Robert Nelson, Tuttle's replacement as agent, complained in March 1862 that the copper mines along Portage Lake had "drained us of men."[127]

By May of 1862 the shortage of manpower had become acute. President W. J. Gordon, after visiting the mines, wrote Mather: "You must manage to send Nelson more men . . . don't spare any effort." He suggested seeking out Germans and Irish, paying a bonus to those who stayed at least three months, and using cheap housing as a selling point.[128] But the labor problem remained serious. In August Nelson reported men were leaving "by the dozen," fearing they would be drafted.[129]

The pivotal year of the war, 1863, brought no relief, and pushed the company to new efforts. Cleveland Iron joined with other iron and copper-mining companies in March 1863 to support the Mining Emigrant Aid Association of Lake Superior to recruit men from Lower Canada.[130] When this yielded no immediate results, William Ferguson, who had returned to the company for a second stint as Marquette agent, traveled to Canada to recruit mining labor, bringing thirty-seven men back in mid-May.[131] On the Cleveland end, Mather tried to help, paying passage from Cleveland to Marquette for dozens of men who agreed to work in the company's mines, but this proved a disaster. In June 1863 Peter White, a director of the company and a prominent Marquette banker and entrepreneur, wrote Mather that of the thirty-three men he had sent up recently, only two remained at work. He also noted that some miners had been frightened off by the newly instituted military conscription law, enacted to try to fill the ranks of federal armies depleted by the prolonged war.[132]

The situation worsened in 1864. Ferguson complained that the region had only half the labor it needed and that men were "running from one locality to another" expecting to benefit by pitting one company against another. Moreover, he added, the men available were inexperienced, "hardly one of them understanding drilling."[133] Only with the end of the war in April 1865 and the sudden slackening of demand for ore did this problem end.

Labor shortage, however, was just one aspect of the labor problem. The company also faced a much more militant labor force, one unwilling to accept without a struggle the periodic pay reductions imposed by the company to coincide with seasonal and market downturns and one knowledgeable enough about the growing need for ore to periodically demand increased wages. In March 1860 and again in May 1861, the company's miners—at this point largely Irish immigrants—struck over pay issues. The strikes failed, largely because at the time of the walkouts the supply of labor was relatively large because demand for ore was low. The company's Upper Peninsula management discharged the "ring leaders."[134]

As labor shortages grew acute during the war years, however, Cleveland Iron and other iron-mining companies found they could no longer prevail in labor disputes so easily. This became clear in 1864. By then wages for miners had risen from $1 per day to nearly $1.75 as companies, including CIMC, voluntarily raised wages to attract and hold increasingly scarce labor. In May 1864 men at several mines, including the Cleveland, struck, demanding $2.00 per day, a ten-hour work day instead of eleven, and the right to quit work at 4:00 on Saturday instead of 7:00. The companies, anxious to keep operating amid high wartime demand and prices, willingly granted the wage demand, but resisted shorter working hours since this would have impacted production. In response, the labor force soon became, in the words of company agent Ferguson, "disposed to be somewhat riotous."[135] To shut down all area ore-related operations, 300 to 400 miners marched from mine to mine and then to nearby blast furnaces and to the docks at Marquette. In a bind, with shipping contracts fixed and demand high, the mining companies settled with their miners, compromising on a 6:00 p.m. closing time on Saturday, but conceding both the $2 wage and the ten-hour day.

By then, however, the strike had spread from the miners to dockworkers, whose work was more sporadic, depending on vessels being in the harbor. Paid 37.5¢ per hour, they now demanded 50¢. On the docks things quickly got out of hand, as striking dockworkers forcibly prevented anyone from loading or unloading vessels, including sailors from incoming vessels in need of supplies. Cleveland Iron officials, expecting a favorable hearing due to wartime need for iron, traveled south to persuade Michigan's governor to send a force to restore order and received a gunboat (the USS *Michigan*) and a body of troops. The combination quickly restored order; several miners were arrested for rioting. Mine company officials talked about creating a "home guard" but did not follow up.[136]

Worse was yet to come. When the war ended in the spring of 1865, demand for iron ore dropped. Companies reduced their labor forces, and the region's labor short-age suddenly became a labor surplus. Several regional mining companies, including the Cleveland, reduced wages in concert from around $2.25 to $2.50 to around $1.75 a day on July 1. At the time, Marquette harbor was full of vessels, so when workers on the dock resisted the magnitude of the reduction and struck for $2, the companies con-ceded. Miners, who had initially accepted the reductions, now felt betrayed and struck. When the companies hesitated, the strike turned violent. Mills, the company's mining superintendent, reported on July 11, 1865, that the "state of things is frightful," with 150 club-armed men, largely Irish, passing by his office on the way to a nearby mine to prevent work there.[137] Miners, many apparently drunk, probably joined by hundreds of laid-off workers, threatened to march on Marquette and loot the town. They burned some company property and cut a telegraph line or two.

Coincidentally, the USS *Michigan* arrived at Marquette shortly after the miners' strike began. Persuaded to intervene by fearful local officials, the *Michigan*'s officers mounted two small cannon on a flatbed railcar, moved on the 1,500 to 2,000 striking miners at the inland mines, and threatened to shoot first and ask questions later if they did not disband and return to work. Intimidated, the miners returned to work, only to go out again shortly after the *Michigan* left port, forcing the *Michigan*'s commander to repeat the process a few weeks later, this time with support from a Volunteer Reserve Corps shipped up by boat and rail from Chicago. With company property protected from violence and demand for iron ore low, the strike collapsed; miners returned to work at the reduced wages or left the region.[138] From 1866 on the Irish, who had hith-erto dominated the company's labor force, were steadily replaced by Scandinavian and Cornish miners. In 1874 one observer estimated the workforce at the Cleveland mine as

The USS *Michigan*. This gunboat, which patrolled the Great Lakes during the Civil War, played an instrumental role in suppressing labor problems on the Marquette Range in both 1864 and 1865. *(Courtesy of Historical Collections of the Great Lakes, Bowling Green State University, Bowling Green, OH.)*

only a quarter Irish, down from the nearly all-Irish workforce of the early 1860s. Scandinavians now made up half the workforce, and the Cornish a quarter.[139]

In addition to having to struggle with its ordinary labor, CIMC also had to struggle in the 1860s to keep its skilled management personnel. William Ferguson, for example, served the company in several capacities, including shipping and financial agent in the late 1850s and again in the early 1860s. In April 1864, frustrated that his children had no school in the isolated and primitive area around the mines and tired of interacting with the "class of people that we now have to deal with," presumably unskilled, immigrant labor, Ferguson was ready to quit. A few weeks later he commented that he was so discouraged by the way work was dragging that he was "almost tempted to run away out of sight & hearing of [the] whole thing."[140] Frank Mills, who was to serve as the company's mine superintendent for almost two decades, briefly left in 1862, declaring: "I am heartely [*sic*] sick of this place . . . & as soon as you can lit [*sic*] me go I want to leave for Home."[141]

Mather and his associates, having come to recognize the critical importance of good local management in the mid-1850s, attempted to correct the problems and even anticipate others. For instance, in 1867 company officials appointed a clerk to assist Mills at the mine to permit Mills, who hated paperwork, to focus on mining alone.[142] Mills had long lived in a house so close to one of the active pits that when the company blasted,

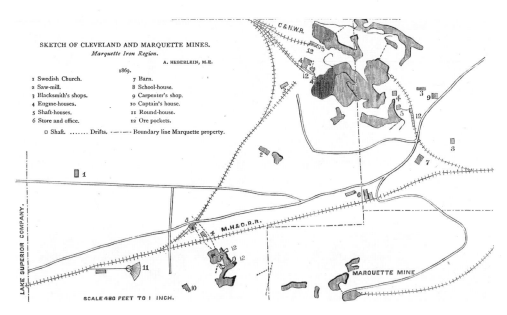

SKETCH OF CLEVELAND AND MARQUETTE MINES.
Marquette Iron Region.

A. HEBERLEIN, M.E.

1869.

1 Swedish Church.   7 Barn.
2 Saw-mill.   8 School-house.
3 Blacksmith's shops.   9 Carpenter's shop.
4 Engine-houses.   10 Captain's house.
5 Shaft-houses.   11 Round-house.
6 Store and office.   12 Ore pockets.

□ Shaft. .... Drifts. -·—·- Boundary line Marquette property.

SCALE 480 FEET TO 1 INCH.

Map of the Cleveland and Marquette mines, 1869. The Marquette mine of the Marquette Iron Company was a Civil War venture owned and operated independently by a number of the officers of the Cleveland Iron Mining Company. The Cleveland Iron Mining Company's pits are at the upper right and lower center, served by two railroads: the Marquette, Houghton, and Ontonagon and the Chicago and North Western. Note also that several of the pits have engine houses (labeled with 4s), indicating the growing use of steam engines to hoist materials out of ever-deeper mine pits. *(From T. B. Brooks et al.,* Geological Survey of Michigan: Upper Peninsula, *1869–1873, vol. 1 [New York: Julius Bien, 1873], pl. 2, facing p. 142.)*

four times each day, his wife and family had to go into the basement for safety. In 1871 the company built Mills a new house farther from the pits, with Mather instructing the company's local agent in Marquette, Jay Morse, "Make it convenient & comfortable for Mr. Mills & his family in every respect." Mills was to remain with the company until his retirement in 1879.[143]

While the Cleveland Iron Mining Company struggled with labor shortages, strikes, and assorted other matters during the Civil War, it was clear by mid-1862 that, as far as profits were concerned, the war's impact would be profoundly positive. Cleveland Iron paid its first dividend of $1.40 a share in late 1862, a sizable amount since the early investors had paid in only around $14 a share. Peter White noted in August 1863 that he thought the company would make a clear profit of $125,000 that year and that the few shares of Cleveland Iron he had for sale were priced at $30. In fact, the company paid out two dividends of $5 a share each in 1863, and regular, large dividends followed.[144] The *Lake Superior Mining Journal* reported on June 17, 1864, that Cleveland Iron had already paid out more in dividends than had been paid in to capitalize the company and that the stock was now worth six times its original cost. The company's mining operations, which had shipped an average of only 12,000 tons of ore between 1857 and 1859, averaged shipping nearly 45,000 tons annually between 1862 and 1864.[145] The profits to be made mining ore on the Marquette Range had not gone unnoticed. The high prices that iron commanded during the Civil War prompted the establishment of a host of new mining and furnace companies in the region, including the Iron Cliffs Company. In 1864 alone thirteen new mines began operations.[146]

By the close of the Civil War in 1865 the Cleveland Iron Mining Company was on firm foundations, paying regular dividends and making significant profits. Moreover, the accelerated demand for ore during the war helped transform the Lake Superior iron district from a peripheral segment of the American industrial scene to a central feature. By mid-1862 some furnace companies in Pennsylvania and Ohio reported they were "*suffering* for [want of] Lake Superior Ore."[147] In early 1863 the *Lake Superior News and Journal* commented that if double the amount of iron ore had been produced the previous year—already a record high—it would have been sold.[148] By the end of the Civil War, the demand for ore had outstripped supplies available from many small, local iron mines in western Pennsylvania, New York, Ohio, and Indiana. Iron furnaces in the Ohio Valley had become dependent, instead, on Lake Superior ores. The availability of these rich and abundant ores through steadily improving water and rail transport networks had moved the center of the American iron and steel industry from the eastern to the western side of the Allegheny Mountains, helping to guarantee the company's future prospects.[149]

Growing output did not come without a price. Wartime production had begun to exhaust the easily accessible ores of the Cleveland Mountain. The low-cost mining techniques on which the company had been founded—better described as quarrying rather than "real" mining—were no longer viable. Shifting to more technologically intensive and capital-intensive mining would now become a priority.

2

TECHNOLOGY, TRANSPORT, AND TRANSFORMATION, 1865–1891

In 1864 a journalist declared that the day was "forever past" when iron works east of the Alleghenies would furnish the West with iron. The future, he asserted, belonged to iron works west of the Alleghenies, those supplied with Lake Superior ore.[1] Indeed, this transition was well under way when he wrote. In 1860 Lake Superior mines had produced less than 5 percent of American iron ore. By 1870 the proportion had tripled to over 16 percent. By 1880 the figure was 30 percent, with Michigan on the verge of becoming the nation's leading iron ore–producing state, surpassing the traditional leader, Pennsylvania. By 1890 the Lake Superior district was producing over 50 percent of American iron ore; by 1900 it was producing 75 percent and still climbing.

During the 1870s and 1880s, as a result of the growing dependence on Lake Superior iron ores, local ores produced from small, widely scattered mines and supplied to small local furnaces ceased to be a significant factor in the American iron and steel industry.[2] Increasingly, large blast furnaces located in a handful of strategic areas that could use the country's rail and water transportation systems economically to combine high-quality coking coal and high-quality ores dominated production. By 1900 most of these furnaces were located in places, like Pittsburgh, Youngstown, Cleveland, and Chicago, that had available to them Lake Superior ores. The previously dominant iron industry of the East Coast slipped to secondary status, often having to draw on imported iron ore.

The Cleveland Iron Mining Company, one of the pioneering producers in the Lake Superior district, not only participated in the district's growth but by 1890 was one of the largest iron ore companies in the country and one of the best managed. In 1862, 1863,

Table 2-1. Employment and production at the Lake Superior iron mines, c. 1872

Company	Tons of ore produced	Approximate employment
Lake Superior Iron Co.	185,070	450
Cleveland Iron Mining Co.	151,724	375
Jackson Iron Co.	118,842	180
Champion Iron Co.	70,538	195
New York Mining Co.	68,950	270
Pittsburgh & Lake Angeline Mining Co.	61,247	?
Iron Cliffs Co.	57,545	325
Washington Iron Co.	38,841	180
20+ other mining companies	200,000	?

Source: Data compiled from Thomas B. Brooks et al., Geological Survey of Michigan: Upper Peninsula, 1869–1873 *(New York: Julius Bien, 1873), atlas, tables. Thomas Dunlap, ed.,* Wiley's American Iron Trade Manual . . . *(New York: John Wiley, 1874), 479, gives significantly higher figures for the number of men employed (600 for the Cleveland Iron Mining Co., for example). It is likely that Brooks, who worked in the Upper Peninsula, had the more accurate figures.*

and 1864, the central years of the American Civil War, Cleveland Iron had produced and shipped nearly 45,000 tons of iron ore annually, triple its output in the late 1850s. By 1872 the company had again tripled its output, producing over 150,000 tons and employing around 375 men.

To put these figures in context, the United States had 420 iron-mining firms in 1870. The average production was only around 8,000 tons annually and the average workforce around 35 men.[3] No single source provides production and employment information for individual iron ore companies nationally, but scattered data indicates that only the very largest iron ore companies in the traditional ore-mining states (Pennsylvania, New York, New Jersey, and Connecticut) or in the significant iron ore–producing western states (Missouri, Michigan) were larger than Cleveland Iron. In Pennsylvania, the traditional leader, the famous Cornwall ore banks, produced about 200,000 tons annually by the 1870s, but it had no close in-state rivals; perhaps the next-largest Pennsylvania ore miner was the Cambria Iron Company, which in 1875 dug out only around 60,000 tons.[4] In New York, the giant was the new Port Henry Iron Ore Company, which produced 270,000 tons in 1873, but the next largest New York company only mined 60,000 and the remainder much less. Lackawanna Iron and Coal, New Jersey's largest iron ore company in the mid-1860s, employed 350 but dug less than 60,000 tons per year.[5] In Connecticut, the largest company was the Brook Pit Company, which in 1872 employed 50 men and 20 horses and produced only around 25,000 tons.[6] Missouri's Iron Mountain Company of St. Louis had a labor force of around 1,200 and an output of 371,000 tons in 1872, but the handful of other Missouri companies were much smaller, unearthing 50,000 tons or less annually.[7] These figures suggest that Cleveland Iron was, perhaps, the fifth-largest iron ore producer by the early 1870s.

The Cleveland Iron Mining Company's growth, however, did not come without a price: to maintain its position it had to embrace more capital- and technology-intensive mining methods. The focus of the company's energies between 1850 and 1865 had been on establishing a viable operation in the remote wilderness of Upper Peninsula Michigan and finding a market for its ores hundreds of miles away in the river valleys of western Pennsylvania and eastern Ohio; in the decade following the conclusion of the Civil War that focus shifted. To remain competitive, CIMC now had to transform itself from a simple, labor-intensive, aboveground quarry mining company to a much more machine-intensive operation; it had to more economically and reliably ship its now-large production to furnaces that had become heavily dependent on those shipments; it had to initiate underground mining; it had to embrace difficult-to-mine ores it had previously regarded as waste; and it had to become much more technologically aggressive.

From Open Quarry to Open Pit

The key figure in the company's transition to greater use of mechanical power sources and to underground operations was not a Cleveland officer like Gordon or Mather but Frank P. Mills, the company's local mining superintendent. Born in New York in 1827, Mills worked on railroads early in his career. He came to the Upper Peninsula in 1858 as an assistant to Robert Nelson, who had just been appointed mine superintendent for the company. Mills and Nelson left Cleveland Iron in 1861 to try their luck, instead, in the local mercantile trade, operating out of the port of Marquette. After the death of his first wife in 1864, Mills returned to the community growing up near Cleveland Iron's mines. In May 1865 he took charge of the company's mining operations. During his nearly fifteen-year tenure the company's workforce grew from around 100 men to over 800. He aggressively tackled the company's growing need for ore, opening new pits and mines, installing the appropriate technology to operate them effectively, and guiding the company through the transition from surface to subground operations. Active in the community, in 1873 he became the first mayor of Ishpeming, the town organized adjacent to the Cleveland, New York, and Lake Superior company mines. He remained in charge of the company's mining operations until he left the region in 1879 to farm in Indiana.[8] He was succeeded as mining superintendent by his assistant, Don Bacon, who, in turn, was succeeded in 1886 by Frank Mills's son, Frank P. Mills Jr.

The ore deposit on which the Cleveland mine had been founded was large. Its axis ran east and west, but pitched—that is, declined—toward the west. At the higher east-

Frank P. Mills Sr. (1827–1910). Mills served as general superintendent in charge of the Cleveland Iron Mining Company's mining operations through most of the 1860s and 1870s, guiding the company's transition into mechanized pit operations and then into underground mining. He also served as the first mayor of Ishpeming when that town, adjacent to the CIMC's workings, incorporated as a city in 1873. *(From* Ishpeming Centennial, *July 25–31, 1954 [Ishpeming: Ishpeming Centennial Committee], 1954, 31.)*

Early "quarry" mining on the Marquette Range. Taken around 1860, this iconic photo from the Jackson mine, around three miles east of the Cleveland mine, illustrates quarry mining operations. Note the miners operating on the bluff in the center of the photo. Once loosened, the ore tumbled to the bottom of the cliff, where it was loaded into horse carts and carried to railcars (*bottom*) for transport to Marquette. No similar photos survive of Cleveland mining operations, but they were carried out on roughly the same lines. *(Courtesy of Cliffs Natural Resources, Inc.)*

ward end, glacial action had left ore exposed at several spots around the original Cleveland Mountain. Here "mining" had first begun from outcrops on the hill's side. Early mining on the Marquette Range was more like quarrying. Ore simply had to be drilled and blasted from the side of the hill, then broken up further by sledgehammer and loaded on railcars.

Early above-the-terrain quarrying activities had evolved by the 1860s to below-the-terrain open-pit work. As company crews exhausted supplies above ground level, they followed the thick ore vein as it began to pitch into the earth. As early as 1859 CIMC was below terrain level in some areas.[9] By mid-1863, as demand accelerated during the Civil War, the company's No. 3 mine had been compelled to go "under rock," that is, follow an inclined ore deposit underneath rock, without removing the overburden, because the overhead rock was extremely hard. This left large rock ledges overhanging giant trenches leading at an angle into the earth. While the resulting rock cap provided miners with welcome protection from storms, sun, and cold, mining now became a more expensive, dangerous, and labor-intensive operation.[10] By 1870 company crews under Mills's direction had penetrated the Cleveland Mountain from both sides, working the deposits from the sides downward following the inclined ore lens. Because the rock was hard, CIMC crews left large pillars of ore (often twenty to thirty feet thick and fifty or more feet high) standing to support the overhanging ceilings, rather than use timber supports. In 1873 an observer commented that in a few more years the company would have a mountain of rock sitting on iron ore columns, creating a "honey-combed mountain sitting upon spider-legged supports."[11]

As company crews followed ore beds downward below the water table, water removal became a problem. In April 1864 runoff from melting snow forced abandonment of work at the bottom of one pit.[12] Recognizing that continuation of mining required capital investment on a scale the company had not hitherto faced, Cleveland Iron purchased its first steam-powered pumping engine in 1865, expanding the plant with two more pumping engines in 1867.[13] That November the company's Marquette agent commented that the new Cornish pumps had to be run night and day and consumed an enormous amount of fuel, but they kept the place "nice & dry."[14] As the company's diggings went steadily downward, CIMC raced against the water problem. In early 1873 Mills commented that he was "losing sight of almost every thing" but the pumps. "Our salvation," he declared, "depends on these."[15] Indeed, a later observer estimated that two tons of water had to be pumped for every ton of ore mined.[16] The company made the investment necessary to stay in operation. In May 1867 CIMC had only one steam-powered pump in its inventory; six years later it had ten.[17] By using a portion of its profits from wartime and postwar prosperity to install capital-intensive pumping equipment, Cleveland Iron remained one of the region's top three iron producers.

Pumping was not the only area where the company had to embrace new technology. The geology of its deposit prompted it to be aggressive in adopting new drilling and

Large pillar in the Cleveland Iron Mining Company's Incline mine, 1875, left standing to support the roof as mining activities penetrated underground in the 1870s. Note figures standing in the left opening for the scale of the pillars. *(Courtesy of Cliffs Natural Resources, Inc.)*

explosives technologies. The Cleveland Mountain contained not only very hard specular hematite but a high proportion of a very hard rock commonly called jasper or jaspilite, an impure quartz with a 20–35 percent iron content. The higher proportion of hard ores and hard waste rock in Cleveland Iron's deposit than in those of other local mining companies helps explain why the Cleveland mine was among the earliest in the district to experiment with both mechanical drilling and the new nitroglycerine-based high explosives.

Perhaps the most labor-intensive process in iron mining was drilling. Because of the geology of the formation on which it was working, drilling was a particular problem for CIMC. Working as it was on one of the hardest deposits in the region, its mining costs were higher than those of its local rivals, whose deposits had less of the troublesome jaspilite.[18] Traditionally, miners drilled by hand. One miner held a steel rod, typically 1.25–1.50 inches in diameter with a 2-inch-diameter chisel bit at the end. The holder rotated the drill as one or two other miners with eight-pound sledgehammers hit the end. Typically, a team drilled nine to eleven feet of hole per day, and multiple holes had to be drilled before explosives could be inserted into the holes for blasting the rock down. This was a slow job under the best of conditions. It was even slower at the Cleveland Mountain.

Power drills had been successfully introduced in the United States in the late 1860s during construction of the Hoosac railroad tunnel in Massachusetts.[19] In these machines steam or compressed air moved pistons to which a drill bit was attached up and down,

Hand drilling at the Jackson mine, c. 1870. Until the introduction of the first reliable compressed air drills, all drilling in Michigan iron mines was carried out in this manner, with one miner holding a long drill shaft with a forged cutting bit on the tip and rotating it by hand as one or two other miners struck the end of the shaft with sledgehammers. The holes drilled by this process would be packed with explosives and ignited to break the ore from surrounding rock. No surviving photo shows this process at a Cleveland Iron Mining Company mine. *(Courtesy of Cliffs Natural Resources, Inc.)*

providing the impact and rotating motion previously imparted by hand. As early as 1859, long before mechanized drills had proven a commercial success, Robert Nelson, then Cleveland Iron's mining superintendent, had traveled to Niagara to investigate steam drills.[20] However, CIMC, as did other mining companies on Michigan's iron ranges, stayed with hand drills at this point, because the early power drills could not stand up to the very hard rock of the Marquette Range. In the nearby Michigan copper district and in American mining generally, the experience was the same. Early mechanical drills could simply not stand up to hard rock mining.[21]

Once miners drilled a number of adjacent holes, they packed them with explosives. The explosives blasted the rock to fragments, which would then be sorted and loaded on carts for transfer to railroad cars, although larger pieces might have to be drilled and blasted a second time. Black powder, the standard early explosive, was very vulnerable to moisture and frequently did not fire completely or evenly. In 1863 Alfred Nobel applied for an American patent on the use of nitroglycerine, less susceptible to water and significantly more powerful in fragmenting rock. Its successful use in liquid form on the

Hoosac Tunnel attracted attention, but liquid nitro was extremely volatile and dangerous. By 1867 Nobel had discovered that he could mix the volatile nitroglycerine liquid with inert absorbents, creating a much more stable solid explosive, soon called dynamite or "giant powder."[22] The hard rock of the Marquette iron range made nitroglycerine very attractive, even it its early liquid form. Local mining companies quickly tried it, one company as early as December 1867.[23] Local interest was sufficient for the Lake Shore Nitroglycerine Company in 1869 to erect a plant near Marquette—soon moved closer to Ishpeming—to furnish mines with the new explosive.[24] Nonetheless, nitroglycerine explosives remained on "trial" for a period, especially after an explosion killed the "nitroglycerine man" at the nearby Jackson mine, an incident that reinforced miners' initial opposition to the substance.[25] Mills, CIMC's mining superintendent, was among those initially skeptical of the new explosive.

However, Mills had changed his mind by 1870, before most of the region's other mining superintendents, convinced that the substance was reasonably safe to handle if sufficient precautions were taken and essential to efficient mining. He had good reason to be open to nitroglycerine's advantages. Cleveland ore was so hard that even when blasted it came down in large chunks, requiring further drilling and blasting to get the ore to manageable size for loading onto carts and transportation to rail sidings.[26] Not only did blasting with nitroglycerine produce smaller boulders and chunks than with traditional black power, but there was another savings. With nitroglycerine blasts, holes did not have to be bored as deeply. Mills saw that the new explosive would keep the Cleveland Iron Mining Company's operations competitive with companies with easier-to-drill deposits by reducing drilling costs. He began using it extensively, declaring by 1871 that he "would not like to be compelled to work . . . without it."[27] Dealing with harder rock, Michigan's iron ore ranges would embrace high explosives almost a decade ahead of the state's nearby copper-mining district.[28]

Once miners blasted rock lose, others sorted it, loading the good ore on wheelbarrows or horse-drawn carts for transportation to nearby storage bins by railroad sidings. Animal-powered whims, or hoists, drew the ore up inclined rail lines from the company's steadily deepening pits. By 1870 CIMC had forty-four horses, two mules, and two oxen assisting these processes.[29] But the progressively steeper and longer inclines up which ore had to be pulled by animal-powered whims as mining pits and inclines went deeper encouraged further mechanization.

While Michigan iron mines adopted high explosives earlier than did copper mines, they were distinctly slower in adopting steam power for pumping and hoisting. With deposits that could be exploited with quarry techniques and with investors with shallower pockets than those backing many of the Keweenaw copper mines, good reasons existed for steam power's slower adoption. By the time steam-powered pumps and hoists first appeared at the Cleveland Iron Mining Company's works, Michigan's deep-shaft copper mines had already installed impressive steam-power plants.[30]

Cleveland mine, No. 3 pit, 1879, following a rock collapse. The inclined ramps were used to hoist ore from the workings in the bottom of the pit to the surface for dumping into railcars. Early on the skips (or small railcars) that operated on the inclines would have been hoisted by horse whims; by this time steam-powered drawing engines were in use. *(Courtesy of the Marquette County Historical Museum, Marquette.)*

Nonetheless, once the company's mines began to shift from quarrying ore from bluffs to operating in pits below the terrain, the adoption of steam power occurred rapidly. Recognizing that the steam engines he had installed for pumping could be adapted to hoisting, Mills typically used them for both operations beginning with the company's first pumping engine in 1865.[31] In ensuing years, as its operations went belowground and began to face the same problems in pumping and hoisting that Michigan copper miners—underground right from the beginning—had faced a decade or two earlier, Cleveland Iron steadily expanded its steam capacity. The two 50–80 hp engines purchased and installed in 1867 worked splendidly as both pumps and hoists. Jay Malone, then the company's local business agent, reported that with these engines the Cleveland Iron Mining Company could finally pull ore out of its mines faster than it was able to mine it.[32] By 1867 the company's hoisting installation was probably the most advanced in the district. Mills counterbalanced the skips (ore cars) so that as the hoisting engine lifted a skip full of ore, an empty one descended. This reduced hoisting costs and speeded loading. At the

Engine house at Cleveland Incline mine, 1879. Ore cars from the Marquette, Houghton, and Ontonagon Railroad that carried the ore to Marquette stand in the foreground. On the left edge of the photo note the inclined ramps going down into the open pit for hoisting skips to storage compartments located above the rail lines. The company's No. 4 engine house, which contained the steam engine that pulled skips up the ramps, is on the right. *(Courtesy of the Marquette County Historical Museum, Marquette.)*

top, he designed the rails and skips so that the skips dumped their loads automatically into storage pockets located above rail sidings, reducing handling costs.[33] The expansion of the company's steam hoisting capacity paralleled in rapidity the expansion of its pumping plants. By 1869 CIMC had begun to reduce its herd of workhorses.[34] By 1873 it had also added a small locomotive to assist in moving loaded railcars on its premises.[35]

Struggling with Transportation

While Cleveland Iron's mining operations improved steadily in output and efficiency under Mills in the decade after the Civil War, transportation issues became progressively more troublesome to the company's Cleveland-based officers. They recognized that efficient bulk transportation was central to the profitability of any iron-mining operation, since iron ore was a low-cost, bulk commodity. For an iron-mining operation located hundreds of miles from most of the blast furnaces that used its ores, the problem was even more critical.

Although the opening in 1857 of Ely's railroad between the mines and the port of Marquette alleviated the most serious land transportation woes, they did not completely disappear. The company's Marquette agent constantly feuded with the rail line over rates and the availability of ore cars. The arrival in early 1865 of a second railroad, the Peninsula Railroad, affiliated with the Chicago and North Western, reduced these problems. Running from Escanaba on Lake Michigan northward sixty miles into the iron ore district, the Peninsula Railroad not only forced Ely's railroad to lower rates but offered CIMC an alternative route and an alternative port for getting its ores out, especially late in the season, when Lake Superior began to freeze over but when open water still existed at Escanaba to the south.

Water transportation, however, was a bigger problem than land transportation. In the 1860s and early 1870s lake transportation comprised around 40 percent of the company's ore costs, more than mining itself.[36] The primary function of the company's secretary—Samuel L. Mather until 1869—was chartering vessels and coordinating waterborne ore shipments with the company's Marquette business agent. CIMC early handled lake transport by contracting with independent vessel or fleet owners to carry ores from Marquette or, after 1865, from Escanaba to lower Great Lakes ports. Vessel owners received a set price per ton for individual loads.

From the beginning, Lake Superior iron-mining companies had serious problems coordinating ore deliveries on the docks at Marquette or Escanaba with vessel arrivals. If ore was not waiting when a contracted vessel put into port, the financial consequences were potentially serious. Vessels under contract and not loaded within a specified time were entitled to bill the ore companies demurrage (loading delay charges). Especially in the company's early years, this happened frequently, and the cumulative costs were serious enough that the Cleveland office regularly warned its Upper Peninsula agents to have sufficient ore at dock to avoid such charges.[37] Delayed vessels and coordination problems also had potentially broader repercussions: loss of customers. In 1860 George W. Tifft, owner of the Buffalo Furnace and Machine Shop, responded to loading delays in Marquette by telling Tuttle, then acting as the company's Marquette agent, that he did not care to send vessels for ore "and be 8 days in loading there. The iron business is *too poor* to take the chances on that. . . . There is no money in iron made from LS [Lake Superior] ore at $3 per ton . . . and if I did not have my own coal would not buy it at all."[38]

The causes of the coordination problem were inherent in the system. Before telegraph lines reached the region in 1865, delays in communications between Cleveland and Marquette were long. Mather, in Cleveland, wrote to his Marquette agent J. J. St. Clair on June 21, 1854, noting that he had "just this moment" received St. Clair's letter of June 10, sent out eleven days earlier.[39] Because of the time delay, problems at the mines could interrupt production at the same time that Mather, ignorant of these developments, continued to charter vessels to pick up ore that would not be there. Making matters worse, the railroads serving the mines did not always provide sufficient ore cars to meet the capacity of incoming vessels.[40] Other times, delays due to weather conditions caused the mix of sail and steam vessels serving the area to arrive "in a heap." In June 1860 Cleveland Iron's representative in Marquette reported that a "calm on the lake" had resulted in a "fleet of vessels" arriving almost simultaneously: eight in twenty-four hours.[41] Such mishaps caused delays in loading and demurrage charges. Sometimes Cleveland officials got too enthusiastic about low vessel rates and chartered too many vessels. In August 1866 Mills complained that the company was "a bit to [sic] fast in taking orders." He was doing all he could possibly do to mine sufficient ore, he asserted, but six vessels waited for ore at Marquette and six at Escanaba, meaning demurrage charges were likely.[42] Such coordination problems created tensions between the company's Cleveland chartering agents and its Upper Peninsula mine superintendents and business agents.[43]

While the difficulties of coordinating ore delivery with shipping contractors were serious enough, the agendas of independent boat or fleet owners complicated matters further. Independent boat owners on short notice shifted to more lucrative and desirable shipping options, leaving Cleveland Iron high and dry. In July 1860, for instance, the owner of *Ironsides* informed Mather that grain shipments were paying so well that he did not care to make any new arrangements to carry ores to Cleveland.[44] In addition to paying more, grain shipments, of course, were much cleaner than dirty, dusty iron ore, providing a further excuse. During the Civil War, when the need for shipping was high, the grain trade once again lured independent lake vessels away from the dirtier and more dangerous ore trade.

A steady supply of ore was a very serious issue for the blast-furnace operators who were Cleveland Iron's primary customers. Furnaces were normally run night and day for months until they had to be shut down for repair—usually relining. They were fed, during this process, a continuous stream of fuel, ore, and limestone. Shutting down a blast furnace for lack of ore and having to restart it was a huge problem, one that furnacemen avoided at all costs. Restarting a blast furnace and getting it back up to steady output was a process that took, at minimum, several days, and it could take several weeks before the furnace settled into normal operation. In 1862 one of Cleveland Iron's customers—a furnace owner who had previously contracted on his own with vessel operators to purchase ore in Marquette—commented that it was "impossible to get vessels to go for ore" and that he would, in consequence, rather pay $4 per ton for other ores than risk having to blow out his furnaces for want of Lake Superior ores.[45] The unreliability of independent vessel operators cropped up again late in the 1865 and 1872 shipping seasons. In both instances demand for ore was high, Cleveland Iron could sell all it could get down the lake, but vessels had shifted to the more lucrative and cleaner grain trade. Mather wrote his agent in 1872: "I am a good deal exercised about it [the lack of vessels]."[46]

Beginning in 1867, in an attempt to get some control over water transport, CIMC

The *George Sherman*, the Cleveland Iron Mining Company's first venture into owner-ship of a vessel to transport ore, c. 1870. CIMC had a half interest in the *Sherman* from 1867 to 1872. The *Sherman* was typical of the nonspecialized vessels converted from the general merchandise trade to iron ore transport in the period from the 1850s to c. 1880. *(Courtesy of Cliffs Natural Resources, Inc.)*

flirted with moving away from its focus on producing iron ores by integrating verti-cally into lake shipping. It purchased half interest in the 550-ton capacity barque *George Sherman.*[47] Five years later Cleveland Iron officials sold their interest in the *George Sher-man,* shifting their attention to a more ambitious project: the Cleveland Transportation Company, a firm created specifically to build and operate vessels for transporting Lake Superior iron ores.[48]

The formation of Cleveland Transportation in the fall of 1872 reinforced the compa-ny's emerging practice of seeking partners for capital-intensive projects. Leading officials of the Cleveland Iron Mining Company provided a bit over 50 percent of the capital for the new company. They worked with another group of Cleveland businessmen headed by Marcus Hanna, who served as general manager of the new company because he had more experience in lake shipping. By 1874 Cleveland Transportation had four steamers (capacity 1,000 tons each) devoted to the ore trade. These each towed a schooner-barge (capacity 1,100 tons each), doubling the steamers' carrying capability. Cleveland Trans-portation often held its annual meetings in the office of the Cleveland Iron Mining Com-pany, and Cleveland Transportation from 1873 to the late 1880s was the leading contrac-tor for transporting Cleveland Iron Mining Company ores.[49] In the midst of problems securing sufficient vessels to move ore in the boom year of 1872, Mather rejoiced that his company would, at last, have its own "steam barges" next season.[50]

By 1873 the Cleveland Iron Mining Company had enjoyed well over a decade of profitable operations. The future seemed rosy. The company had not faced a significant labor dispute since 1866. It had aggressively and successfully responded to the technological challenges of mining in progressively deeper pits, increasing production from 40,000 to over 150,000 tons annually while keeping its mining costs under control.[51] With the creation of the Cleveland Transportation Company, it also seemed to have its water transport problems under control. In 1874 Thomas Dunlap, editor of *Wiley's Iron Trade Manual,* noted: "Having their own ore docks at Marquette . . . and their own vessels and transports," the Cleveland Iron Mining Company is "enabled to handle this vast product [150,000 tons] with perfect facility."[52] In addition, the demand for the company's iron ores was growing rapidly, as the center of the nation's iron industry continued to shift from east to west of the Alleghenies to tap into Lake Superior iron ores. As one observer noted in 1873, Pittsburgh furnaces (on the west side of the Alleghenies) were "universally" using Lake Superior ores, without which they could not have matched the quality of imported iron.[53]

The Panic of 1873

Early in 1873, however, disquieting notes began to sound. By this time a half dozen large iron ore–mining firms, including Cleveland Iron, dominated the Lake Superior iron district, supplying hundreds of independently operated blast furnaces in the lower Great Lake regions. As these furnaces become increasingly dependent on Lake Superior ores, mining firms like the Cleveland Iron Mining Company enjoyed a superior bargaining position. In January frustrated furnace operators in Pennsylvania, in the face of a decline in demand for their product, met to protest the high ore prices still demanded by Cleveland and St. Louis iron ore houses.[54] In May Mather, who regularly met with the company's customers, ominously reported that the market for pig iron was so dull that he was concerned that his contracted sales for the coming season, which had been at high prices, would not stick and that furnace operators were looking for any excuse to break contracts.[55]

On September 18, 1873, the nationally important banking house of Jay Cooke closed its doors, largely due to overexpansion and overspeculation in railroads, setting off a major financial panic. Other banks soon floundered. Nearly a quarter of the nation's railroads went bankrupt, and within a few years the unemployment rate was well into double digits.[56] Investors pulled back, terminating planned railroad and construction projects. The demand for iron ore, already tenuous, dried up. Many of Cleveland Iron's customers, unable to sell pig iron and resentful of past high ore prices, defaulted on payments; others quit ordering iron ore.

On September 22, a mere four days after the panic began, Mather wrote to his Marquette agent Jay Morse that it was "impossible to get any currency here now—our banks have all shut down—You must manage in some way. . . . We are passing through a severe crisis & many must go under." By early October the Cleveland Iron Mining Company was itself in crisis mode. Fearing it, too, might be among those to go under, Mather wrote Morse: "Put off payments for hay, feed, coal, nitro glycerin [sic], powder etc. . . . We will provide pay for all labor as that cant [sic] wait—but other debts must wait awhile." Not a dollar, he noted, was coming into company coffers. A few days later he told Morse: "We must shut down somewhere & unless there is a change soon . . . we may have to stop mining everywhere. . . . I tell you it is fearful." On October 25, a month after the panic began, Mather confessed: "I am nearly worn out with care & anxiety." In November he wrote: "You have no idea what a crisis we are passing through. . . . It will be years before before [sic] we recover."[57]

Matters only got worse. In 1874 several important furnace operations that owed the company money failed.[58] Moreover, furnaces that had once readily bought all the ores they could get their hands on now became particular. The company's laxity in quality control in the years before 1873, when any ore sold, now came home to roost. Several furnaces refused to pay for ores already delivered, complaining that they contained impurities, leading Mather to note in April 1874, "[W]e are generally in bad odor all around." In May he commented that Cleveland ore had lost "its reputation" and that now even his best friends were going elsewhere to get their ore, even though "[t]hey don't like to tell me the reason." To cap matters off, one furnace simply referred to bad ores generically as "Cleveland" ores. Mather noted in July 1874: "I work harder to sell 500 tons this year that I used to in selling 100,000 tons."[59] As the shipping season of 1874 approached an end, Mather was glad things were closing down. The company needed no more ore. What it had was not selling; heavy payments were due for freight. In November, as a means of reducing costs, the company considered giving away the horses on which it had spent thousands of dollars to avoid the cost of feeding them though the winter. As to whether the company could pull through the winter, Mather declared: "All is uncertainty."[60]

As Mather struggled in Cleveland to keep the company afloat, the situation on the mining ranges deteriorated. By the end of October 1873, Mills and Morse had reduced the labor force by a quarter, lowered wages, and still did not have sufficient funds to pay workers. The company resorted to paying them, as in the Panic of 1857, with a form of "iron money": sixty-day IOUs. Mather instructed Morse to tell the company's miners that if they objected to payment in this form, there was no alternative but to discharge them and close the mines.[61] Miners reluctantly accepted the IOUs, but their acceptance merely pushed problems further up the food chain. Peter White, a member of the company's board of directors but also a prominent local banker, complained in December that merchants, who had accepted the iron money as tender for goods, were now pressing him with the iron money certificates for redemption. "I am almost crazed," he wrote, "one more week will drive me mad. *I cannot stand* it longer than that without relief."[62]

With layoffs and reduced wages, Morse predicted in late 1873 that the local poorhouses would be full before spring.[63] Unable to support themselves and with the region's unforgiving winter approaching, miners left the region in "droves."[64] The region's blast furnaces, one by one, began to shut down. By fall 1875 only nine of twenty-one Upper Peninsula blast furnaces remained in operation.[65] The number of mines shipping ore dropped from forty to twenty-eight between 1873 and 1876.[66] In August 1874 Marquette's *Mining Journal* editorialized: "All we can do is to continue whistling to keep up our courage."[67]

If the depression were not enough to threaten Cleveland Iron, two additional factors complicated its struggles: the revival of labor problems and the company's recent solution to its transportation problems. Just as wage cuts prompted by the rapid drop in demand for iron ore following the Civil War had led to labor disturbances, wage cuts in 1873 and 1874 ended the labor peace enjoyed by CIMC since 1866. In July 1874 miners at the Lake Superior mine, led by immigrant Swedes, now the largest ethnic group in the region, walked off their jobs, discontent that wages had dropped from $2.25–$2.50 per day to $1.30–$1.50 in the past year. The group moved on to six area mines, including the Cleveland, demanding that they shut down or that management restore a portion of the wage cut. At several mine locations, notably at the Iron Cliffs Company's Barnum mine, strikers stoned nonstriking miners out of the mine pits and forcibly shut mines down.

Threats by some strikers to gut stores and businesses panicked the local sheriff, who appealed to Michigan's governor for the state's militia. Struggling to keep the company afloat, Mather pondered shutting the mines down for the rest of the year in response to the strike, but refrained because the company's vessel contracts tied it into shipping ore. At one point he told his agent to "pay them [striking miners] off and discharge the whole set." But then, recognizing that the local agent probably knew better how to handle the situation, he backed away: "I won't advise you what to do . . . be governed by circumstances and whatever you do, we will back you in it."[68] Eventually Michigan's governor dispatched two militia companies from Detroit. These units, reinforced by a newly created Marquette militia company, protected mine properties and arrested half a dozen strikers on riot charges. Relatively quickly order returned. Within two weeks the strike had collapsed. The Swedes who had led the strike left the area "in squads, by every train."[69]

Paralleling the revival of labor problems, CIMC discovered that its recent solution to shipping uncertainties—its joint venture with Marcus Hanna and his associates—was, in times of depression, a burden rather than a help. When the Cleveland Transportation Company came on line in 1872 with a fleet of vessels dedicated to the iron ore trade and majority-owned by Cleveland Iron management, CIMC signed a long-term contract. Now, Cleveland Iron had to meet its contractual obligations to the other partners in

Cleveland Transportation. Mather soon recognized that this had serious implications. In March 1874, with few orders for iron ore coming in, he declared, as he would again a few months later during the miners' strike: "Were it not for our vessel contracts—I should say, shut down the mine and wait for better days."[70] With contractual obligations to ship thousands of tons of ore using Cleveland Transportation vessels, shutting down the mines—even in the face of the labor problems, shortage of cash for payrolls, and the absence of any demand for ore—was an option that could be taken only as a very last resort. In April 1874 Mather complained: "We are in for it with our vessels & I must give them employment. . . . Produce all the ore you can with the least expense, as we must have ore for our vessels."[71] To reduce the total annual expense of transporting iron that the company could not sell, Mather secretly hoped for a short shipping season and even suggested that his Marquette agent go slowly in loading vessels without appearing to do it on purpose.[72] Four years later, as the depression dragged on, Mather commented: "Our great misfortune is, that we have vessel contracts—as it is, our ore *must* come down and it *must* be sold."[73]

To make matters worse, because Cleveland Iron was tied by contract to the Cleveland Transportation Company, it was unable to take advantage of the rock-bottom shipping rates available from independent shippers desperate for any cargo during the depression years. This gave other ore companies—without captive fleets—the advantage in securing cheap transportation.[74] In 1876 Mather observed that vessels could not operate at a profit charging less than about $1.40 or $1.50 a ton for transporting ore from Marquette to Cleveland. Yet, so desperate were vessel owners that some accepted charters at as low as $1.10 a ton. Mather wrote Morse: "It is a hard thing to refuse vessels at $1.10." He eventually gave in and chartered several to bring down ore that the company did not really need.[75]

Discovering the downside of having a captive fleet of ore vessels during the depression that followed the Panic of 1873 led CIMC to back away from what would have been the logical next step: outright purchase of the Cleveland Transportation Company and formal vertical integration into lake shipping. Short of cash and struggling to stay afloat, Cleveland Iron once again—as it had following the fire at its forge in 1853 and in selling its horse-drawn railroad in the Panic of 1857—backed away from diversification to focus on its mining operations. Specialization rather than diversification in times of crisis was becoming a company tradition.

Changing Conditions: New Ranges, New Problems, New Ores

For five years after 1873 iron and iron ore prices dropped. Weaker mining companies collapsed. Survivors like Cleveland Iron had to reinvent themselves in a new environment that included permanently lower ore prices, higher operating costs, and more particular customers with more demanding standards for ores. In 1873 grade-one hard specular ore sold for as high as $12.50 to $15 a ton at Cleveland. By 1878 the same ore sold for $5.50

a ton.[76] These declining ore prices coincided with rising operating costs as the exhaustion of deposits easily accessible from the surface pushed Marquette Range mines toward more expensive and complicated shaft-mining operations. Finally, changes in the iron and steel industry, notably the emergence of Bessemer steel, which replaced both cast and wrought iron as the focus of the industry and required ores low in phosphorus, made iron ore customers more particular and meant that securing previously neglected and harder-to-mine low-phosphorus soft hematite ores was a necessity for iron ore companies.

In 1877 the first significant shipments of iron ore from a new Upper Peninsula iron range, the Menominee, began to hit the market. The influx of these new ores kept prices from making a quick recovery from the panic and ensured that they would never again reach their 1873 levels. Deposits of ore had been noticed north of the Menominee River, the border between Wisconsin and the Upper Peninsula, as early as the late 1840s, but development awaited effective transportation. The deposits lay forty to fifty miles from navigable water instead of the twelve to fifteen miles that separated the early Marquette Range mines from Marquette. Only in the mid-1870s, after long delays due to the Panic of 1873, did the Chicago and North Western Railroad push a branch westward from Escanaba into the new district. Unfortunately for Marquette Range ore producers, the new Menominee Range ores reinforced the impact of the 1873 depression and simply drove prices further downward. In January 1878 Mather complained that if rumors were true that 60,000 to 70,000 tons of Menominee Range ores would enter the market next year, "it will break the price & all will be losers." In August the secretary of the Lake Superior Mining Company suggested to Morse that the new range could "trouble us for some time to come." "New discoveries," he added, "in the hand of irresponsible people who will work for a year or two on comparatively cheap ore or give it away to introduce it, are the worst class of competition."[77] Making matters worse, the new mines on the Menominee Range had significant competitive advantages. Menominee Range ores were more friable and, not having previously been tapped, lay closer to the surface, making them cheaper to mine. The region had a milder climate, lay nearer to markets, and had a transportation advantage in the longer shipping season of its port, Escanaba, on Lake Michigan.[78]

When yet another new iron range, the Gogebic, opened in the western Upper Peninsula and adjacent Wisconsin in 1885, ore prices received another push downward. William G. Mather, Samuel's son, who had entered the company and begun to take over some of his father's responsibilities in the 1880s, commented that the large influx of low phosphorus ores from the Gogebic region expected in 1886 would make company sales more difficult.[79] The first of the Minnesota iron ranges, the Vermilion, began shipping ore the same year as the Gogebic, with similar expected results.[80]

As ore prices dropped in the 1870s and 1880s, other forces drove the cost of mining up. In 1873 only one mine in the Marquette area operated entirely underground, but by 1880, for new operations and for many of the old ones as well, underground mines had

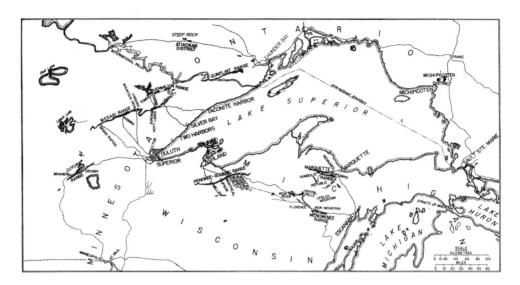

Map of the Lake Superior iron ore district indicating individual iron ore ranges in Michigan, Wisconsin, Minnesota, and Canada. The Menominee Range is located almost directly south of the Marquette Range. Escanaba, on Lake Michigan, was its chief shipping port. *(From Wm. T. Lankford Jr. et al, eds.,* The Making, Shaping and Treating of Steel, *10th ed. [Pittsburgh: Association of Iron and Steel Engineers, 1985], 259. Courtesy of the Association of Iron and Steel Technology.)*

become the rule rather than the exception.[81] From the 1850s to the mid-1870s, initiating mining operations on the Marquette Range had required only modest capital. Ore was discovered in large, aboveground outcrops. Only slowly did operations move to open pits, and even then companies made every attempt to keep the workings open to daylight, even as they angled into the earth, often creating wide, gigantic, sloping caverns, referred to as stopes. This approach avoided or reduced costs normally associated with underground mining: sinking shafts, drifting, timbering, and use of skilled labor.[82]

The cheapness of early mining on the Marquette Range, when combined with the high price of ores, had contributed to inefficiency in mining operations, such as the continued use of horses and carts to move ore to railroad sidings even when long hauls and steep grades called for mechanization. Company directors and shipping agents pushed mine superintendents to produce ore to take advantage of high prices, often neglecting planning and long-term improvements. As geologist and mining engineer T. B. Brooks noted in 1873: "Mine owners did not . . . want surveys, nor machinery, nor tunnels, nor anything that had reference to the future; they only wanted ore, nor did they care much what it cost, nor what the quality was: it was ore, ore, ore!"[83] By the late 1870s, with prices declining, competition from the Menominee threatening, and near-surface deposits approaching exhaustion, surveys, machinery, tunnels, planning for the future, cost controls, and concern for quality had to become more of a way of life.

Making these elements more essential, regional mining companies also had to reconsider an ore they had hitherto ignored: soft hematite. Soft hematite did not usually outcrop on the surface like the hard specular hematite first discovered and long the major product of the Marquette Range; it occurred underground. It was more costly to mine not only because of its location but because it was a much softer rock. All passages through it required extensive support, unlike in hard-ore mines, where only occasional pillars had to be left to support the overhang. In the 1850s and 1860s mine operators often regarded soft hematite as waste. Its iron content was lower than hard specular hematite. Because of its more friable nature, it was not only more expensive to mine but much more messy to store and transport, leaving a red dust over everything in the vicinity. Blast-furnace operators feared that it would clog their furnaces.[84] In 1871 Mather noted that "literally no demand" existed for soft hematite ore. "No one will take it if he can get any other kind of L.S. [Lake Superior] ore."[85] When ore demand was high, Mather and his peers at other regional ore companies often insisted customers take some soft hematite along with the hard specular hematite they really wanted as a condition of the transaction.[86] Charcoal blast furnaces in the Upper Peninsula, however, took soft hematite because they could get it cheaply from local mines, which were glad to have any customers for the stuff. But regional furnacemen soon discovered that, when mixed with hard specular hematite, the soft hematite produced a very high-quality pig iron and was less hard on furnace linings.

In the late 1870s the long-maligned soft hematite came into its own. In the 1850s Henry Bessemer in Britain and William Kelly in the United States independently discovered that tons of molten pig iron from blast furnaces could be converted directly into steel in less than an hour by blowing air through it in large "converters." Previously steel had been almost a semiprecious metal because it could be fabricated only in very small batches of a few pounds using processes that took days. Initial problems with iron ore selection, however, slowed adoption of the Bessemer method, but by 1870 the keys to successful operation had been worked out. One key was using only low-phosphorus (less than 0.1 percent and preferably lower) pig iron. In the mid-1870s steel began rapidly to replace wrought iron as the material of choice for plates and rails, and Andrew Carnegie, among others, erected large mills to produce Bessemer steel.[87]

Producing low-phosphorus pig iron for use in Bessemer converters, however, required that blast-furnace operators use low-phosphorus iron ores (soon to be called "Bessemer-grade ores"). Unfortunately, many of the hard ores of the Marquette Range were non-Bessemer ores (that is, they contained a bit too much phosphorus). Ironically, many of the range's long-neglected soft hematites were Bessemer-grade (low-phosphorous) ores. The early ore discoveries on the Menominee and Gogebic ranges were soft, low-phosphorus ores as well. When blast-furnace operators on the lower Great Lakes discovered that they could mix Michigan hard ores (with their slightly high-phosphorus content) with low-phosphorus Michigan soft hematite ores and simultaneously eliminate concerns about furnace clogs *and* produce high-quality, Bessemer-grade pig, the hematite ores suddenly came into heavy demand.[88] Furnace operators could now produce the low-phosphorus pig iron needed as a feed for Bessemer converters.

The rise in demand for low-phosphorus hematite ores for producing Bessemer-grade pig was extremely rapid. In 1871, as noted above, Mather had commented that there was literally no demand for hematite ores. By 1875 the situation had changed dramatically. Mather now noted that the "great demand at present is for ore that will make Bessemer Iron." One of his customers had commented that if he (the customer) could not make Bessemer iron he would blow out, since "no other irons will sell now."[89]

Reinventing the Company

Thus, in the aftermath of the Panic of 1873, the Cleveland Iron Mining Company faced a whole new world, one characterized by declining iron ore prices and increased operating expenses brought on by the need to mine underground and mine soft, low-phosphorus hematite ores. Mather quickly recognized the necessity of cutting costs. In 1875 he praised Jay Morse, his agent in Marquette, for looking "so carefully" at the cost of mining. If costs could not be cut, he noted, "we shall be obliged to shut down."[90] Later Mather wrote: "I am afraid Ore will soon be where Pig Iron is, & only they that can mine cheaply & have light expenses will be able to keep alive."[91] In the mid-1870s, in response to the rapidly changing environment in the iron and steel industry, the company began the process of reinventing itself, undertaking four major transitions almost simultaneously: shifting from surface to underground mining, developing the soft hematite ores it had previously neglected, embracing advanced mining technologies, and focusing systematically on cost reduction.

Cleveland Iron's first attempt at underground mining occurred in 1866, about the time that the first of the region's completely underground mines began operation.[92] By 1876 at least one of CIMC's mines, the School House, was an underground operation.[93] That year Morse reported that the company's transition from open-pit to shaft mining had enjoyed "a much greater degree of success than we had expected . . . due to the fact that we were perfectly well aware that we knew nothing of that method of mining, and consequently went to work with great caution and care."[94] For a few years, open-pit mining continued to be the dominant mode of operations, but severe rock falls in the company's inclined pits in 1879 accelerated the transition.[95] This shift, of course, entailed additional expenses, especially for timbering. A local newspaper commented in October 1879 that the volume of timber disappearing daily into the Cleveland mine was "astounding."[96]

After commencing underground operations, CIMC quickly learned how to deal with the soft hematite deposits on its lands. It mined some hematite as part of existing hard-ore operations. There the soft hematite tended to appear under the hard ore, closer to the footwall of the deposit. Other soft-ore deposits, geologically created by slow leaching of the area's iron-bearing formations, were independent of hard-ore deposits. The company opened these as independent soft-ore mines. This was the case with the Cleveland Hematite mine, opened a mile north of the company's hard-ore mine in 1881. As Bessemer-grade soft hematite grew steadily more valuable in the 1880s, the company began to mix it with its traditional non-Bessemer hard ores in order to create a Bessemer-grade ore mix.[97] Thus, ironically, within a decade the company had switched from using its hard ores to force sales of its soft ores to the reverse.

Soft hematite deposits tended to come in high chimneys, or vertically situated, lens-shaped deposits, located some distance underground and requiring huge amounts of timbering to support the hanging wall, or roof, as miners removed the soft ore. In the early 1880s CIMC personnel introduced a new mining technique called sublevel caving in its Cleveland Hematite mine. In this system miners removed the ore from the top of the lens, instead of the bottom, allowing the overlaying burden of soil and rock to cave in a controlled manner as the mine went deeper, rather than supporting it with timber. This, of course, left a widening chasm on the surface, not then regarded as a problem in the sparsely populated region. The caving technique sharply reduced the amount of costly-to-install timbering required in the mine since it did not require shoring up an underground cavity that steadily grew in size as ore was removed from the top down. Safer and more economical in the use of timber, the technique spread from the Cleveland Hematite to other soft-ore mines in the district and eventually to other Lake Superior iron ore districts as well.[98]

Cleveland Iron simultaneously sought to reduce mining costs through new laborsaving technology. As surface-level deposits disappeared and as demand rose for soft hematites, which were usually some distance beneath the surface, one of the critical elements of local mining became exploration. Digging shafts and tunnels through barren rock in a blind search for new ores was expensive. The key new technology was the diamond drill, invented in 1863 in Europe and patented in the United States shortly thereafter. In a diamond drill cutting edges studded with industrial diamonds—cheaper than jewel-quality diamonds but still expensive—were set in the end of a cylinder linked to a drill rod. Rotated by machine, the diamond-studded cylinder would slowly cut through rock, leaving a rock core in the cylinder to be brought to the surface for inspection. The diamonds needed for the cutting bit were expensive, but, as one contemporary noted, the harder the rock the greater the savings in labor the diamond drill delivered.[99] The 1869 annual report of Cleveland Iron suggests the company was awaiting delivery of a diamond drill to determine the size of the deposit at its water-plagued School House mine.[100] By 1870 several mines in the region were using the device.[101]

The diamond drill came into regular use in the aftermath of the Panic of 1873, as Cleveland Iron began to go underground and had to look more systematically for ore deposits, especially the low-phosphorus, soft hematite ores now in heavy demand. In 1877 CIMC, concerned about depletion, contracted with an independent drill operator,

Diamond-drill outfit at work on a Michigan iron range. Capable of producing a core sample of the ground through which it drilled, the diamond drill was introduced onto the Marquette Range in the 1870s and had become a regular fixture by 1880. *(Courtesy of Superior View, Marquette.)*

hoping to find additional ore lodes around its School House mine. In December Mather wrote Morse that the stockholders had heard so much about the diamond drill and "the expense thus far attending it, that more than a usual interest is felt in its success."[102] The following year the company bought a diamond drill of its own.[103] It soon became the core of the company's exploratory operations both on the surface and beneath it.[104]

Once ore deposits had been identified by diamond drills, they had to be reached by drilling and blasting operations. As noted previously, Cleveland ore had the reputation of being so hard that it required "more labor in mining than at any mine in the district."[105] For this reason, as early as 1859 the company had considered steam-powered rock drills.[106] But early mechanized rock drills were expensive and unreliable, so the company turned, instead, to nitroglycerine to reduce mining costs. In 1870 Cleveland Iron's drilling was still done entirely by hand. The transition to underground mining changed this. Around 1870 the nearby Lake Superior mine began to experiment with compressed-air-powered rock drills, after proof of their success in the late 1860s in boring the Hoosac Tunnel. Other mines in the district soon followed suit, with varied success.[107] In both the Marquette mining district and the nearby Keweenaw copper district, the weight and unreliability of the early drills soon quenched early enthusiasm. As late as 1883, the superintendent of the Michigamme Company complained that his mine had used seventeen Rand drills and never had one that worked sufficiently.[108]

Frank Mills, Cleveland Iron's mine superintendent, began to take a new look at mechanized rock drills in 1875–76 as he contemplated more underground work, but he was initially disappointed, probably because of the frequency of mechanical failure but possibly because of some resistance from the miners, which was certainly the case in the nearby copper district.[109] Convinced, nonetheless, that mechanized drilling was essential to reducing mining costs, CIMC continued to test them, often working closely with drill manufacturers, eventually finding one that would work. A representative of the Rand Drill Company would comment in 1880 to the company's Marquette agent: "[Y]ou had many severe trials before you got a drill that gave any thing at all like a satisfactory result."[110] In 1879 CIMC began regularly to use Rand compressed-air drills.[111] By 1880 the company had five at its Incline mine with little, if any, drilling done there by hand. They quickly became so essential that, according to one observer, that mine's managers would almost prefer to forego the use of explosives as be without the Rand drill.[112] CIMC's success in finding, finally, a reliable drill paralleled the experience in Michigan's copper region. That district's two largest companies—Quincy and Calumet & Hecla— also embraced the Rand between 1878 and 1880.[113]

In 1871 it had cost the company around $100 a foot to drive a drift, or tunnel; by the 1890s the introduction of high explosives and, especially, the rock drill had reduced that cost to $16 a foot.[114] The growing use of compressed-air drills reduced mining costs, but it was not without a negative side. Hand drilling was a skilled art. As the company switched to mechanized drilling, hand drillers left the company for other locales. When three Rand drills broke at once in 1889, the company could not find any men to take the place of the machines while they were being repaired.[115]

Cleveland Iron was among the earlier companies in the Lake Superior iron district to adopt both diamond drills and mechanized rock drills. The same can be said of electric lighting, something that penetrated the mining industry generally only very slowly. A mining engineer in the mid-1890s would comment that "little [had been] done in the way of introducing the electric light into mines."[116] The Cleveland Iron Mining Company, however, installed its first electric lighting system in 1880. Charles Brush, a Clevelander, introduced one of the first commercially successful arc lighting systems in the United States in 1878. Hoping to find a new market for the system, his company in 1879 approached three leading iron ore companies, including Cleveland Iron, about giving the Brush arc light a trial under conditions as "*dark & black* as possible."[117] In 1880 the Brush Company installed one of the first electric lighting plants located at any mine in the country at CIMC operations. The system used sixteen arc lights to illuminate several open-pit workings as well as some underground locations.[118] CIMC found the Brush arc

Rand rock drill as depicted in *Scientific American*, Dec. 25, 1880, being used to sink a shaft in hard rock. By this date the Cleveland Iron Mining Company was increasingly shifting to drills of this type and turning away from traditional hand drilling.

lights worked well for lighting the company's open-pit mines, its ore stockpiles, and other surface operations, such as ore sorting. It found them especially valuable during the short days of winter. Parallel attempts to use them belowground were less successful. The arc lamps proved "too bulky and heavy for practical use," were too frequently injured by the impact of underground blasts, and too often short-circuited due to the damp environment.[119] After permanently adopting arc lighting in 1880 for its surface and open-pit workings, through not underground, CIMC remained committed to electrification. By 1890 the electric lighting system at the Cleveland mine was called "perhaps the most complete" of that of any of the regional companies. By then it consisted of twenty-four arc and fifty-six incandescent lights and was used in the company's shops, offices, shaft houses, and ore pockets.[120] The telephone made its first public appearance in 1876; by mid-1881 CIMC had linked its hematite mine with the company's main office in Ishpeming by telephone.[121]

To reduce mining costs further, CIMC also sought to improve material handling. Local mining journalist A. P. Swineford in 1880 characterized the company's power plant as "the neatest and most substantial, if not the most elaborate, . . . to be found in the iron region." It consisted of twelve boilers that supplied steam to nine engines that operated multiple air compressors and fifteen hoisting drums.[122] To reduce the costs of materials handling once ore reached the surface, Cleveland Iron began looking at steam shovels after company officials saw one in operation on the Menominee Range. One mining expert estimated that the steam shovel multiplied the efficiency of stockpile labor on the surface by six times and reduced the cost of loading ore on railroad cars by half.[123] By 1889 Cleveland Iron had purchased one.[124]

The company's efforts to cut costs and increase production went beyond the adoption of mechanized-drilling and material-handling equipment and the installation of electric lighting. Company officials encouraged mining superintendents to visit other mining districts for cost-cutting ideas. In 1881 Don Bacon toured coal mines in the east, and in 1890 Frank Mills Jr. visited mines on the Menominee Range.[125] Company mine superintendents tested the comparative quality of coals being used to power the steam engines at mining operations and then compared coal to petroleum as a boiler fuel.[126] Not all cost-cutting measures worked, of course. Attempts to operate a blacksmith shop underground to cut time in sharpening and repairing tools failed, since the shop filled the mine with so much gas that the men got sick.[127]

As mining became more technologically sophisticated, management in Cleveland conceded even more autonomy and authority to its Upper Peninsula managers. Three examples suffice as illustrations. In 1876 Mather wrote his agent Morse in Marquette that he would leave the matter of testing and purchasing mechanized drills "entirely" up to him: "Whenever you are satisfied, & wish to use them in our mines, you have full authority to buy such & as many as can be used to advantage." A few years later Mather wrote Morse regarding lubricants: "You know better than I do what we need & I wish you to judge for yourself & I shall be satisfied." Finally, with regard to labor, Mather told Morse in 1881: "You can . . . judge better than I, what to do with *our* men."[128]

Succession

By the early 1880s Samuel L. Mather had been a keystone in the Cleveland Iron Mining Company's edifice for three decades, serving as secretary and treasurer from 1853 and as president and treasurer since 1869. Well into his sixties and with his health beginning to fail, he now had to worry about who would succeed him.

Initially his oldest son—born in 1851 and also named Samuel ("Sam")—seemed the obvious choice. Sam accompanied his father on trips to the mines in his teens. At age eighteen, in the summer of 1869, after completing preparatory school in Massachusetts with the intention of enrolling in Harvard in the fall, he took a summer job as a time-keeper at the Cleveland mine to learn more of the business. Interested in the mining operations themselves, he suffered a serious injury from a rock fall following an explosion. He spent the next two years convalescing locally, reporting to his father on mining operations to make himself useful.[129] He never attended Harvard. In 1873 Sam resumed

LEFT: Samuel "Sam" Mather (1851–1931), 1895. The eldest son of Samuel L. Mather and his first wife (Georgiana Woolson), Sam Mather began working in Cleveland Iron Mining Company operations in 1869 and continued working with his father until 1883, when he left CIMC to form an iron-marketing firm—Pickands, Mather and Company—with two friends. Pickands Mather would become a major player in the industry by the twentieth century and would eventually be purchased by Cleveland-Cliffs in 1987. *(Courtesy of Cliffs Natural Resources, Inc.)*

RIGHT: Colonel James Pickands (1839–96), friend of Sam Mather, brother-in-law of Jay C. Morse (both married daughters of one of CIMC's original stockholders, John C. Outhwaite), and the third principal of Pickands, Mather and Company. *(From Samuel P. Orth,* A History of Cleveland, Ohio: Biographical *[Chicago and Cleveland: S. J. Clarke, 1910], 2:591.)*

work in his father's firm, this time in the Cleveland office. In 1881 he married an Ohio woman, Flora Stone, daughter of wealthy Cleveland industrialist Amasa Stone, and in 1883 he left his father's firm to form Pickands, Mather and Company with James Pickands and Cleveland Iron's able and long-serving Marquette business agent Jay Morse.[130]

Born in 1839, son of a Presbyterian minister and his wife, James Pickands was the oldest of the three partners. After moving to Cleveland from Akron, he had enlisted in the first year of the Civil War and served through the bulk of the conflict, eventually attaining the rank of colonel. In 1867, after a brief stint working for an iron ore merchant firm in Cleveland, Colonel Pickands moved to Marquette, where he opened a hardware

Jay C. Morse (1838–1906), shipping agent for the Cleveland Iron Mining Company in Marquette from the 1860s through the early 1880s and active in other regional mining activities, some in conjunction with other CIMC officers. In 1883 he left the company to form Pickands, Mather and Company with Sam Mather and Colonel James Pickands. *(Courtesy of the Western Reserve Historical Society, Cleveland.)*

store to serve the region's iron-mining companies. In 1870 he entered the coal business as well, supplying that product both to local mining companies, including Cleveland Iron, and to homes and businesses in Marquette. He achieved sufficient local prominence to win election as mayor of Marquette in 1875. When his wife of twelve years suddenly died in 1882, Pickands left Marquette and moved back to Cleveland.[131]

Jay C. Morse, the second partner and also a Civil War veteran, had been employed after discharge from the army by the Cleveland Iron Mining Company as its general agent or business manager in Marquette and Ishpeming. When Pickands moved to Marquette, the two had quickly become close friends. Morse was a partner in Pickands's hardware business and later a sales agent for his brother Henry Pickands, who managed several local charcoal furnaces. Morse invested widely in iron ore properties, participating in several local enterprises with fellow Cleveland Iron officers, including the Marquette Iron Company, the McComber Iron Company, and the Bancroft Iron Company. Morse and Pickands's ties became even closer when Morse, like Pickands before him, married a daughter of John Outhwaite, one of the Cleveland Iron Mining Company's original officers and stockholders, and when the two formed a company to develop the Boston mine and Taylor mine on the western extension of the Marquette Range in 1880.[132]

In 1882, when Sam Mather approached Colonel Pickands and Jay Morse about starting a new iron ore partnership, Pickands, who had just lost his wife, was ready to start a new life. Morse was too. After more than a decade and a half of working as an agent for others, primarily the Cleveland Iron Mining Company, he was ready to strike out on his own, as was Sam Mather. In 1883 the three founded the firm of Pickands, Mather and Company as dealers in pig iron and iron ore. The new organization was more than simply an agency that sold iron and ore on commission for others, although that was its primary function. Morse and Pickands brought to the firm the two small mines on the western end of the Marquette iron range in which they had secured ownership. The new firm would soon acquire part or complete ownership of other mining properties through leases, including a large one—the Colby mine—on the new Gogebic Range. While his father's firm focused on its Marquette Range operations, Sam Mather and his partners' firm rapidly expanded in multiple directions, investing heavily in the late 1880s and early 1890s in the newer Michigan and Minnesota iron ranges (the Gogebic, Menominee, Vermilion, and Mesabi), in coal mines, in vessels to transport ore, and in furnaces and rolling mills in Ohio and Illinois, as well as taking on sales agencies for other companies.

Sam Mather's departure to form Pickands Mather eliminated Samuel L. Mather's most obvious successor. But Sam's departure was apparently amicable. The new company, in fact, conducted business in its first few months from the offices of the Cleveland Iron Mining Company.[133] Even after Pickands Mather became one of Cleveland Iron's major competitors, relations remained good between father and son, as well as between Sam Mather and his half brother William G. Mather. From 1890 Sam Mather would serve regularly on the board of CIMC's successor, the Cleveland-Cliffs Iron Mining Company.

With Sam Mather's departure from Cleveland Iron, his half brother William G. Mather became heir apparent to Samuel L. Mather's position. Samuel L. Mather's first wife, Georgiana Pomeroy Woolson (m. 1850), mother of Sam Mather, died in 1853. Three years later Samuel L. Mather married Elisabeth Lucy Gwinn. They had a single child—William G. Mather—in 1858. Like his older half brother, William G. Mather attended preparatory school in the East. He then matriculated at and in 1877 graduated from Trinity College in Hartford, Connecticut, with numerous honors. In 1878 he joined the Cleveland Iron Mining Company as a clerk, working initially under his older brother Sam's tutelage.

William G. Mather was more of a scholar than his brother. He would become a major collector of the published works of American Puritans (his family forebearers) and a collector of both rare books and rare art. He would be deeply involved in charitable and church affairs, to the point of being called the "leading layman" of the Episcopalian Church in America. He was charming, gentlemanly, and cultured, traveling widely in Europe. But he shared with his father and brother a devotion to hard work and the iron ore business. He would not marry until 1929, at age seventy-one. From 1878 until 1929 the Cleveland Iron Mining Company and its successor the Cleveland-Cliffs Iron Company served as surrogates for wife and family.[134]

By the mid-1880s Samuel L. Mather was suffering from rheumatism and deteriorating eyesight.[135] As his father's health worsened, William assumed progressively more responsibilities. In 1886, for example, Samuel wrote to Don Bacon, then serving as the company's agent, that he wished he could go to the mines, but couldn't. "[M]ust leave that to Will. He has done well here during my absence & has sold as much ore as we dare

William Gwinn Mather (1857–1951), c. 1885. The second son of Samuel L. Mather and half brother of Samuel (Sam) Mather of Pickands, Mather and Company, William G. began working for the Cleveland Iron Mining Company in 1878, becoming his father's heir apparent by the mid-1880s. *(Courtesy of the Western Reserve Historical Society, Cleveland.)*

to sell at present."[136] One of the early areas in which William G. Mather began to play an important role was in handling the company's growing transportation needs during the boom years of the late 1880s.

Integrating into Transportation

On land the Cleveland Iron Mining Company had long had problems with the two railroads serving its mines, in terms of both price and service, and these problems continued into the 1880s.[137] Thus, late in that decade, the company seriously considered constructing a railroad of its own from Ishpeming to Marquette bay. In keeping with previous practice, the company sought to leverage its capital by finding a partner or partners for a joint venture. In 1888 and 1889 it secretly began talks with the Pittsburgh & Lake Angeline Mining Company, the Champion Iron Mining Company, and outside capitalists to establish the line. However, the outside parties demanded more guaranteed business for the projected line than Cleveland Iron and associated mining companies could provide; the project went into hibernation.[138]

At almost the same time CIMC began to reconsider integrating vertically into waterborne transportation. By the late 1880s the company faced pressure from two directions. In the mid-1880s two new iron ranges began shipping iron ore: the Gogebic in the western Upper Peninsula and adjacent Wisconsin and the Vermilion in Minnesota. Because the ores from these new iron ranges were initially closer to the surface and easier to extract than the long-exploited deposits on the Marquette Range, they contributed to keeping ore prices low, while simultaneously driving transportation prices higher by creating a shortfall in shipping capacity. William Mather, already playing an increased role in the company, wrote to the company's Marquette agent Don Bacon in November 1885 noting that the large influx of ores from the Gogebic would make sales more difficult the coming year. In April 1886 he commented: "[T]he competition from these cheaply mined ores is heavy for us & we must find some way to meet it."[139]

By 1886 the Cleveland Iron Mining Company was suffering increased difficulty getting vessels, and when it did freight costs were high. Samuel L. Mather commented that vessel men "are getting crazy" and freights "getting wild."[140] Later that year William, who had begun to take responsibility for chartering vessels, complained that vessel men were "rampant" and that the outlook for the following year, as far as lake freights went, was grim. The new Gogebic Range anticipated shipping as much as 1.25 million tons of ore, and demand for grain was likely to be high, pulling vessels out of the ore trade.[141] Several other ore companies, experiencing the same frustrations, had purchased their own ore carriers. One was the Republic Iron Company. In 1886 William Mather attended the Republic Iron Company's stockholders' meeting and found that Republic was making a profit on its ores *and* on its vessels.[142] After turning away from direct control over vessels in the aftermath of the Panic of 1873, in 1888 the company reversed course. After several years of problems securing shipping at reasonable rates, CIMC directors decided to contract for two steam vessels, even though it involved mortgaging the company's mining properties.[143] The company's two new, carefully designed, state-of-the-art ore carriers, named *Frontenac* and *Pontiac,* the first steel (as opposed to iron) ore freighters on the Great Lakes, could carry ore from Marquette to Cleveland for $1.22 a ton; the older and now-obsolete Cleveland Transportation Company vessels they replaced charged $3.50 a ton.[144] By 1890 the company's new vessels owned records for cargo size and speed of transit.[145]

The most important step that the Cleveland Iron Mining Company took not only to survive but to remain competitive in the new era of lower ore prices, however, was not the adoption of new mining machinery and techniques or the purchase of a fleet of vessels to lower transportation costs. It was an acquisition. In 1890 the Cleveland Iron Mining Company acquired controlling interest in the Iron Cliffs Company, a leading fixture on the Marquette Range since 1864.

Iron Cliffs: Organization and Early Years

The Iron Cliffs Company had its origins in attempts to construct a railroad into the Marquette ore district from the south. On June 3, 1856, Congress awarded land grants to several proposed railroad lines, including one that would start at Little Bay de Noquet, modern Escanaba, on Lake Michigan, and run directly north to the iron district and on to Marquette.[146] Although ore shipped on this line had to traverse sixty to sixty-five miles, rather than the twelve to fifteen miles to Marquette on the Ely railroad, the port on

The SS *Pontiac*, one of the modern specialized steel vessels constructed for transporting iron ore by the Cleveland Iron Mining Company in 1888–89. *(Courtesy of Historical Collections of the Great Lakes, Bowling Green State University, Bowling Green, OH.)*

Lake Michigan linked to the new line would be closer to lower Great Lakes ore customers and free from ice for several weeks longer than Marquette in both fall and spring, thus extending the shipping season. Chicago entrepreneurs, headed by William B. Ogden, saw the land grant as a major opportunity and began to plan extending their railroad—the Chicago, St. Paul and Fond du Lac Railroad—through Wisconsin toward the iron districts of the Upper Peninsula.[147]

The Panic of 1857 and opposition from the existing Ely railroad interests set plans back. Ogden's railroad went bankrupt in 1859. Ogden's lawyer, Samuel J. Tilden of New York, was appointed receiver and soon took control of the line, now renamed the Chicago and North Western Railroad.[148] Tilden, a wealthy New York investor, was soon to become governor of New York and eventually the nominee of the Democratic Party for president of the United States, losing to Rutherford B. Hayes by a single electoral vote in the hotly disputed 1876 election. In February 1862, a group affiliated with Tilden's Chicago and North Western Railroad, including Charles Harvey (builder of the canal at Sault Ste. Marie), William Ogden, and Perry Smith, chartered a new company—the Peninsula Railroad—to build into the iron ore region from the south.[149] This company inherited the old land grant and moved rapidly because of Civil War demand for iron ore, building the new port of Escanaba on Lake Michigan with a modern ore dock while pushing rails northward. They reached the iron ore mines of Negaunee and Ishpeming in December 1864 and began shipping ore early the following year.[150]

To guarantee that the new road would have something to ship once it reached the

Samuel J. Tilden (1814–86), founder and first president of the Iron Cliffs Company. Tilden also controlled the New York mine, which operated on forty acres immediately adjacent to the Cleveland Iron Mining Company's early workings. *(Courtesy of the Western Reserve Historical Society, Cleveland.)*

ore fields, some of the railroad's investors sought interest in ore mines. Tilden himself organized the New York Iron Mining Company on March 14, 1863, as majority stockholder.[151] Even more significant, he and several other Chicago and North Western investors purchased 38,000 acres of mineral land in Marquette County from the St. Mary's Mineral Land Company, inheritors of the land grant that had funded construction of the Sault canal. With this purchase they formed the Iron Cliffs Company in September 1864.[152] The company quickly opened two mines: the Ogden and the Tilden.[153]

Iron Cliffs early on took a path contrary to that being pursued by the Cleveland Iron Mining Company. Perhaps because the ores in its first two mines turned out to be of lower than normal grade or because pig iron was selling at nearly record-high levels in 1866, Iron Cliffs turned from its original intent of shipping iron ore over the Peninsula Railroad and declared that it would focus, instead, "chiefly [on] making pig iron, instead of selling ore."[154] In pursuit of this objective Iron Cliffs negotiated at first to lease and then to purchase the Pioneer Iron Company, which had established the region's first successful charcoal-fueled blast furnace in 1858 in Negaunee, but had since floundered.[155] It also began to plan a completely new set of charcoal-fueled blast furnaces south of Negaunee near its new Foster mine.[156] Iron Cliffs diversified further by establishing a sawmill, brickyard, store, and small foundry in conjunction with its charcoal works, mines, and furnaces.[157] This was a modestly aggressive agenda for a new company, but the stockholders went ahead, rejecting a local director's suggestion of using a partnership to ease the capital burden of opening a company store. Board policy was "to have no partnerships in any department of their business," quite the opposite of the policy pursued by Cleveland Iron.[158]

Iron Cliffs' decisions initially appeared to be paying off. In 1866 its profits from making iron exceeded $15,000, while it made less than $5,500 on ore sales.[159] The store, likewise, did well, sales hitting $200,000 in 1870, generating $20,000 in profits.[160] The store initially had no serious rival, and the company's miners often had to trade with the

The Pioneer furnace of the Iron Cliffs in Negaunee as it appeared in the 1880s. The two-stack, charcoal-fueled Pioneer furnace, whose first stack was completed in 1858, was the first blast furnace to operate in the iron-mining district of Lake Superior. Purchased by Iron Cliffs in 1866, it burned in 1877, but was rebuilt. The furnace was the heart of the company's operations. *(Courtesy of Cliffs Natural Resources, Inc.)*

store in lieu of pay during the winter months when Iron Cliffs, like many other regional companies, was short of cash.[161]

Iron Cliffs, however, made serious errors in judgment early on. The Pioneer furnaces, acquired in the purchase of the Pioneer Iron Company, were in worse condition than expected, and locally "not one man in the companies [*sic*] employ knew anything about iron making."[162] The decision to build a new furnace (called Cliffs) at the Foster mine was also a disaster. In 1867 Iron Cliffs began excavation and had almost completed the plant before suspending construction for winter. The suspension lasted until 1873 as the company hesitated, perhaps because of problems securing sufficient charcoal even for its Pioneer furnaces.[163] Iron Cliffs finally completed the new Cliffs furnace in 1874, just in time for the collapse of the iron trade following the panic of 1873, operated it off and on for several years, and then abandoned it.[164]

Iron Cliffs' large and diversified fixed-capital investments in widely scattered operations—land, sawmill, brickyard, mines, foundry, charcoal kilns, and blast furnaces—meant it was undermanned and undercapitalized. Its managers initially had to spread their attention "over a hundred structures from a log house to a furnace scattered over fifty square miles."[165] Problems between headquarters in New York and the company's Upper Peninsula managers began early. T. B. Brooks, the first of its managers, a Civil War veteran and a respected local geologist, wrote the company's executive committee in December 1866 complaining that the committee's instruction had come so late in the season that it was impossible to get the elaborate estimates for building repairs it desired, that high late-season prices had to be paid for supplies, and that by the time work began "every competent mechanic" in the district was fully occupied, resulting in higher prices for both labor and materials.[166] In July 1868 Brooks told Charles Canda, Iron Cliffs' secretary, that the reason Iron Cliffs had made less pig iron than other companies the previous year was that he had followed "the spirit of the instructions" given him on charcoal purchases by the New York office. These instructions, he asserted, "had their origin in an entire misconception of the then future of the iron market . . . and . . . a total misunderstanding of the nature of coal making here."[167] Problems between the New York office and Upper Peninsula agents remained endemic. In early 1877, for instance, T. J. Houston, Brooks's successor, had to respond to accusations by Canda of fraud and intent to deceive in his accounting.[168]

Besides its furnace operations, the company also made errors in its mining operations. Its first two mining operations—the Tilden and Ogden mines—required substantial funds to open, yet produced ores "hardly up to shipping standards."[169] The third mine it opened, the Gilmore in 1868, was a total flop.[170]

Iron Cliffs: The Barnum Years

One of the key figures in Iron Cliffs was William H. Barnum, a leading iron manufacturer from western Connecticut who served extended periods in the U.S. House of Representatives, the U.S. Senate, and as Democratic national chairman.[171] His experience in charcoal iron production may have been the critical factor in Iron Cliffs' turn from shipping ore to producing charcoal iron. It was to Barnum that Canda and the other directors often turned for guidance, but Barnum was deeply involved in Democratic politics and often unavailable in the early years of the company's operation. In late 1868 Canda commented to E. B. Isham, then working as the company's agent, that he hoped Barnum would become Iron Cliffs' president so that when advice was needed "we will have the *right* to call for it and not wait upon it till all political meetings and conventions are over."[172]

Canda got his wish. In 1869 Barnum replaced Tilden, deeply involved in New York state politics by this time, as president of Iron Cliffs. Barnum would direct the company until 1886. But problems remained. Barnum tried to manage affairs from afar, insisting on long, detailed reports from local agents and complaining that previous reports had not lived up to his expectations.[173]

In his first few years Barnum enjoyed some success, partly as a result of a rising market for pig iron.[174] Iron Cliffs was probably on the verge of paying dividends when the Panic of 1873 hit. By November Iron Cliffs was in serious trouble, and the company's employees were becoming "clamorous" because they had not received pay since August.[175] By 1876 the company had begun to cut back on all operations.[176] Stockholders would not see the company's first dividend, a small one of 2 percent, until early 1879, nearly a decade and a half after the company's formation.[177]

In some ways, Barnum's knowledge of charcoal iron making under eastern conditions may have hurt rather than helped the company. It made the other directors deferential toward him even when that was probably not appropriate and even when his absorption in political activities made it advisable to find someone more devoted to the enterprise. Barnum was the dominant figure in Iron Cliffs and insisted that Upper Peninsula agents do things his way. In 1875, for instance, Barnum, concerned about getting the phosphorus levels in Pioneer pig iron down to Bessemer-grade levels, interfered with new company agent T. J. Houston's attempts to reduce production costs by using a proportion of cheaper soft hematite ore in the Pioneer furnace, something that Houston had found prolonged furnace linings and would, in fact, have helped get phosphorus levels down. Barnum ordered Houston to quit and to use strictly hard ores and no hematites.[178] In 1878 the dispute over the proportion of local hematites to use in the furnace flared up again.[179] Both Houston and the company's furnace manager complained that the company's eastern management simply did not understand the specific difficulties of operating furnaces in the Lake Superior region.[180]

By the 1880s the local view was that Iron Cliffs was "badly handled from the main office," that the local agents were top quality but not permitted to "exercise their judgment in the management of affairs."[181] The accumulating discontent led to Barnum being eased out as president in 1886, but he remained a major shareholder and important figure in company operations.

Iron Cliffs and the Cleveland Iron Mining Company

While Iron Cliffs struggled, the Cleveland Iron Mining Company prospered. By the early 1880s Samuel L. Mather's company had acquired the reputation of being one of the best-managed companies in the Lake Superior mineral district. In 1883, for instance, Michigan's commissioner of mineral statistics commented that Cleveland Iron's affairs "have always been consistently and ably managed."[182] In 1889 Charles D. Lawton, then the commissioner of mineral statistics, noted that no other iron-mining company "more completely systematized" details and no other mining location was kept "neater and in better trim than the Cleveland."[183]

About the time Barnum lost control of Iron Cliffs, the generation of major shareholders who had founded that company began to die off. Tilden died in 1886. Edmund Miller, a large stockholder and prominent director, died in 1887. Charles A. Rapallo, another major stockholder, died shortly after. Their replacements were less committed to the company's traditions. In 1887 Iron Cliffs' board closed the twenty-year-old company store. When reports suggested some of the company's mines were approaching exhaustion and explorations to find cheaper, readily accessible ore deposits failed, the board opened the company estate, by now over 50,000 acres, to exploration. This initiative yielded nothing, and in April 1889 Iron Cliffs' board appointed a committee to consider the condition of the company and "the possibilities of selling the shares."[184] This was the opportunity for which Samuel L. Mather had been waiting.

It is not completely clear what prompted Mather to go after Iron Cliffs. Fear of the exhaustion of Cleveland Iron's own properties was certainly a major factor. CIMC owned only about 2,000 acres of mineral land. In the early 1880s correspondence between

Don H. Bacon (185?–1922), assistant to Frank
P. Mills as Cleveland Iron Mining Company's
Upper Peninsula mine superintendent beginning in 1872. Bacon succeeded Mills as mine
superintendent in 1879 and directed the
company's operations until 1886, especially its
transition to an underground mining company.
Bacon later headed the Minnesota Iron Company and the Tennessee Coal and Iron Company. *(Courtesy of Cliffs Natural Resources, Inc.)*

Mather and his Upper Peninsula agent Don H. Bacon suggests growing concern about
exhaustion. In March 1882, for example, as output from CIMC's No. 3 mine declined,
Mather wrote: "Is it possible . . . that that mine is going to play out entirely! Why, it
would be dreadful, & makes me sweat even to think of it." He told Bacon to keep the
company's diamond drill going: "We must not holdup in any direction. Can't afford
to."[185] By 1886 Mather was particularly concerned about the company having insufficient
low-phosphorus, soft hematite (Bessemer-grade) ores, commenting to Bacon: "Bessemer
ores are all the rage. . . . My great anxiety is about our Hematite." He instructed Bacon
to push exploration for hematites at all costs, even if it meant no dividends.[186] In 1889
Mather wrote Morgan Hewitt in Marquette: "Our ore is becoming more of a second
class ore & sells slowly & at a lower price."[187] Mather may also have been concerned
because even though his company's ore output had increased by 60 to 70 percent since
1870, its share of Lake Superior ore production had declined from a bit over 15 percent
in 1870 to under 10 percent by 1885, partly the result of the opening of the Menominee,
Gogebic, and Vermilion iron ranges. In 1872 the Cleveland Iron Mining Company was
probably the fifth-largest iron ore company in the country in terms of production. The
comprehensive study of the iron ore resources of the United States conducted as part
of the 1880 census indicates that the Cleveland Iron Mining Company had dropped
to ninth.[188] Although no comparable data was gathered for the 1890 census, CIMC's
ranking certainly had not improved, for in 1880 the Menominee Range had just begun
to ship large tonnages, and the Gogebic Range in Michigan and Wisconsin and the Vermilion in Minnesota did not began shipping large tonnages until the mid-1880s. New
companies opened these ranges; Cleveland Iron had no holdings in the 1880s on any
ranges save the Marquette.

Jeptha H. Wade Sr. (1811–90). Initially an itinerant portrait painter, in the 1840s Wade became a major contractor for telegraph line construction and eventually became president of the Western Union Telegraph Company. Wade joined the board of the Cleveland Iron Mining Company in 1867, became a major shareholder in Iron Cliffs in the late 1880s, and was a decisive figure in organizing the syndicate that acquired control of Iron Cliffs for CIMC in 1889–90. *(Courtesy of the Western Reserve Historical Society, Cleveland.)*

With fears of ore exhaustion growing and the company's relative position in the Lake Superior ore district in decline, Samuel L. Mather began vigorously to pursue acquisition of new mineral properties as a survival mechanism. William G. Mather wrote Bacon in April 1886: "Father and I want to get our directors interested in pushing the company to the front. . . . We have got the richest board of directors in this city . . . , and I think we can get their backing in any promising scheme."[189] A few months later Samuel asked Bacon to keep his eye out for properties to buy or lease, prophetically adding: "The time may come when we may be able to pick up some good ore property at little cash."[190] Among the properties Mather had in mind was certainly Iron Cliffs. Iron Cliffs had mines near the Cleveland Iron properties in Ishpeming, proven ore underneath its newly opened "A" and "B" shafts (the new Cliffs Shaft mine), and over 50,000 acres, much of it in the Marquette Range mineral belt. Its acquisition would relieve fears of exhaustion and make the Cleveland Iron Mining Company the dominant firm on the Marquette Range.

When Samuel Tilden, the founder of Iron Cliffs, died in August 1886 after a long illness, Mather immediately began to investigate acquiring the company. In October he wrote Bacon: "I wish we could control . . . Iron Cliffs, we are pegging away on it but don't meet with any success so far."[191] In November Mather traveled to New York and visited the executor of the Tilden and Ogden estates. He had no luck. The executors relied on William Barnum for direction. "He is the Law & Gospel to them on all Mining matters," Mather complained, "& they will do nothing without consulting him." Barnum apparently opposed the sale.[192]

The Iron Cliffs' board's decision on April 15, 1889, to investigate the possible sale of the company, coupled with Barnum's death two weeks later, changed matters entirely. Jeptha Wade, a director of the Cleveland Iron Mining Company since 1867 who had made his fortune building and then presiding over the Western Union Telegraph Company, headed a syndicate of Cleveland investors whose goal was acquisition of controlling interest in Iron Cliffs, apparently (and secretly), for Cleveland Iron. Samuel L. Mather, William G. Mather, Selah Chamberlain, and others associated with Cleveland Iron participated in the syndicate. By February 1890, to the surprise of outside observers, the syndicate had secured over 14,000 of the 20,000 outstanding shares of Iron Cliffs, and at Iron Cliffs' April 1890 meeting the syndicate took control, placing Cleveland Iron Mining Company officers in charge of Iron Cliffs.[193]

Before the Cleveland Iron Mining Company could digest its acquisition, Samuel L. Mather and Jeptha Wade died. In May 1891 their successors formed the Cleveland-Cliffs Iron Mining Company as a holding company controlling majority stock in both Cleveland Iron and Iron Cliffs, adopting this organization because of restrictions on property ownership and capitalization under existing Michigan law.[194] Only in the 1910s would the Cleveland Iron Mining Company and Iron Cliffs pass out of paper existence.

The acquisition of Iron Cliffs by CIMC stockholders and the subsequent formation of Cleveland-Cliffs put the Cleveland Iron Mining Company, in its new form, in a solid position as the largest holder of mineral acreage and known ore reserves on the Marquette iron range. As the table below demonstrates, it also became the largest ore producer on the range.

Table 2-2. Leading companies on the Marquette iron range, 1890 and 1892

Rank/Company, 1890	Shipments in 1,000s of tons, 1890	Rank/Company, 1892	Shipments in 1,000s of tons, 1892
Queen Group	388	Cleveland-Cliffs Iron Co.c	590
Cleveland Iron Mining Co.a	338	Queen Group	380
Lake Superior Iron Co.	318	Lake Superior Iron Co.	367
Iron Cliffs Groupb	297	Pittsburgh & Lake Angeline Mining Co.	288
Pittsburgh & Lake Angeline Mining Co.	262	Republic Iron Co.	168
Total range shipments, all companies	3,001	Total range shipments, all companies	2,665

Source: Calculated from Lake Superior Iron Ore Association, Lake Superior Iron Ores, 1938 (Cleveland: Lake Superior Iron Ore Association, 1938), 130–31.
aIncludes Cleveland Lake and Moro mines.
bIncludes Barnum, Cliffs Shaft, Foster, and Salisbury mines.
cIncludes Cleveland Lake, Moro, Barnum, Cliffs Shaft, Foster, and Salisbury mines.

Because Cleveland Iron and Iron Cliffs properties were contiguous, the same management team could operate both, saving administrative costs. The larger organization could undertake more capital-intensive mining and negotiate lower costs for transportation and supplies.[195] The new company—Cleveland-Cliffs—would need all of these advantages and more in the decade to come, for, in the words of one near-contemporary author, "[A] storm was gathering over the Lake Superior ranges compared to which all previous ones were but summer breezes, and the region's companies were about to be subjected to a test before which all but the very strongest were to go down to ruin."[196]

3

CRISES, DIVERSIFICATION, AND INTEGRATION, 1891–1930

On October 8, 1890, Samuel Livingston Mather died in Cleveland. On the day of his funeral—the 11th—over 500 miles away on the Marquette iron range, "all work of every kind" ceased on Cleveland Iron Mining Company properties "out of respect to his memory."[1] Mather had served as an officer in the company from 1853 and as president from 1869 to 1890. He had become the heart and soul of the corporation, devoting most of his energies from the mid-1850s onward to overseeing the company's agents and mining superintendents, to chartering vessels, to finding purchasers for its ores, and to directing its expansion. The succession was seamless. His second son, William G. Mather, assumed the helm of first the Cleveland Iron Mining Company and then Cleveland-Cliffs when the holding company was formed to combine Cleveland Iron Mining Company and Iron Cliffs in May 1891. Like his father, William G. Mather would dedicate his life to the company: between them they devoted nearly a century of leadership, giving Cleveland Iron and Cleveland-Cliffs a continuity that set it apart from virtually every other iron ore company in the country.

One of William G. Mather's earliest undertakings was construction of a cottage in Ishpeming, built on a hill overlooking the southeast side of the city near the soon-to-be-abandoned Moro mine. Designed by local architects Charlton, Gilbert and Demar of Marquette and landscaped by Warren Manning, a nationally known landscape architect, the cottage became Mather's headquarters on his very frequent visits to the company's mining operations. With additions in 1903, the cottage came to be used increasingly for business purposes, including meetings between Mather and local mine mangers and meetings of Cleveland-Cliffs' board of directors when the board visited the region.[2]

Cliffs Cottage, c. 1900, on the hill in the center background, served as William G. Mather's headquarters during his visits to the company's primary mining district. The local mining agent's residence is the two-story house in the center foreground. *(Courtesy of Cliffs Natural Resources, Inc.)*

Cleveland-Cliffs' board of directors posed in front of Cliffs Cottage, 1902. *Back row, left to right:* Peter White, Austin Farrell, William G. Mather, H. R. Harris, and F. A. Morse. *Front row, left to right:* M. M. Duncan, R. C. Mann, Samuel Redfern, and J. H. Sheadle. *(Courtesy of Cliffs Natural Resources, Inc.)*

In 1891, with the absorption of Iron Cliffs, William Mather and his company had every reason to look forward to a prosperous future. Immediate concerns about ore reserves had been put to rest. The discovery of ore under Lake Angeline and the acquisition of Iron Cliffs had halted and then reversed the declining relative position of the company in the Lake Superior iron ore district (see table below).

Table 3-1. Cleveland Iron Mining Co./Cleveland-Cliffs Iron Co. ore shipments, 1870–1910

Year	Tons of iron ore shipped by all Lake Superior mines	Tons of ore shipped by CIMC/CCI	% of Lake Superior shipments by CIMC/CCI
1870	856,000	133,000	15.5
1875	899,000	130,000	14.5
1880	1,965,000	213,000	10.8
1885	2,468,000	219,000	8.9
1890	9,011,000	688,000	7.6
1895	10,441,000	899,000	8.6
1900	19,169,000	1,993,587	10.4
1905	34,575,000	2,539,374	7.3
1910	43,630,000	2,669,491	6.2

Sources: Lake Superior ore shipments are from Lake Superior Iron Ore Association, Lake Superior Iron Ores, 1938 (Cleveland: Lake Superior Iron Ore Association, 1938), 308. For CIMC/CCI shipments, the table uses Cleveland-Cliffs Iron Co., The Cleveland-Cliffs Iron Company: An Historical Review of This Company's Development and Resources Issued in Commemoration of Its Seventieth Anniversary, 1850–1920 (Cleveland: Cleveland-Cliffs, 1920), 30, which reports shipments "from the Company's properties."
Note: The post-1890 figures for CCI ore may be a bit high because some of the ore shipped from company properties was owned by partners. Moreover, the figures sometimes do not quite mesh with figures published in company annual reports.

Little did Mather and Cleveland-Cliffs realize that they would very shortly need all their resources and abilities simply to stay afloat, for beginning roughly three years after Samuel L. Mather's death, the company was nearly swamped by a succession of crises: a financial panic in 1893 followed by a long depression, growing labor militancy, the opening of the Mesabi Range and resulting oversupply of iron ore, an invasion of ore mining by steel companies, and dramatically lower ore prices.

This series of events radically altered the nature of the American iron ore industry. The succession of crises between 1893 and 1903 killed many iron-mining companies and left most ore production in the hands of steel and ore merchant firms. When the dust settled, Cleveland-Cliffs had survived. In fact, it was now the most important independent iron ore–mining company remaining in the Lake Superior region, but its relative importance in the iron-mining world—something that had been enhanced by the 1891 absorption of Iron Cliffs—was once again on the decline and the playing field remaining for the few surviving independent ore firms had shrunk sharply.

The Panic of 1893

The first of the crises hit Cleveland-Cliffs in the spring of 1893. Early that year ominous clouds hung over the iron ore industry as talks between iron ore producers and iron ore consumers deadlocked over prices.[3] A few months later the gold reserves backing the national currency dropped. The resulting panic sent the nation's economy into a tailspin. The Philadelphia and Reading Railroad, then the nation's largest, and the National Cordage Company, then the largest publicly traded stock, declared bankruptcy. Panic spread in financial circles, leading to runs on banks; over 500 failed, accelerating financial panic. As investment funds dried up, other major railroads, including the Northern Pacific, Union Pacific, and Santa Fe, declared bankruptcy. Ultimately 15,000 companies went under.[4]

In this environment, demand for iron and steel plummeted, and the ripple effect quickly worked its way down the food chain to the iron ore mines of Upper Michigan. By spring 1894 ore that had sold for $9 a ton in 1882 and had held at $5 a ton in the early 1890s, sold, if it could be sold at all, for $3, and some grades for only $2.10, less than the cost of production.[5] "Few who have not been through it," William Kelly, a regional mining supervisor, would later write, "can appreciate the agony of heart which seized us when . . . we realized what great reductions must be made in cost to insure a continuance of operations and were confronted with the appalling condition that at the same time the output must be reduced enormously."[6]

As demand for ore fell precipitously, ore companies closed mines and reduced workforces. In 1887 ninety-six Michigan mines shipped iron ore; by 1894 the number was fifty-five.[7] Iron mine employment in Michigan dropped from over 17,000 in late 1892 to slightly over 3,500 in late 1893. The few miners still employed faced wage cuts that left them with incomes barely sufficient for survival.[8] Many left the area, hoping for better luck in the nearby copper mines, in the coal mines of Illinois, or—for laid-off Cornish miners—back home in England. Others took municipal work, when available, for bare subsistence wages.[9]

William Mather, facing the first major crisis of his presidency, reacted by reducing the salaries of Cleveland office personnel in June 1893.[10] By July 1 he had closed down most of the company's mines. Responding to reporters, Mather commented: "We shut down simply because we are unable to dispose of the product at cost price."[11] By October miners were begging for any type of work to keep food on their families' tables. Regional mine operators tried to oblige, but struggling to keep rising losses to a minimum, they reduced wages to their lowest point since 1850. Even pro-company papers conceded these were at bare subsistence or even sub-subsistence levels.[12] Things got no better the following year. In 1894 the Cleveland-Cliffs Iron Company (CCI) operations lost over $72,000.[13] Its stock plunged from $100 a share to $25.[14] Cleveland-Cliffs employees were in distress. In 1895 company land agent Samuel Redfern reported that collecting rent from people in company-owned houses was difficult because the occupants had fallen so far into debt.[15]

Under these conditions, labor grew restive. Nationally, labor militancy had begun escalating in the 1880s as bonds between employer and employee weakened with the growing size of industrial enterprises. The number of strikes in the United States rose from around 500 annually in the early 1880s to nearly 1,500 by the 1890s.[16] Although CCI and regional companies had successfully fought off attempts to organize miners in 1890, the long depression following the Panic of 1893 exacerbated problems.[17] On July 13, 1895, frustrated with bare subsistence wages, believing that the depression was easing and that companies could and should now pay more, miners walked out at Negaunee's Lillie mine, demanding restoration of some depression-induced wage cuts. Led by miners of Cornish descent but headed by a council with members of all major immigrant groups, the walkout quickly spread to most major mines on the Marquette Range, including Cleveland-Cliffs'. Striking miners, numbering as many as 3,000, paraded daily between the two chief mining towns on the Marquette Range: Ishpeming and Negaunee. They soon organized an independent local union, formalizing its structure on July 16, 1895. The new union's demands included a standard wage in all regional mines, a wage increase, and union recognition.

In late July CCI and other regional mining companies attempted to end the strike by offering a modest raise without union recognition. Mather traveled to Ishpeming to talk personally to company employees about the company's economic condition and the proposal. Prompted by regional union officials, however, strikers rejected the offer.[18] Contrary to the expectations of local newspapers and mine officials, the strike continued, and relations deteriorated between the miners and the companies. In early August a mob of 300 miners ran an unfriendly reporter out of Ishpeming, prompting local law enforcement officials to request Michigan's governor, John T. Rich, send troops to preserve order. Rich delayed.

Mather, meanwhile, assumed leadership of the local mining companies' resistance to union demands. He coordinated hiring the Pinkerton Detective Agency to recruit outside labor—especially steam shovel operators—from Chicago, Milwaukee, and Minneapolis to load cars from the companies' very large stockpiles of unsold ore.[19] On August 26, Marcus Hanna, whose lake fleet transported ore, and Peter White, a member of CCI's board and a prominent regional businessman and politician, informed Rich that area mining companies intended to load ships from their stockpiles before the end of shipping season using the imported labor. Violence, they asserted, would likely result.[20] Faced with the determination of mining companies to load ore with imported labor if necessary, regardless of the consequences, Rich relented on August 31, ordering units of the state militia to the area. On September 2 five companies of the Michigan National

Michigan militia company at Ishpeming, 1895. Companies of Michigan militia were stationed around mining properties all over the Marquette Range in the closing phases of the 1895 strike to protect property from real or imagined threats from striking miners. Their presence was a key factor in the failure of the 1895 strike. *(Courtesy of Superior View, Marquette.)*

Guard took up positions around area mines and ore stockpiles. Protected by troops, imported steam shovel operators and a handful of nonstriking personnel loaded ore onto trains from company stockpiles.

The companies' attempts to bring in outside labor, including their partial success in securing steam shovel operators, their success in getting state militia sent to the area to protect their properties, and the growing desperation of the striking miners transformed what had been a generally calm and peaceful strike into the most serous labor-management confrontation in the Marquette Range's history to date. At one point a mob started out with the intention of hanging one of the local mine agents. Police, many of whom half sympathized with the strikers, were helpless to bring order. Strikers met the trains bringing in Pinkerton-recruited labor from Chicago, Milwaukee, and Minneapolis, persuading, gently or not so gently, most to abandon employment plans. Crowds of miners jeered those who went to work anyway. They had to be protected by militia or special deputies to reach work sites and then sleep in train cars on company property. Eventually, Cleveland-Cliffs and the other regional companies secured—with great difficulty—just enough labor to set their steam shovels into operation.

As the striking miners grew more desperate, violence increased. Beginning on the 10th of September massed groups of strikers (referred to as "mobs" by company officials) boarded every train entering the region to search for "scabs." Several men were severely beaten, either because strikers suspected that they were regional miners attempting to return to work or because strikers suspected they were imported "scabs." One man had to be hospitalized. In one case strikers discovered, too late, that those they had attacked were not looking for mining employment at all, but looking for work, instead, in area lumber operations. By early September large numbers of miners had begun to leave the area for other mining fields. By the second week of September calmer elements among the striking miners and local business and professional men, thoroughly frightened by how close the situation had come to escalating completely out of control, began to clamp down on the local toughs who had sought to provoke violent confrontations. When workers were given some assurance that the advance in pay offered earlier still held and that another raise was likely to occur by December with the coming of higher ore prices, support for the strike crumbled. Reading the handwriting on the wall, the union called the nine-week strike off on September 20, even though the companies still refused to recognize its existence.[21]

While Cleveland-Cliffs emerged as the leader of regional corporate opposition to the strike during the long confrontation, coordinating the use of Pinkerton agents to bring in strikebreakers, it was also the one company that local labor union leaders commended for the way it handled returning workers. In the strike's immediate aftermath, the union noted, "The Cleveland Cliffs Company treated their men in a manner worthy to be praised," apparently making no attempt to screen out those involved in the union.[22]

The 1895 strike was simply the beginning of the company's labor problems. The Cornish-led union created during the strike grew rapidly, reaching a membership of 2,600. It spread from the Marquette to the Menominee and Gogebic ranges, "having at one time, within a year of its organization, a considerable portion of the labor in the Michigan iron mines enrolled in its ranks."[23] In August 1897 the union attempted to strengthen its position by having its members refuse to work with nonmembers, hoping to force the latter to join the union.[24] Regional mining companies again worked together to stymie union plans, while this time labor support for the union split along ethnic lines.[25] The effort quickly faltered, but labor problems continued.

In early 1899, as economic conditions further improved, the region's mining superintendents recommended advancing wages without pressure from labor to forestall problems. Mather—in keeping with the company's policy of allowing its Upper Michigan superintendents considerable autonomy—wrote M. M. Duncan, Frank Mills Jr.'s successor as mining superintendent: "You men on the ground know what is best to do, and I approve of the decision to which you have come."[26] The miners' union, however, decided to try once more. On March 11, 1899, it demanded that nonmembers join the union by April 1. The union offered a low reinstatement rate and threatened other actions if nonunion miners did not join. On April 2, 700 men walked off their jobs, seeking to compel regional companies to hire only union men. Again the strike failed, with ethnic differ-

NOTICE!

To Every Mine Worker in the Mines of Ishpeming and Winthrop Location:

GENTLEMEN:--

At a mass meeting of Union Mine Workers held on March 11th, 1899, it was decided by a unanimous vote that an invitation be extended to all mine workers in the places above mentioned, including surface and under ground men to join their respective unions before the first of April, 1899, at which time, if any mine worker is outside of his organization, the above unions will take other steps to accomplish this end.

The reinstatement fee, until the first of April, is one dollar ($1.00) which, on and after that date, will be three dollars. The Initiation fee will remain at one dollar.

MINE WORKERS UNION NO'S 7 and 8.

MATT. WASLEY, Sec'y.

Union poster from 1899. The Northern Mineral Mineworkers Union established in the aftermath of the failed 1895 strike attempted to strengthen its position by compelling all miners to become members, leading to a strike in April 1899. This labor action failed. *(From Burton H. Boyum, "Cliffs Illustrated History," manuscript, c. 1986, 10-6.)*

ences and internal disagreements among the strikers major contributing factors.[27] The Cornish represented the area's most skilled miners and dominated union leadership; the Finns—who had begun to arrive in numbers in the 1880s—had been compelled to take the least desirable and lowest-paying jobs in the mines. Marquette's *Mining Journal* commented that, generally speaking, Cornish miners were the "most ardent supporters of the union," while Finnish miners "almost as a unit" opposed being forced into the union. Early in the strike Finnish miners met, decided to continue work, and asked for protection from authorities to do so.[28] The strike collapsed. But three major labor disturbances in five years and reports of another attempt to create a union for all three Michigan iron ranges clearly signaled to CCI management that labor problems were not simply going to go away.[29]

The Mesabi and Consolidation in the Steel Industry

Besides the impact of the economic downturn of 1893 and increased labor militancy, competition from a major new ore range and the resulting downward pressure on ore prices added to the company's woes. Around 1890 entrepreneurs opened a vast new iron ore supply—the Mesabi Range of Minnesota. The Mesabi hit the established Lake

Superior iron ore industry like a hurricane. The range shipped its first ore only in 1892. By 1895, a scant three years later, the Mesabi shipped more ore than any of the other Lake Superior iron ranges. The tidal wave of Mesabi ore, moreover, hit the market at the very time when ore prices were already depressed by the Panic of 1893. Eight years after its opening, in 1900, the Mesabi produced more than twice as much iron ore as any of the Michigan ore districts; by 1902 it produced more ore than all the other four Lake Superior ore ranges (Marquette, Menominee, Gogebic, and Vermilion) combined; and by 1910 it shipped twice as much ore as all the other ranges combined.

The huge volume of ore coming from the Mesabi at a time when demand and prices were falling due to the panic was problem enough. Making matters worse, Mesabi ore could be mined much more cheaply than CCI's underground Marquette Range ores. Mesabi deposits largely lay horizontally in the earth rather than vertically as in Michigan. Thus many Mesabi mines operated as open-pit mines once the relatively shallow earth overburden had been removed. Mesabi ore was also soft, so it could be dug out and loaded by steam shovel with minimal drilling and blasting. While CCI miners loaded ore cars in increasingly deeper underground mines a few dozen pounds at a time by hand, a Mesabi steam shovel operating in an open pit loaded four to six tons in a single scoop. While CCI underground miners had to push heavy ore cars to a hoist to get the ore to the surface so it could be loaded into railroad cars, Mesabi operators could run railroad tracks down into the pits and load railroad cars directly with steam shovels. Moreover, open-pit mines did not require the expensive timber supports needed in many underground mines. Once overburden had been removed, tracks laid, and steam shovels installed, vast volumes of Mesabi ore could be extracted far more cheaply than CCI's hard or soft-ore underground deposits.

Thus, by the mid-1890s Cleveland-Cliffs had to deal with a major economic depression and competition from abundant, cheaply mined, low-phosphorus Mesabi ores, both of which contributed to sharply falling ore prices. Simultaneously it faced upward pressures on its own costs from a more militant labor force. If things were not bad enough, Cleveland-Cliffs faced another challenge in the 1890s: reduced competition between iron furnaces for iron ore due to growing consolidation within the iron and steel industry and the backward integration of steel firms into iron ore mining.

Before the 1890s over 500 independent iron and steel firms operated in the United States.[30] Very few owned iron mines. Practically all competed on the open market to purchase ore from equally independent mining firms like CCI or from iron ore marketing agencies like Pickands, Mather and Company or Oglebay Norton. The competition for ores—particularly Bessemer-grade ores—by hundreds of users kept ore prices relatively high, even though the opening of new iron ranges beginning with the Menominee in the late 1870s had generally pushed prices downward. Not until 1892 did the largest of the steel companies—Andrew Carnegie's operations—seriously begin trying to secure control over ore supply and ore prices by integrating backward into mine ownership and, initially, Carnegie did so slowly and reluctantly. When the Panic of 1893 pushed weaker mining companies to the wall, John D. Rockefeller set off the scramble by buying up struggling mines and mine leases at bargain prices, especially on the new, highly productive Mesabi Range, hoping to duplicate with iron ore the monopolistic position he had earlier achieved in petroleum. The Oliver Iron Mining Company, by 1896 a virtual subsidiary of Carnegie Steel, entered the lists soon after on a major scale as Carnegie overcame his initial resistance to involvement in ore mining. To the shock of both producers and consumers of ore, Rockefeller and Carnegie reached a tentative accommodation and alliance in late 1896. Meanwhile, Carnegie's Oliver Iron Mining Company continued to buy up major ore properties, even entering into negotiations to buy Cleveland-Cliffs in the fall of 1897.[31]

Desperate to do something to hold the line on declining iron ore prices, a number of the older companies operating mines on Lake Superior attempted to create a price pool for Bessemer-grade ores in 1895, 1896, and 1897.[32] The Carnegie and Rockefeller interests, centered on the new Mesabi Range, however, had little desire to support ore prices since Carnegie's primary focus was on making a profit converting iron ore to steel, not mining and selling iron ore. The pool struggled to hold up prices, failed, and then quickly disintegrated.[33] It was now every man (mine) for himself. *Iron Age* editorialized in early 1897: "Everybody realizes that a fight is imminent. . . . The fittest only will survive."[34]

These events sparked fear in other steel companies. Concerned about the cost advantage that Carnegie might gain through control of his own mines and fearful of being left without an ore supply due to a Rockefeller-Carnegie monopoly, and the aggressive purchase of remaining independent properties by Oliver, they joined the rush to buy iron mines. By 1900 at least nine large iron and steel companies had purchased or leased iron mines, with some owning or leasing literally dozens. Traditional ore-selling agents, in turn, fearing that the expansion of iron and steel companies into mining would leave them in the cold, joined the feeding frenzy. Pickands Mather; M. A. Hanna and Company; Oglebay Norton; and Corrigan, McKinney and Company—all old Cleveland ore-selling agencies—soon became major mine leasers or owners.[35] As competition among iron and steel firms intensified in the late 1890s, acquisitions and mergers became almost daily news. Between 1898 and 1900 the iron and steel industry saw eleven large mergers involving nearly 200 previously independent companies.[36] By 1900 just eight steel companies controlled almost three-quarters of the Lake Superior region's iron ore output.[37]

The largest of the companies created by the merger mania was U.S. Steel, formed in 1901 by J. P. Morgan. Morgan, seeking to dampen competition and to maintain prices on finished iron and steel products, combined the Carnegie and Rockefeller interests with the shipping, mining, and manufacturing interests of a half dozen of the nation's other large iron and steel firms, all themselves the product of previous consolidations or mergers. U.S. Steel was the nation's first billion-dollar corporation. At formation, its mining subsidiary—the Oliver Iron Mining Company—controlled at least half of the iron ore being produced in the Lake Superior region (which was around 80 percent of U.S. production).[38] It sought even more. In 1901 Oliver offered to purchase Cleveland-Cliffs for $8 million. Mather and his associates wanted more, giving Oliver an option for $9.5 million. Oliver considered but then backed away.[39]

In little more than a half decade, between 1895 and 1901, the prevailing pattern of iron ore supply in the United States—sales by independent iron ore companies to a myriad of ore consumers—had been obliterated. Consumers had been consolidated. Moreover, through purchase or lease of mines, many consumers had become self-sufficient in ores. Despite some withdrawal from mining by U.S. Steel—under government antitrust pressure—independent ore-mining companies like Cleveland-Cliffs and ore merchant companies like Pickands Mather in their capacity as mine owners or mining or managing agents for mining companies controlled by 1948 only around 15 percent of the total tonnage of iron produced in the Lake Superior district.[40] By shortly after 1900 the hundreds of potential customers CCI had before 1893 had been reduced to a few dozen, and one company—U.S. Steel, through its iron-mining subsidiary, the Oliver Iron Mining Company—had become the dominant element in the iron ore industry.[41]

Table 3-2. Lake Superior iron ore shipments of the six leading firms, 1902

Company	Tons shipped	% of Lake Superior ore shipped
U.S. Steel Corp. (Oliver Iron Mining Co.)	16,063,179	58.3
Corrigan, McKinney & Co.	2,190,000	7.9
Cleveland-Cliffs Iron Co.	1,720,000	6.2
Pickands, Mather & Co.	1,270,000	4.6
Penn Iron Mining Co.	792,000	2.9
Jones & Laughlin Steel Co.	442,000	1.6

Source: Modified and calculated from table in Henry Raymond Mussey, Combination in the Mining Industry: A Study of Concentration in Lake Superior Iron Ore Production *(New York: Columbia University Press, 1905), 144.*

The reduced competition for iron ore by few consumers reinforced what the onset of Mesabi ores and the Panic of 1893 had begun: it accelerated downward pressure on ore prices. Mather warned CCI investors in early 1898 that the large investments in ore properties being made by Carnegie Steel, Cambria Iron, and other large consumers "much diminished" the number of consumers for company ore and that consequently "prices will doubtless be on a low plane hereafter."[42] Indeed they were. Iron ore (non-Bessemer), which had sold for as high as $8 a ton in Cleveland in 1880 and $5.25 in 1890, by 1898 sold for only $1.85. Average annual prices were only a bit better. In the ten years before 1893, the average price per ton for iron ore was $5.92. In the next six years the average was $3.54, some 40 percent lower.[43] While prices would rise, they would not hit the 1890 level again until the high-demand, inflationary years of World War I.[44] Because net profit per ton was now "extremely small," Mather argued that CCI could make "a fair showing" only by significantly increasing its ore output to take advantage of economies of scale and, of course, by keeping its operating costs as low as possible.[45]

Corporate Paternalism

Mather recognized labor tranquility was one key to survival since labor accounted in the 1890s for 70 percent of the cost of producing ore in underground mines. With large fixed costs, ore prices trending downward, and a very small profit margin per ton, Mather sensed that CCI could secure adequate return on investment only by full and steady operation at increased volume. Strikes and labor turnover militated against full and steady operation.[46]

Many employers in the late nineteenth and early twentieth centuries responded to the growing labor unrest with repression, hiring gangs of spies to obstruct union organization and thugs to break strikes and intimidate labor.[47] William G. Mather was uncomfortable with this approach. In the aftermath of the 1895 strike, it was probably his influence that ensured fair treatment to returning strikers. During the 1899 strike, some regional companies implemented a policy of not employing any union men who had gone out. M. M. Duncan, Mather's own general manager in Ishpeming, saw the failure of the 1899 strike as a great opportunity to "obliterate the union." Mather, however, wrote Duncan: "Theoretically I do not like such an arbitrary stand. It seems to me too despotic to insist that our employes [sic] shall not belong to such organizations as they wish." He argued that a policy of not rehiring union men would arouse "the most determined opposition" and cause much-needed skilled laborers to either leave the country or remain to carry on "most open and antagonistic agitation against our interests."[48]

For companies seeking a middle ground between unionization and heavy-handed repression, corporate paternalism, sometimes called welfare capitalism, offered a more appealing option. Paternalism involved providing employees with amenities not required by law or absolutely needed for operations. Proponents hoped that such amenities would undermine the attractiveness of labor activism and guarantee a more stable and tractable workforce.[49] William Mather had seen paternalism practiced in European mining districts during his wide travels and likely picked up the concept there.[50] He became convinced that a healthy and content labor force was essential to the company's future and that the proper treatment of employees was, moreover, a matter of *noblesse oblige*, the moral obligation of those more favored by inheritance and the circumstances of life to act with honor toward those less favored. Acting on these convictions, Mather made his company a leading—if not the leading—practitioner of paternalism on the Lake Superior iron ranges.[51] Cleveland-Cliffs under Mather initiated an ambitious corporate welfare program that embraced housing, improved communities, aesthetics, recreation, medical care, rest homes, pensions, workplace safety, education, and even agriculture.

Like other mining companies in frontier regions, the Cleveland Iron Mining Company had adopted some elements of paternalism—particularly housing—very early, out of necessity. It was the only way to attract and hold a labor force in a rugged, frontier area where such amenities were not available from established local communities. Cleveland-Cliffs' predecessor companies had long provided housing to employees at modest rental. CCI continued this policy. By 1916 CCI owned or had under construction 660 houses.[52] But it now also pushed programs to entice employees to own their own homes by building houses and selling them at cost or by selling company-owned homes with extended payment terms to employees who had long rented them.[53] In 1906 at Gwinn, in the southern part of the Marquette Range, Cleveland-Cliffs went further, constructing a complete model company town, laid out by Warren J. Manning, one of the best-known landscape architects of the period. Cleveland-Cliffs provided the town with a variety of different housing styles, modern water and sewer systems, hospital, school, recreation center, swimming pool, railroad station, and business blocks.[54] A smaller company settlement at North Lake, west of Ishpeming, while not as elaborate, included some of the same facilities.[55]

To improve the aesthetics of its mining locations, in 1894 Mather began awarding cash prizes annually for the best-kept premises and vegetable gardens. The company provided galvanized garbage cans and initiated garbage pickup at company expense in mining locations where regular garbage collection was not rendered by city or village.[56] To entice employees away from saloons, Cleveland-Cliffs erected recreation centers in its primary mining districts. Those erected at Gwinn to serve the Swanzy district (south of Ishpeming) and near the Lloyd mine to serve the North Lake district (west of Ishpeming) had bowling alleys, billiard tables, reading rooms, meeting rooms, and baths. In the more "urbanized" areas of Ishpeming and Munising, CCI helped finance YMCAs for the same purpose.[57] Mather often paid very close attention to what was going on in Cleveland-Cliffs communities in the Upper Peninsula. For example, in March 1899 he gave Austin Farrell, the head of the company's furnace department, a very detailed critique of his annual report regarding the company's new furnace location at Gladstone, Michigan, near Escanaba. He concluded: "I want to see a marked improvement this year in the appearance of our furnace location. I hope you will succeed in getting the township or city authorities . . . to ditch the road and prevent water from washing on to our lots, and also to encourage in some way . . . so that these lots will be cleaned up and put into a state of cultivation. I want to see this done by the close of this year sure."[58]

The CCI clubhouse constructed for employees at the model town of Gwinn, 1913. The company hospital is the building at the right; the Methodist church is on the left. The trees—part of Warren Manning's landscaping scheme for Gwinn—are still small saplings at this point. *(Courtesy of Cliffs Natural Resources, Inc.)*

Austin location in winter, c. 1915. The housing near the Austin mine in the Gwinn region was typical of that in many of CCI's mining locations, with its row of nearly identical houses located close to the mine shaft (*left background*). *(Courtesy of Superior View, Marquette.)*

J. Ferrero's home, Austin location, September 12, 1907, winner of second prize in the vegetable garden category. For decades William Mather awarded prizes in each of the company's locations for best-kept premises and best vegetable garden in an attempt to improve the appearance of company properties and the morale of those living in them. *(Courtesy of Cliffs Natural Resources, Inc.)*

To further ensure healthy and happy employees, Cleveland-Cliffs in May 1908 introduced a visiting nurse program, becoming the first iron-mining company to do so. Within four years the company had placed visiting nurses in Ishpeming, Negaunee, and Gwinn and, a bit later, jointly sponsored with other companies, in Iron River on the Menominee Range. Working under the direction of company-employed physicians, the nurses visited families of employees free of charge, instructing them in care of the sick, hygiene, and pre- and postnatal care. Some years Cleveland-Cliffs' visiting nurses made almost 10,000 professional visits.[59] During the 1918–19 influenza epidemic that killed millions across the world, CCI nurses and physicians provided free inoculations.[60] At the recommendation of its visiting nurses, in 1909 CCI established a rest home for overworked mothers, since many had large families and had never had a vacation.[61] In 1914 alone this facility took in 107 guests; each stayed an average of 8.5 days.[62] For older, long-term employees, Cleveland-Cliffs initiated a pension system in 1909, giving retirees, in the period before Social Security existed, at least a miniscule income.[63] In addition, the company carried out relief work for needy families in its communities. In 1938 it provided payments to fifty-four families and donated additional funds to incapacitated miners or widows of mine employees.[64]

Cleveland-Cliffs also began to pay increased attention to conditions in and around its mines. CCI used Manning to landscape mine sites as well as its model town at Gwinn,

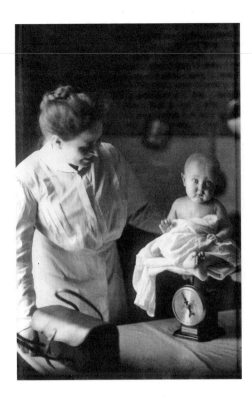

Cleveland-Cliffs visiting nurse weighing baby during home visit. CCI initiated the program in 1908, eventually employing three full-time visiting nurses. Providing the wives of miners with advice about children's health was one of their primary tasks. *(Courtesy of Cliffs Natural Resources, Inc.)*

blending neat lawns accented by shrubbery with well-kept roads around mine offices and shaft houses.[65] In 1900 none of the company's changing houses, or drys, where miners changed from mining clothes to street clothes and vice versa, had bathing facilities. Miners often had to walk home in wet, dirty clothes in the dead of winter, tracking reddish slime into their houses. In 1901 the company began to install showers in its old drys and build new dry houses with hot and cold water, good ventilation, drying racks for wet clothes, lockers, and showers. The dry erected at Cliffs Shaft mine in 1901 was one of the first modern fireproof drys in the region.[66]

Perhaps the most dramatic step taken to improve working conditions came in the area of safety. In the late nineteenth century injuries and death on the job were viewed as almost routine, with blame usually placed on worker carelessness.[67] Underground iron mining was an inherently dangerous occupation. In the late nineteenth century the fatality rate in Michigan's iron mines was a "very high" 4.5 per 1,000 men per year, exceeding that of the more notoriously dangerous bituminous coal mines.[68]

The prevalent view among most historians is that the passage of workmen's compensation laws and the rising cost of liability suits, which shifted the burden of safety from workers to management, were the forces behind the drive for greater workplace safety.[69] Cleveland-Cliffs was certainly interested in securing what it regarded as a tolerable workmen's compensation bill and sought to influence state legislation in that direction.[70] Nonetheless, some evidence suggests that William G. Mather saw safety as essential to

Morris mine, North Lake district west of Ishpeming, 1913. The mine's shops are at the far left; its dry, where miners changed clothes before and after work, is in the center with the mine's headframe behind it. On the far right is the North Lake mining district's main office. The power poles indicate that this mine has electrified its operations, while the nicely landscaped area around the dry and headquarters building are indicative of Manning's work for CCI. *(Courtesy of Cliffs Natural Resources, Inc.)*

keeping an experienced and contented labor force at least a decade before the passage of Michigan's workmen's compensation law in 1912 and a half dozen years before U.S. Steel initiated the modern "safety first" movement in 1906. In early 1900, for instance, he congratulated his general manager for reducing the number of fatal accidents in company mines and noted that the company needed some system to "incite" foremen and bosses to push safe practices, perhaps by awards or publication of accident records.[71] Comments of the company's master mechanic O. D. McClure in various reports likewise suggest that an interest in safety in CCI had emerged before workmen's compensation and the rapid growth of liability suits.[72]

Perhaps influenced by U.S. Steel, which inaugurated its safety first campaign in 1906 and had extended it to its iron ore–mining subsidiary the Oliver Iron Mining Company by 1910, Mather in 1911 hired a full-time professional safety director, William Conibear, making CCI one of the first American mining companies to have such a position. Under Conibear the company became one of the strongest promoters of the safety first movement, establishing a comprehensive set of safety regulations, local and central mine safety committees, and regular mine inspections. Among other things, Conibear replaced

First class of Cleveland-Cliffs employees to undergo first-aid training, Lake mine, Ishpeming, 1910, as part of the company's increased emphasis on safety. Note those demonstrating proper bandaging techniques on the right. Within a few years a substantial portion of the company's employees had received first-aid training. *(Courtesy of Cliffs Natural Resources, Inc.)*

miners' candles with carbide lamps, required a telephone to be installed in each mine office, and specified that a canary be kept at each mine, a traditional means of detecting poor air quality.[73] Since only 20 percent of the company's workforce was of either American or English origin, Conibear printed safety circulars in multiple languages.[74] He required supervisors to prepare detailed reports on all injuries and eventually made maintenance of a safe working environment an element in the remuneration of supervisory personnel, shifting responsibility for safety from the miner to the company.[75] The U.S. Bureau of Mines early saw promotion of rescue stations and first-aid training as an essential element of its mission.[76] Conibear promptly embraced the idea, establishing five fully equipped mine rescue stations and pushing first-aid instruction, training more than 1,000 company employees within a few years.[77]

Cleveland-Cliffs' safety program succeeded in multiple respects. In 1913 the superintendent of the Morris-Lloyd mine reported that the miners were "impressed with the efforts being made by the Company to decrease accidents."[78] More important, the fatal accident rate dropped dramatically, from 5.0 per 1,000 employees annually in the decade preceding Conibear's arrival to 2.3 between 1911 and 1945—even lower (about 1.6) if CCI's one major mine disaster is excluded from calculations.[79] Within eighteen months of the creation of the safety department under Conibear, minor accidents dropped 16 to 20 percent.[80] In 1932 Daniel Harring of the U.S. Bureau of Mines wrote that Cleveland-

Cliffs had "achieved a standing in safety in mining second to no mining organization in the United States."[81] By 1940 the manger of the company's mines could report that "no organization in the Lake Superior district, employing a similar number of men, . . . can show such results," referring to the company's safety record.[82]

When accidents did occur, company-financed hospitals provided improved care for injured miners. By 1915 Cleveland-Cliffs operated or was affiliated with hospitals in one Minnesota and four Michigan towns. In 1918 CCI erected a new modern hospital in Ishpeming, the center of its operations.[83] The company managed the hospital, opening it to the entire community, but keeping two beds always open in case of mine accidents.[84]

To provide employees with the opportunity to advance within the company, Cleveland-Cliffs developed education programs. In February 1912 it initiated classes to train electricians. Forty enrolled. The following year CCI began a "Shift Boss School." Its two-year night course provided instruction in geology, mathematics, mine analysis, leadership, safety practices, explosives, mining systems, and related matters. The *Engineering and Mining Journal* reported that 130 enrolled in August 1912. On completion, the better students entered a pool from which the company selected future shift bosses. By 1919 most of the men in charge of electrical work at CCI and an increasing number of its shift bosses were miners who had taken the courses.[85] In addition, CCI sponsored classes in English for foreign-born miners and their families.[86]

In a move somewhat unusual for corporate paternalism anywhere, CCI also launched a wide-ranging, even if ultimately unsuccessful, program to promote commercial gardening and agriculture in the region. The somewhat disjointed program included a greenhouse and experimental garden to determine what food plants might grow best in the cold climate of the Upper Peninsula and to nurse the appropriate seedlings. It included attempts to start an agricultural fertilizer business and to introduce sheep ranching into the region. It saw the creation of three different demonstration farms and involved hiring a graduate of the Michigan Agricultural College to run the farms and advise those who purchased land from the company for farming. Mather saw agriculture as providing miners with an alternative occupation when mines closed in hard economic times and, perhaps, also as a means of keeping local food prices, and hence wages, under control, while providing a safety valve for discontented workers with no alternative local occupation.[87]

This relatively comprehensive set of paternalistic programs helped the company manage its labor problem and made Ishpeming, the heart of the company's operations, into something of an exemplary mining town. After two members of the U.S. Commission on Industrial Relations visited the region in 1914, they wrote W. H. Moulton, the head of CCI's "pension department," that Ishpeming had become one of the standards by which they measured other company towns. CCI, they declared, was in the "vanguard" of "sensible and practical welfare work."[88]

A final element in CCI paternalism was the active involvement of its general managers and mine superintendents in local community affairs. Frequently mine superintendents occupied local political offices. Often their prominence in a local community was sufficient to secure election without much effort. Sometimes, however, they used techniques that placed their adversaries at a disadvantage, such as drawing on company employees or equipment to assist their campaigns. Once in office, however, most tried to improve the communities they served. One of the best was Charles J. Stakel who, as mine superintendent in several Marquette-area mines and then as general manager, frequently ran for political office and once there sought to assist the communities in which he resided, improving water supply and sewage, bringing financial order, bettering roads, and securing new facilities such as airports.[89]

In the late nineteenth and early twentieth centuries, paternalism was widely practiced throughout the Lake Superior region, both by other iron-mining companies and by copper-mining companies. Attempted earlier and better capitalized, comprehensive paternalistic programs emerged in the copper district several decades before they began in the iron region. The comprehensive nature of paternalism practiced in the copper district, especially by the giant firm of that region—Calumet & Hecla—attracted considerable attention in the contemporary press and has been noted subsequently by historians.[90] CCI's expanded system of paternalism emerged later, but included most of the elements found in Michigan's copper country plus some additions. Cleveland-Cliffs' extensive visiting nurse program, its attempts to beautify its mining sites with professional landscaping, its efforts to promote agriculture, and its construction of several model communities from scratch were elements of paternalism that were either not practiced at all or practiced to a much smaller degree in the more frequently cited copper-mining region. In fact, the comprehensiveness of Cleveland-Cliffs' program compares with only a handful of other practitioners of welfare capitalism, generally much larger firms operating in the nation's manufacturing core.[91]

The paternalistic carrot, however, was sometimes balanced by subtle use of the stick. CCI rarely used the rather heavy-handed measures of some other companies in the Lake Superior iron district.[92] William G. Mather advised M. M. Duncan, his mining superintendent, during the 1907 recession that he should discourage attempts by mining captains and foremen to "get even" with the men under them for "annoyances" that had occurred during the previous year when labor was scarce. This, he said, was "incompatible with the dignity of the company."[93] By the mid-1910s Mather's philosophy had been accepted by many of CCI's managers on the Marquette Range, for in a joint letter to Mather in 1915 the company's key department heads noted: "We have always to remember that this is a Non-Union district, and unless the companies treat the men fairly and liberally, it will result in their organizing."[94] Nonetheless, when demand was slack and the company laid off employees, some of its mine managers clearly rid themselves of personnel they regarded as problems. In 1913 the manager of the Negaunee mine noted: "We still have quite a number of undesirable men and are constantly making every effort to weed them out," especially "young Socialistic Finns." When demand for ore slackened in

1914, he used the necessary force reduction to rid the mine of "every man whom we had any reason to think was not desirable."[95]

The rather restrained nature of CCI's dealings with labor and its comprehensive paternalism inhibited labor organization, for it made it difficult for labor organizers to paint Cleveland-Cliffs as a completely heartless, exploitive, and remote corporation. CCI paternalism contributed to the relative labor peace it enjoyed for almost half a century after 1900. The company experienced no strike of any significance between 1899 and 1946.

Expanding Operations and Increasing Production

Maintaining steady production through labor peace was only one element of the equation that had to be solved if CCI was to survive as an independent ore-producing firm in the changing landscape of the Lake Superior iron ore industry after 1893. In the midst of the crises hitting the company in the 1890s Mather informed shareholders that the company would have to use its earnings not only to pay them regular dividends, but also to strengthen the company's position through expansion of production facilities and acquisition of key resources.[96] He saw that lower per-unit profits for ore could be offset only by increased volume.

To increase its ore output Cleveland-Cliffs had to expand its operations. In 1899 Mather tried to purchase or merge with an old rival: the Lake Superior Iron Company, one of the first three companies to operate on the Marquette Range. But other firms— notably the Oliver Iron Mining Company, affiliated with Andrew Carnegie's growing empire—were also interested. To avoid a bidding war, Mather and Oliver considered several alternatives, including merging CCI into Oliver, with CCI as a subsidiary company in charge of all of Oliver's Marquette Range properties. In the end the two agreed to have Oliver buy three-fourths of the property and CCI one-fourth.[97] In 1899 CCI bought half the stock of the Arctic Iron Company and the Regent Iron Company, plus interest in the Webster and Imperial mines. It also leased the Michigamme mine. Mather recognized that the grade of ore in most of these properties was not high, but with rising demand he hoped to sell some at profit.[98] In 1905 CCI acquired the dwindling assets of its old partner and rival, the Jackson Iron Company.[99] The Jackson had ceased active mining in 1897, largely because it had exhausted its high-grade ores and was having difficulties selling the low-grade ores it had left.[100] Cleveland-Cliffs purchased Jackson for the potential value of its mineral lands.[101] Between 1913 and 1915 CCI also purchased two other old rivals: the Republic and Lake Angeline mines. While the mines acquired in this process were largely played out of high-grade ore deposits, the acquisitions increased the mineral land holdings that the company controlled in fee and further consolidated its position as the leading mining company on the Marquette Range.

Meanwhile, Cleveland-Cliffs did not forsake its tradition of seeking partners for

The Negaunee mine, 1904, looking southwest from the #2 shaft toward the #1 shaft. Note the absence of trees around the mine. Much of the once-dense forests around the mining towns of Ishpeming and Negaunee had been consumed by 1904 to feed charcoal-fired blast furnaces, like the Pioneer in Negaunee, to provide fuel to the steam engines that pumped and hoisted material from the mines, or to supply support timbers for the mines themselves. Mather would, a few years later, attempt to beautify mine locations like this one with professional landscaping. *(Courtesy of Cliffs Natural Resources, Inc.)*

capital-intensive ventures. When Oliver's lease on the Negaunee mine expired in 1903, CCI acquired a thirty-year lease on the property. The cost was high—$1.5 million—so CCI sought a partner, sharing costs with the Lackawanna Steel Company, each owning half, but CCI handling operations.[102] In 1905 CCI explorations revealed a very deep ore deposit on company land east of Negaunee. Because of high development expenses, CCI again went the joint-venture route. In 1912 the company, with partners, opened two mines in the area, the Lucky Star and the Athens (the latter working with Pickands Mather), with CCI holding, respectively, a 25 percent and a 46.46 percent interest and operating the mines.[103] By 1905 Cleveland-Cliffs owned nine operating mines, leased six others, and partly owned and controlled three more, for a total of eighteen.[104] In 1893 the company had operated only six.

CCI complemented its purchases and joint ventures by "quite vigorously" expanding its exploration program, not only on company-owned property but also upon leaseholds, "so that the steady exhaustion" of older mines would be counterbalanced by the discovery of new ones.[105] In 1901 the aggressive exploration policy disclosed a deep deposit near Negaunee, which the company developed through a lease; it became the Maas mine.[106] In 1903 Cleveland-Cliffs' explorations in the Swanzy district, twenty miles south of Ishpeming, discovered moderate ore bodies sufficient to justify purchasing the Princeton mine and leasing adjacent properties to open up additional mines, beginning with the Austin. CCI erected the model community of Gwinn to lure labor to the new area.[107] In 1907 discoveries west of Ishpeming led to the opening of the North Lake district—named after an adjacent lake—in 1910.[108]

These properties were all on the Marquette Range. After 1900 CCI also sought to expand its operations outside that area. On the Menominee Range Cleveland-Cliffs acquired half interest in the Michigan Mineral Land Company in 1908 because that company controlled 50,000 acres of potential mineral lands near Crystal Falls.[109] Although little came of this investment, CCI's extensive exploration program uncovered a moderate deposit of ore near Iron River/Caspian. Although the ore was high in phosphorous, World War I demand prompted the company to begin developing the Spies mine on the property.[110] On the Gogebic Range CCI acquired a foothold in 1901 by leasing the Ashland mine in Ironwood. This lease was particularly attractive because the Ashland was a Bessemer ore property, an ore type in which the company held too few reserves. In a few years the company had rejuvenated a mine poorly treated by the previous leasers, transforming it into "one of the finest" in the Lake Superior region.[111] In 1905 CCI added the Iron Belt mine to its leased Gogebic properties, while continuing explorations on that range.[112]

In January 1902 Cleveland-Cliffs decided to try its luck on the booming Mesabi Range. It sent mining engineer J. E. Jopling to carry out exploration work and retained E. J. Longyear, a prominent regional mining expert, as its chief consultant. It was a major venture. The company's mine superintendent M. M. Duncan had no experience with the district and had never even visited the region. When he arrived he found conditions different and more frantic than on the older Michigan iron ranges. "From the first visit," Duncan noted, "a constant stream of people interested in exploring came to offer me lands, which they held under option, or thought they could get from friends." Separating the frauds from the honest men proved impossible. Moreover, exploration was difficult because the ground was so hard that drilling went slowly.[113] Longyear had intimated to Cleveland-Cliffs that the region was already so well explored that important new finds were unlikely. He proved right. At the end of CCI's second year of exploration, Duncan confessed that results were "disappointing" and the chance of finding easily accessible ore bodies was "so small" that he did not recommend further work.[114]

This left the option of purchasing leases on previously explored property. Duncan purchased several, including the Crosby mine of the East Itasca Mining Company. In 1905, after two years of preparatory work, the Crosby became Cleveland-Cliffs' first mine on the Mesabi to ship ore.[115] In the aftermath of World War I, CCI expanded its holdings on the Mesabi Range largely through joint ventures much like those it had initiated on the Marquette Range. CCI and its partners shared the costs of leasing and developing a property by establishing a jointly owned subsidiary mining company. CCI then operated the property, making a profit both from its ownership share and from compensation for management. In 1918, for example, CCI acquired two mine leases jointly with the Struthers Furnace Company and began operating the mines.[116] Then in 1920 CCI worked with several iron and steel companies to secure five additional mining leases under the corporate name of The Mesaba-Cliffs Iron Mining Company. CCI owned 25 percent of Mesaba-Cliffs and served as operating agent.[117] In 1918, reflecting the company's growing presence on the Mesabi, Cleveland-Cliffs established a district office at Hibbing, Minnesota.[118] In 1926, in conjunction with several large steel companies, CCI leased and began operating the Holman-Cliffs mine on the Mesabi and the Clark mine on the Cuyuna, a smaller Minnesota iron ore range south of the Mesabi.[119] In 1929, again in conjunction with partners, CCI leased the Canisteo mine on the Mesabi.[120]

While Cleveland-Cliffs used joint ventures earlier and more heavily than most regional mining companies, partially because it did not have the capital resources on its own to compete with the large, integrated steel firms that had come to dominate the field, it was, of course, not the only company that resorted to them to undertake risky and expensive mining operations. As the scale of the American iron and steel industry grew in the late nineteenth and early twentieth centuries, other firms faced similar problems and took the same route. Bethlehem Steel, for instance, partnered with Pennsylvania Steel to undertake ore explorations in New York and the development of mines in Cuba as early as the 1880s.[121] And—as the scale of mining steadily increased between 1880 and 1920—joint ventures became steadily more common.[122] Besides the usual reasons for joint ventures—sharing risks and undertaking capital-intensive projects—Cleveland-Cliffs had an additional reason to embrace the strategy. The backward integration of iron and steel companies into iron ore mining to secure their own ore supplies and the consolidation of the industry had sharply reduced the number of potential customers available to an independent ore-mining company like Cleveland-Cliffs. With nearly 50 percent of the country's steel-production capacity, U.S. Steel was self-sufficient in ore. And many of U.S. Steel's chief rivals, like Bethlehem and Jones & Laughlin, had purchased sufficient ore deposits to become nearly self-sufficient as well. To find outlets for its ore, Cleveland-Cliffs had to cultivate the smaller steel companies. Joint ventures with these firms not only helped distribute risks and raise capital for ventures that would have stretched the limited resources of CCI, but they helped establish or maintain vital personal relationships with the smaller iron and steel producers and thus ensured a market for the company's ores.

By 1920 CCI owned or operated twenty-nine different mines in Michigan and Minnesota, producing all grades of ore, a far cry from the six mines it had operated when the Panic of 1893 broke out.[123] The acquisition of new operating properties through

Stripping overburden at the Boeing mine, Hibbing, Minnesota, c. 1920. The Boeing mine, operated by the Mesaba-Cliffs Company, was an example of CCI's expansion to the Mesabi Range of Minnesota. CCI operated the Boeing as an open-pit mine from 1919 to 1928. *(Courtesy of Cliffs Natural Resources, Inc.)*

exploration, lease, purchase, or partnership enabled Cleveland-Cliffs to increase its ore output from around 700,000 tons annually in the early 1890s to between 2 and 4 million tons annually from 1905 onward. While Cleveland-Cliffs was not a large corporation compared to fully integrated iron and steel firms (CCI ranked 105th in assets among American companies in 1917; U.S. Steel ranked no. 1; Bethlehem Steel no. 3; Midvale Steel no. 6), it remained the largest independent iron ore–mining firm.[124]

Reducing Costs

Labor peace and expanded output were two legs of the stool on which CCI's survival rested in the post-1893 world. The third was reduced production costs—a central task since most Cleveland-Cliffs mines were labor-intensive, underground operations. In early 1911 Mather reiterated what the company's directors and stockholders already knew: the tendency toward decreased prices for iron ore due to reduction in tariffs on ore imports and competition among ore producers compelled the company "to constantly strive to improve our methods and reduce our costs."[125] Beginning in the 1890s, CCI had begun already to systematically reduce its costs on a variety of fronts: transportation, energy, surface operations, underground operations, and management.

As prices for ore dropped after 1893, CCI quickly took major steps to cut transportation costs. It was an obvious area. Most of the furnaces that consumed Lake Superior ore lay 600 to 800 miles from company mines, making transportation a significant portion of total ore cost. Cleveland-Cliffs had built two of the largest and fastest vessels on the Great Lakes in 1889, and these had even made a modest profit in 1893, when the company's mining operations lost money.[126] Faced with the decline of independent vessel operators—another casualty of the consolidation of the iron and steel industry between 1895 and 1905—and the growing fleets of steel companies and ore agents like Pickands Mather and Oglebay Norton, CCI began to expand its modest fleet when economic prosperity returned. In 1899 CCI purchased a schooner (*Chattanooga*) to be pulled by its steamer *Pioneer;* the following year it acquired two steel steamers from the Lake Superior Iron Company.[127] In 1904 and 1905 the company ordered three additional steamers.[128] By 1916 Cleveland-Cliffs controlled, either directly or through captive companies like the Grand Island Steamship Company, a fleet of twenty-three vessels.[129] The tradition of joint ventures established in its mining operations extended to shipping as well. In 1923 CCI partnered with four other companies to create the Progress Steamship Company and formed another partnership—the Leathem Smith–Cliffs Company—to purchase an old steamer, with each partner putting up half the capital. An element of these latter two partnerships involved converting vessels to use the self-unloading patents of Leathem Smith, potentially an additional cost-reduction measure.[130]

To complement reduced costs gained by controlling its own shipping, CCI also entered the business of land transport. The Cleveland Iron Mining Company had regularly feuded over rates and service with the rail companies serving the Marquette Range, but in earlier years it had insufficient capital to undertake a line of its own. In 1888 and 1889 CCI's attempt to work with several local companies and outside capitalists to establish a rail line from Ishpeming to Marquette harbor failed. The crises of the 1890s, however, drove ore prices so low that bringing land transportation costs under control became urgent. CCI again looked to partnerships to leverage its limited capital. Mather renewed negotiations with both the Pittsburgh & Lake Angeline and Lake Superior Iron companies. Lake Superior balked at the last minute, but Lake Angeline and CCI pushed ahead.[131] Mather commented in early 1896 that even if the resulting railroad company made no profit, the project would still be "of great advantage" because it protected the company against being "financially injured" by railroad rates, which would have risen without the creation of the new company.[132] Ore trains began rolling over the new Lake Superior and Ishpeming Railway, or LS&I for short, on August 12, 1896, unloading ore at a new wooden ore dock of the latest design erected at the north end of Marquette

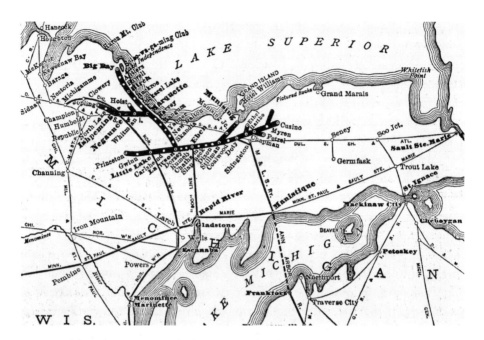

Map of the Lake Superior and Ishpeming Railroad (LS&I) and its connections, 1929. The railroad initially ran only from company mines near Ishpeming to a new, modern ore-shipping dock on the northern edge of Marquette, but it soon expanded through construction and the purchase of small local railroads so that it could also haul timber to supply the company's new charcoal-fired blast furnace and wood chemical plant at Marquette. *(Courtesy of Cliffs Natural Resources, Inc.)*

harbor. "There is no doubt," Mather declared, "but that this road prevented a raise in railroad rates to that in force a few years ago, viz, 40¢ and 65¢ to Marquette and Escanaba, respectively, as compared with 32¢ and 52¢, the present rates."[133] When its partner in the venture encountered financial problems after 1910, Cleveland-Cliffs increased its investment in LS&I from 50 percent to 65 percent and eventually to 75 percent.[134] In 1912, to further speed up operations and reduce costs, Cleveland-Cliffs replaced the 1896 wooden ore dock with a new steel and concrete dock capable of storing 50,000 tons (the old dock had stored 32,000) and of discharging that ore into vessels in one-fifth the time.[135]

As it cut transportation costs, Cleveland-Cliffs simultaneously reduced its energy and mining costs. Early mechanization in iron mining had relied on steam power. Steam engines required coal. And the coal had to be imported at considerable expense since the Upper Peninsula had no good deposits. Moreover, some tasks underground could not be mechanized using steam. For example, the labor-intensive task of loading ore once it had been broken from an ore deposit and moving it to a shaft where it could be hoisted could only be handled by man or animal since steam engines could not operate effectively underground. Electrification held the promise of more complete mechanization at lower cost than steam. Cleveland-Cliffs had experimented with electric lights at its mines

Hand trammers loading ore in the Cliffs Shaft mine, c. 1900. Almost universally before the 1890s ore was moved from the place where it was basted down to a hoisting shaft underground by hand or animal. In this photo, the trammers—still using only candles for underground lighting—are loading the hard specular hematite of the Cliffs Shaft mine into a mine car; they would then push it by hand perhaps hundreds of yards to the hoisting shaft. *(Courtesy of Cliffs Natural Resources, Inc.)*

as early as 1880. In the following decade Edison, Westinghouse, Sprague, and others had expanded the applications of electricity from lights to traction and motors, opening up new possibilities.

The ideal place for Cleveland-Cliffs to try expanded electrification proved to be a new mine. In the winter of 1886–87 a Cleveland Iron Mining Company diamond-drill rig operating on the thick ice overlaying Lake Angeline, immediately south of Ishpeming, uncovered a rich deposit of non-Bessemer ore beneath the lake. CCI decided to drain the lake to exploit the deposit. CCI officials, however, recognized that with non-Bessemer ores the company could make a profit only "by reducing the cost of mining to the lowest possible point." The large deposits and the long distances between the points where ore would be blasted down (under the old lake bed) and where it would be hoisted (a shaft house near the old shore line) offered a particularly good opportunity for mechanical haulage. To do this, CCI installed the first underground electrical locomotives in the district.[136] Electric trams reduced the cost of moving ore within the mine from 29¢ per ton to less than 4¢. CCI's general superintendent noted in late 1894 that electric haulage was "certainly a great success" and that the company could not profitably have gotten along with hand tramming in the mine.[137]

The success of electric haulage at the Lake mine encouraged further electrification.

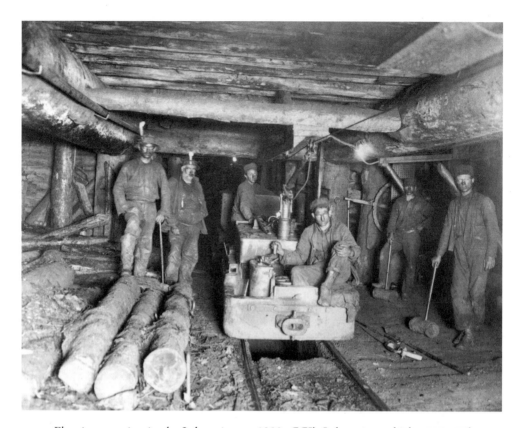

Electric tramming in the Lake mine, c. 1900. CCI's Lake mine, which penetrated under the drained-out bed of Lake Angeline south of Ishpeming, was the earliest mine in the Lake Superior district to replace hand and animal tramming with electric loco-motives, drastically reducing the cost of moving ore within the mine. Note the heavy timbering required of a soft-ore mine like the Lake compared to the hard-ore Cliffs Shaft mine, which required relatively little. *(Courtesy of Superior View, Marquette.)*

By 1900 CCI's Salisbury mine had five electrically powered locomotives for hauling ore, rock, mud, and timber.[138] When CCI opened the Negaunee mine in 1905, it put in an electric haulage system duplicating the one at the Lake mine, largely because it provided significant labor savings by enabling Cleveland-Cliffs to dispense with trammers, the men who pushed ore cars from the mine face to the hoisting shaft. The replacement of trammers had an additional bonus from the company's point of view. They had long been the most difficult element of the labor force with which to deal.[139]

The success of electrifying haulage underground encouraged CCI to push electri-fication aboveground. This was especially encouraged by William G. Mather who, in the 1890s, had visited Swedish mines and observed their use of hydroelectric power.[140] When CCI opened mines around Gwinn and west of Ishpeming in the so-called North Lake district between 1907 and 1910, it used electrical equipment not only for haulage underground but for hoisting engines and pumps.[141] The company simultaneously made plans for retrofitting older mines with electric motors as existing steam machinery wore out. The transition went well. In 1910 O. D. McClure, the company's mechanical engineer, noted that electric motors had proven "reliable and efficient" and that the shift from steam to electricity caused "very little trouble."[142]

While electrifying hoists, pumps, and underground haulage improved the efficiency of these operations, the company's dependence on expensive imported coal continued since the steam turbines installed to produce the electricity still required coal. To make full use of electrification as a cost-cutting measure CCI needed to move away from dependence on coal. In 1901 and 1902, apparently at Mather's direction, Cleveland-Cliffs began quietly acquiring waterpower sites in the Upper Peninsula and the watersheds that fed them.[143] In 1908 mine superintendent M. M. Duncan and chief mechanical engineer O. D. McClure visited hydroelectric stations in both Europe and the United States. With this experience, CCI in 1910 initiated two hydroelectric projects. The first was on a small scale, more in the nature of a pilot plant. CCI dammed the Au Train River around twelve miles from Munising, installed water turbines capable of developing 1,000 hp, and delivered the power to the Munising Paper Mill, a large, partly owned customer of the company's timber holdings. When this project was completed, CCI engineers turned to a more ambitious one: a plant on the Carp River between Negaunee and Marquette. Completed by 1912, it provided hydroelectricity to the company's mining operations.[144] When CCI began to electrify its mining operations, virtually all mines in the Lake Superior district and all iron mines in Michigan still operated by steam, save one or two using compressed air produced by waterpower. The company's shift to electricity was, according to a local newspaper, initially greeted by criticism "from engineers which amounted almost to ridicule."[145]

While as late as 1911 Cleveland-Cliffs' electric power was entirely generated by steam, the company transitioned rapidly. By 1913 CCI depended primarily on its hydroelectric plants for power, using steam units only as backup.[146] By 1918 the company's electrical system had 536 miles of transmission lines and 27,000 hp of electric motors, compared to just over 3,000 hp a scant eight years earlier.[147] By 1920 CCI owned five hydroelectric plants. These supplied 99 percent of the company's power needs, eliminating 140,000 tons of coal purchases annually. Moreover, the cost of the company's hydropower had remained static, while the cost of coal had jumped by 250 percent between 1914 and 1920.[148]

Eventually the company's hydroelectric system did more than simply lower the operating costs of its mines, becoming a profit center in itself. In 1926 CCI incorporated its hydroelectric system as a wholly owned public utility (Cliffs Power and Light) because the system's capacity exceeded what CCI mines needed. This enabled the company to sell power on the public market and realize additional revenue.[149] In the 1930s, when mines went into hibernation during the Great Depression, Cliffs Power would sell nearly half of its output to outside parties.

Besides reducing costs through expanded control of transportation and power, CCI's

CCI's McClure hydroelectric plant on the Dead River, west of Marquette. Named after O. D. McClure, the company's chief mechanical engineer, and erected between 1917 and 1919, this plant was one of a series of hydro plants built by CCI to reduce its dependence on steam power. A penstock 7 feet in diameter and 13,600 feet long linked the powerhouse shown here with a 500-foot-long dam across the Dead River. The powerhouse contained two 5,000-kVA generators. *(From Alexander N. Winchell,* Handbook of Mining in the Lake Superior Region *[Minneapolis: Byron & Learned, 1920], 65.)*

Upper Peninsula and Minnesota managers and superintendents scrutinized every other element of ore production to keep the company competitive with open-pit producers and the vertically integrated giants that dominated the American steel industry. A few examples will suffice to illustrate. To reduce assay costs, Cleveland-Cliffs established a chemical laboratory in July 1895 in Ishpeming; it had previously contracted out ore analyses.[150] To reduce stockpiling costs, CCI built at the Negaunee mine the first permanent steel and concrete ore-stocking trestle in the region. Previously crews had constructed temporary elevated wood trestles every fall for winter stockpiling operations. These trestles carried the ore cars above stockpiling locations for ease of dumping. In the spring crews dismantled the trestles to clear the grounds for steam shovels, which would load the ore for shipment in railcars.[151] Cleveland-Cliffs was not the first to erect a steel shaft house in the Lake Superior district to reduce maintenance costs and fire hazards, but it was among the first. By 1907 it had one at the Maas mine that Cliffs' master mechanic called "in every way the most satisfactory shaft house which we have."[152] CCI also had an early steel shaft house at the Negaunee mine, where it also experimented with concrete for lining the mine's shaft. When the company put in a shaft at its Athens mine between 1913 and 1917, it also adopted steel construction on the surface and concrete and steel in the shaft.[153] Both materials were cost-effective since they did not have to be replaced as frequently as wood.

In 1913 O. D. McClure, CCI's chief mechanical engineer, redesigned the skips—the cars that brought ore out of the mines. Soft hematite ore often stuck to the sides and bottom of the standard skip when it dumped its contents on stockpiles or into railcars, requiring a team of four men using picks or bars to clean sticky ore from the corners. McClure's new design speeded up dumping by reducing sticking and decreased significantly the number of men needed for skip dumping.[154]

Cleveland-Cliffs had introduced the caving system of mining to the Lake Superior iron ranges in the 1880s to reduce labor and timbering costs. Attempting to cut expenses every way possible after 1893, it applied the system to any other mine whose underground conditions merited its use (soft ore, weak overhanging wall), such as its new Cleveland Lake mine under Lake Angeline.[155] Drilling, too, underwent improvement. In 1897 J. George Leyner in Colorado developed a hammer drill as an alternative to the then-usual piston drill. In the piston drill the drill bit was attached rigidly to the piston rod. In the hammer drill the bit remained against the rock and was hit, like a hammer, by a reciprocating mechanism. Piston drills usually required two men for operation; by 1908 hammer drills had been made light enough for a single person. The Cliffs Shaft mine had thirty Leyner-Ingersoll one-man hammer drills in use by 1915.[156]

Not all cost-cutting measures were technological. Some were administrative. CCI adopted detailed cost accounting in its operations after 1900, imitating a technique introduced by Andrew Carnegie in his iron and steel business a decade or more earlier. H. R. Harris, superintendent of the company's railroads, commented to Mather 1901 that the detailed reports commenced that year were "unusual so far as my knowledge is concerned" but of benefit for overseeing operations.[157] By 1910, in an effort to scrutinize every source of expense for possible savings, mine superintendents' annual reports contained a detailed analysis of ore-production expenses in over thirty categories, ranging from insurance to pumping and hoisting, from safety to drilling, from ventilation to ore storage. In addition, superintendents had to provide a list of all delays, their causes, and estimates of tons of production lost.[158]

Operating Advantages

Securing labor tranquility, expanding production, and reducing its costs for transportation and ore production enabled Cleveland-Cliffs to survive the crises of the 1890s and the ensuing merger mania and emerge as the only surviving large independent iron ore–mining firm in America by 1910. Technological, economic, and geographical factors, fortunately, reinforced CCI's efforts and contributed to this survival. Technologically, the replacement of the Bessemer by the open-hearth process of steel making was of critical importance. In 1890 only 12 percent of American steel had been produced by open hearth; by 1910 the figure was over 63 percent and rising. The standard Bessemer process required Bessemer-grade iron ores (ores containing less than 0.07 percent phosphorus). The open-hearth furnace did not. It was capable of removing phosphorus from most ores.[159]

This was important for Cleveland-Cliffs. Most of its mines produced non-Bessemer ores. While it had uncovered some Bessemer ores in the 1890s and leased properties to increase the proportion of its output that was Bessemer grade, concerns remained. The rapid replacement of Bessemer steel by open-hearth steel between 1890 and 1910 rescued the company from this conundrum. Even more helpful, the open-hearth steel-making process simultaneously created a demand for hard, or lump, ores, like the specular hematite that had early been the company's primary product, but had been steadily displaced by soft hematite. In open-hearth furnaces, an iron oxide scum formed on the pool of molten steel shortly before it was time to pour the steel off into ingots. This scum represented as much as a 10 percent loss. Hearth operators found that if they dumped lump ore into the pool at the appropriate time, the lump ore dissipated the iron oxide film and sharply reduced iron oxide losses. Thus, steel companies required at least some lump ores for efficient operation. The Mesabi with its soft ores could not provide it, but a handful of producers operating on the Marquette and Vermilion ranges mined lump ore, including CCI at its Cliffs Shaft mine. CCI sales agents took advantage of the need for lump ores—which soon sold at a premium over other Lake Superior ore grades—by using it as a lever to extract parallel orders of its non-Bessemer ores.[160] By 1930 Cliffs Shaft produced more lump and crushed specular hematite ore and sold it to more different customers than any mine in the United States.[161] Finally, so-called old-range ores commanded a slight price premium on the ore market because they were less fine than the Mesabi ores and hence less likely to clog blast furnaces or be carried out the top of their stacks by updrafts. For this reason most furnace operators preferred to mix at least some old-range ore with fine Mesabi ores.

Geography also contributed to CCI's survival. While companies operating the Mesabi Range's open-pit mines had significant cost advantages in ore extraction, the location of Cleveland-Cliffs' properties offset some of these advantages. CCI's Marquette Range mines were significantly closer to Lake Superior, so its land shipping costs were lower. CCI's chief port for exporting its ores—Marquette—was several hundred miles closer to ore users, giving it lower lake shipping costs. Michigan, moreover, kept its taxes on iron ore mines lower than those of Minnesota so that Michigan underground mines could remain competitive with Minnesota open-pit mines.[162] And, most significant, Cleveland-Cliffs owned most of its ore-bearing lands and hence did not have to pay leasing fees. This was critical. Most companies operating on the Mesabi did not own the property on which they mined. They leased the properties and had to pay royalties, typically 25¢ to 50¢ a ton, to the property owners.

Thus, a combination of factors explain Cleveland-Cliffs' survival in the two decades after 1890 when dropping ore prices, the rise of the Mesabi Range, and the entrance of steel companies into ore mining led most other independent iron ore–mining companies to shut down or sell out to vertically integrated steel companies or former iron ore merchant dealers. The combination included the geographical advantages that the company's primary holdings on the Marquette Range had over mines on the Minnesota ranges, the continued need for at least small amounts of hard ores, lighter Michigan taxes, and, especially, the company's ownership of the land on which its mines were located. A handful of other companies on the "old ranges" had many of these same advantages but could not or did not survive. Cleveland-Cliffs survived because it supplemented the advantages given it by changing technology, geography, transport economics, politics, and land ownership. Under Mather's leadership it took active steps to secure labor tranquility, increase output to compensate for lower unit prices, and reduce operating costs through integration into transportation, electrification, and systematic cost accounting. And it enjoyed the solid, devoted leadership of William G. Mather, who headed the company for over four decades and provided continuity in management that other companies often did not have.

Charcoal Iron and Wood Chemicals

Besides simply keeping the company's ore business alive and independent in the post-1893 environment, Mather and Cleveland-Cliffs also pursued two related strategies: diversification and forward integration into iron and steel production. Designed to ensure multiple sources of revenue and a guaranteed market for its ores in the face of steel firms integrating backward into mining, both strategies succeeded in the short term. Ultimately, however, both encountered problems due to the onset of the Great Depression of 1929. But, in the case of forward integration, CCI was one of only very few iron-mining companies to defy the prevailing trend—backward integration by steel firms or ore-merchant firms into mining—by successfully integrating forward into iron and steel production.

Forward integration and diversification were unusual initiatives for the company. Samuel L. Mather, partly out of the virtue of necessity, had kept the old Cleveland Iron Mining Company largely focused on ore mining. After the destruction of its Marquette forge by fire in 1853, Cleveland Iron had turned its back on vertical integration into iron production, focusing instead on mining, letting other companies bear the heavy capital expenses of providing transportation for the ore and converting it into pig iron or, later, steel. In the 1870s, frustrated with periodic shortages of transport, Mather and his associates had briefly flirted with ownership of ore vessels, but backed away there, too. Only in the late 1880s, with the company's rivals beginning to purchase vessels, had Mather integrated forward into shipping. William Mather, facing different circumstances and having different predilections, eagerly embraced what his father had generally avoided: forward integration and diversification.

The Panic of 1893 provided William Mather and CCI an opportunity to reinforce the company's traditional focus on mining. By 1893 the charcoal-fueled Pioneer furnaces at Negaunee that the acquisition of Iron Cliffs had brought into Cleveland-Cliffs were dilapidated, giving the company a good excuse to abandon the production of charcoal iron. In past crises, the company had responded to economic stress by backing away from such peripheral enterprises. This time the company took the opposite tack: it took steps to consolidate the foothold that the Iron Cliffs acquisition had given it in iron production. This decision may have been prompted by the company's large supplies of non-Bessemer ores, which before the replacement of Bessemer converters by open-hearth furnaces were barely profitable.[163] In part, also, it may simply have been the spirit of the age. As we have seen, the period from 1895 to 1905 saw the general consolidation of numerous small and medium iron and steel firms into large, vertically integrated conglomerates. CCI may have simply seen its foothold in charcoal iron production as its response to these trends.

Even so, the decision to stay in the charcoal iron business did not come easily. Mather at first vacillated. In 1891 he acknowledged that he was "unfamiliar" with the business, and he noted that charcoal iron cost too much to make.[164] Samuel Redfern, CCI's land agent, argued that with the price of charcoal iron "continually lowering," the question was not so much whether to stay in the business, but "how to get out of the furnace business as soon as possible with the least loss."[165] In hindsight, Redfern was right. It should have been clear by 1893 that charcoal iron would ultimately be all but completely replaced by coke iron or steel produced from coke iron. In fact, by the 1890s charcoal iron increasingly depended on small, niche markets. Known for its freedom from impurities (particularly silicon) and uniformity, charcoal iron had long had a good reputation, and even when much cheaper coke-produced iron displaced it from many uses in the 1870s and 1880s and the price premium it commanded could no longer make up for its higher production costs, charcoal iron remained important for certain products in which low silicon content was critical, especially the cast-iron wheels used on railroad cars. But manufacturers used charcoal iron also for the jaws of rock crushers, the rollers in iron and steel mills, and some brake shoes. Nonetheless, by the 1890s the handwriting was on the wall. The wood needed to fuel charcoal iron furnaces was rising in price; the newly introduced open-hearth steel-making process was capable of producing steel of very high quality without having to use charcoal iron as a feeder.[166]

In this light, CCI's board of directors may have been a bit surprised in early 1894 when, in the midst of a major depression, Mather suddenly recommended replacing the dilapidated Pioneer furnace with a new one.[167] Furnace manager Austin Farrell's visit to Germany to investigate metallurgical processes in 1893 probably explains this deci-

Charcoal kilns and blast furnace of CCI's plant at Gladstone, Michigan, c. 1920. This plant, when erected in 1896, was the largest and most modern charcoal iron plant in the country. The kilns in the foreground distilled wood to produce charcoal to fuel the blast furnace (*background right*) and chemicals like wood alcohol and lime acetate (extracted from the gaseous by-products of the charcoal production) in the buildings in the left background. (*Courtesy of Cliffs Natural Resources, Inc.*)

sion.[168] Farrell observed that German iron and steel plants had replaced the traditional beehive ovens that produced coke for blast furnaces with special ovens, called retorts, that collected the waste gases from the coking process and extracted various marketable chemicals. Farrell recognized that the processes used in coke ovens to extract by-products could be used in charcoal ovens and might enable charcoal iron facilities to remain competitive by selling both charcoal iron and extracted wood chemical by-products, like acetic acid, methyl alcohol, acetone, formaldehyde, and lime acetate. He apparently sold Mather on the idea. Mather thus proposed building a new, large, modern charcoal-fired blast furnace with an associated wood chemical plant at Gladstone, near Escanaba, in the southern part of Michigan's Upper Peninsula. It was one of the first combination charcoal furnace–wood chemical plants in the country.[169] Billed as "the largest charcoal furnace in the world," it went into operation in 1896.[170] Early on the gamble seemed to work. Of its first year of operation Farrell noted: "The plant started off very nicely and considering our absolute ignorance of this business no very costly mistakes developed."[171] In early 1898 Mather reported that the new furnace had "contributed materially" to profits although prices for charcoal iron were "lower than ever before in the history of the trade."[172]

The Gladstone plant worked well enough to entice CCI to construct another charcoal blast furnace with an associated wood chemical by-product plant at Marquette in 1901. This plant was the "most modern of its kind in existence." Designed to integrate the iron-making and chemical-extraction functions more fully than in any previous plant anywhere, it was fully electrified and equipped with a host of laborsaving devices such as an automatic skip for loading the furnace while minimizing the escape of furnace gases.[173] Now sold on the idea of producing wood chemicals, CCI, through a captive company (the Pioneer Iron Company), acquired control of yet another charcoal furnace: the Carp near Marquette.[174] CCI also partnered in 1912 in the construction of a wood chemical plant (Cliffs Chemical Company) at Goodman, Wisconsin, in conjunction with the large sawmill of the Goodman Lumber Company. CCI had a 72 percent interest in the plant. The new venture utilized waste wood from the lumber company, which was converted into charcoal and wood alcohol. This venture, too, initially produced a profit.[175]

Once committed to charcoal-fired blast furnaces and wood chemicals, with furnaces at three locations (Gladstone, Marquette, and Carp River), CCI had to secure adequate wood reserves. Redfern, CCI's land agent, had noted in 1891 that as long as the company depended on outside parties for its charcoal supply, it would be subject to "fluctuations in the supply caused by the demands of other furnaces." He strongly urged the company to be independent "of all outside influences" in charcoal production.[176] After committing itself to the charcoal iron–wood chemical combination, Cleveland-Cliffs began to seek that independence. In 1900 it purchased the Munising Company and the Munising Railway (80,000 acres and fifty miles of railroad) to ensure a timber supply for its plant at Gladstone and the planned plant at Marquette.[177] The following year it purchased 1,060,000 acres of Upper Peninsula land from the Detroit, Mackinac, and Marquette Railroad for $1.4 million. Since this was a significant investment, Mather, in keeping with CCI tradition, sought a partner. The Upper Peninsula Land Company, half owned by CCI, purchased the lands.[178] CCI then purchased several hundred thousand acres of prime timberland from the captive company to fuel its wood chemical and blast-furnace operations and to provide timber for its mines. Cleveland-Cliffs also purchased an assortment of other large property holdings between 1895 and 1910. By 1914 CCI and its associated companies owned around 1.5 million acres, or 14.5 percent of Michigan's Upper Peninsula.[179]

These enormous landholdings provided adequate reserves for its charcoal iron and wood chemical operations. They also prompted further diversification into wood products. For some years the company struggled in its attempts to get "every cent from the forest products of the Company's lands" by enticing outside industries to locate in the Munising area, around forty miles east of Marquette on the Lake Superior shoreline and near its largest block of timber holdings. High local labor costs and freights and more difficult operating conditions "on account of the long cold seasons" discouraged most prospects.[180] Eventually Cleveland-Cliffs resorted, once again, to a joint venture

1 Cordwood Splitting Mill
2 Retort House and Pre-driers
3 Primary Acetate of Lime House
4 Alcohol Refining Plant
5 Formaldehyde Plant
6 Acetic Acid Plant
7 Acetone Plant
8 Laboratory
9 Sulphuric Acid Plant
10 Blast Furnace House
11 Power Plant
12 Ore Stock Shed
13 Machine Shop
14 Supply House
15 Blacksmith Shop
16 Carpenter Shop
17 Ore Drying Plant
18 Alcohol Barreling House
19 Alcohol Storage Tanks
20 Office

Marquette Plant
The Cleveland-Cliffs Iron Co.

Cleveland-Cliffs' blast furnace and wood chemical complex in Marquette. Erected in 1901, then one of the most modern plants of its type in the country, the chemical portion of the complex operated until the 1960s. The blast furnace, which operated until around 1930, is #10 in the center of the picture; the retorts that destructively distilled wood to produce charcoal for the blast furnace and gasses for the wood chemical plants in the complex are on the right. The facilities for producing alcohol (4), formaldehyde (5), acetic acid (6), acetone (7), and sulphuric acid (9) are also identified. The enormous quantities of wood that would be stacked up in the vicinity of the plant do not appear in this illustration. *(Courtesy of the Marquette County Historical Museum, Marquette.)*

to leverage its limited capital. One facility attracted to the area was the Munising Paper Company, which erected one of the largest pulp and paper mills in the country in 1904. Initially CCI invested $500,000 to finance a 50 percent share.[181] In 1907 and 1908 Mather worked with the Alanson Woodenware Company to establish a joint-venture woodenware company at Munising. CCI again took a 50 percent share.[182] The resulting company—Munising Woodenware—produced wood bowls, spoons, and clothespins. CCI also lured the Piqua Handle Company and a veneer plant to the region and operated a sawmill on its own.[183]

Initially, CCI's diversification into wood products, charcoal iron, and wood chemicals seemed wise. The honeymoon in some areas, however, was short lived, especially charcoal iron. Although CCI's charcoal iron–making capacity did not compare to the large iron and steel firms operating coke-fired blast furnaces on the lower Great Lakes, in 1906 production hit 87,000 tons, making Cleveland-Cliffs the largest producer of charcoal pig iron in the United States.[184] Although charcoal iron formed an increasingly small percentage of total American iron production, it still sold at a premium over coke iron. Its denser structure and greater strength made it preferable for high-wear uses, such as cast-iron wheels for railroad cars, rollers in steel mills, cylinder blocks in engines, and some types of valves, crushers, grinders, and dies.

The proportion of iron smelted with charcoal, however, continued to diminish, falling from 7 percent of American iron in 1890 to 2.8 percent by 1900 to 1.5 percent by 1910.[185] Ironically, the open-hearth process that had helped Cleveland-Cliffs by opening up a wider market for its large reserves of non-Bessemer iron ores and its lump, or hard, ores simultaneously hurt CCI by dooming the company's niche in charcoal iron. The quality of traditional coke-fired iron and Bessemer steel could not match charcoal iron. The open-hearth process, because it was much slower and more controlled, produced steel that could match, and would eventually exceed, the quality of charcoal iron. Open-hearth steel cost more than charcoal iron, but it was stronger and lasted longer. The single leading use of charcoal iron by the 1890s was for casting wheels for railroad cars. As railroads increased operating speeds and loads, the advantages of charcoal iron cast wheels over open-hearth steel wheels gradually dissipated.[186] As early as 1890 a British observer commented that the premium price commanded by charcoal iron did not compensate for the higher cost of production entailed by the use of wood instead of coal.[187] As early as 1897 one of CCI's customers declared that he had found he could make first-class malleable iron castings "from coke iron alone" and that he would "never consider the purchase of charcoal iron again" unless prices were entirely equal.[188] By 1910 open-hearth steel had begun to push into the niche once exclusively filled by charcoal pig iron.

Why did Cleveland-Cliffs invest so heavily in this dying technology? Part of the answer lies in capital availability. An iron-mining company capitalized in the millions of dollars did not have the funds to build and compete with coke-fired furnaces and steel mills whose capitalization was in the tens of millions or, in the case of U.S. Steel, hundreds of millions of dollars, and which had already established footholds in the American iron and steel market. Charcoal iron was the one niche where—thanks to its large landholdings—CCI had an advantage. The belief that profits from wood chemical by-products might enable charcoal iron furnaces to drop their prices closer to those of coke-fired furnaces reinforced the decision. For a few years this strategy worked. In 1901, for example, the company's two operating charcoal furnaces and wood chemical plants made a profit, but by 1904 furnace profits were "unsatisfactory."[189] Losses on charcoal iron continued and began to occasionally exceed the profits made on wood chemicals. In 1912 Farrell noted he was "losing an immense amount of money each year" and that he was almost ready to recommend closing down his own department. Indeed, in 1912 furnace operations lost $64,000, offset only because the associated chemical by-product plants made $106,000.[190] By 1914, however, the company's charcoal iron and wood chemical operations were clearly in trouble. On June 18 the Marquette furnace shut down; CCI suspended all operations except the production of wood chemicals. The Gladstone furnace, under repair the previous year, was blown back in "not because of expected profits, but to keep the organization employed, running losses being practically no greater than 'idle expense.'"[191]

Forward Integration

The desire to have some ensured outlet for its ores and alternate sources of revenue kept CCI in the charcoal iron business. But the heart of the company's business remained ore sales to iron and steel firms on the lower Great Lakes. As the larger iron and steel firms secured their own iron ore supplies between 1895 and 1905, Cleveland-Cliffs sought contracts with smaller companies that owned no mines or insufficient mines. In 1900 the Antrim Iron Company of Grand Rapids, in lower Michigan, agreed to a ten-year contract to buy its ore from CCI, and in 1901 CCI reached long-term contracts with the Lackawanna and Jones & Laughlin steel companies and the Dunbar Iron Company.[192]

With steel mills integrating backward into ore mining, Mather clearly wished to leave open the option of integrating forward into coke iron and steel production, especially as the demand for charcoal iron continued to drop. At a board meeting in 1902 he suggested that the company secure a site "suitable for blast furnaces and steel plant" at a desirable port on Lake Erie for possible future use, since good harbor and dock sites were "getting somewhat scarce on Lake Erie." In 1903 CCI purchased land at Toledo, Ohio, for this purpose.[193]

Because coke iron production involved the conjunction of two primary raw materials, iron ore and coal, Mather also wanted to secure a foothold in coal. In 1909 Cleveland-Cliffs took an option on 2,600 acres of coal land in Greene County, Pennsylvania, "on the theory that it would be well for the company to have its property thoroughly rounded out through the possession of a tract of coking coal, and thus through the possession of the principal raw materials necessary for the production of pig iron and its products, be in a position to engage in such manufactures whenever it saw fit." The price, however, was too high. In keeping with tradition, CCI's executive committee tried to find a partner. It first contacted Pickands Mather, where William Mather's half brother Sam was a key figure. When Pickands Mather declined, William Mather announced he would purchase the property for himself and would sell it to the company later if it wished. Mather's action apparently prompted both companies to reconsider. The result was another joint venture. CCI, the Steel Company of Canada, and Pickands Mather each bought one-third of the property—to be named the Mather Collieries—with Pickands Mather serving as operator of the mine.[194] In 1917 CCI supplemented the Mather Collieries by taking a lease on the Ethel coal mine (3,600 acres) in West Virginia.[195] To provide a market for the coal—since it did not own a blast furnace nearby—CCI diversified further after World War I, purchasing a coal dock at Green Bay in 1922, the first of a number it would acquire.[196] In addition, it acquired part interest in the Low Volatile Coal Company and established in 1926 a vessel-fueling facility at Port Huron, Michigan, capable of refueling vessels with coal faster than any previously established on the Great Lakes. It also began operating as sales agent for other coal producers. These actions furnished an up-bound cargo—coal—for the twenty-two ore vessels owned or operated by the company by the 1920s.[197]

In 1910 Mather again pushed diversification and forward integration by urging CCI's board to purchase controlling interest in the Cleveland Furnace Company, which operated two blast furnaces. CCI made an offer, but the owner declined. In 1915 Cleveland-Cliffs was more successful, purchasing 51 percent of the company's stock. The Cleveland Furnace purchase carried with it a substantial interest in an adjacent by-product coke plant and two mines on Minnesota's Cuyuna Range (the Kennedy and the Meacham).[198] In 1920 CCI pushed still further into coke iron production, undertaking a joint venture with the Trumbull Steel Company. They erected a large blast furnace at Warren, Ohio, under the corporate name of the Trumbull-Cliffs Furnace Company. The Trumbull-Cliffs furnace, operated by Cleveland-Cliffs personnel, was the largest blast furnace in the world, built in record time in 1920–21; it was designed specifically for efficiency.[199] As part of the agreement, Trumbull-Cliffs contracted to purchase a large quantity of its iron ore from Cleveland-Cliffs, and Trumbull Steel, the joint owner, agreed to purchase a large quantity of the pig iron produced by Trumbull-Cliffs for conversion into steel at its mills.[200]

Purchasing equity in iron and steel producers to secure long-term ore contracts became standard operating procedure for Cleveland-Cliffs in the early twentieth century, beginning with Cleveland Furnace and Trumbull Steel. It supplemented the company's traditional policy of using joint ventures for this same reason. In 1919 CCI acquired a large interest in the Otis Steel Company of Cleveland, when that company merged with the Cleveland Furnace Company. In 1923 CCI purchased one-third interest in Central Steel in return for an agreement to purchase 400,000 tons of ore annually for the next ten to fifteen years.[201] That same year Mather proposed acquisition of Federal Furnace property in Chicago, possibly in partnership with Interstate Iron and Steel.[202] The proposal died. But in the mid-1920s Mather did purchase a floundering foundry company in Chicago—Interstate Foundry—to try to get a foothold in that industrial center. He found foundries to be "a dirty business" and got out in 1928 at a loss.[203] In 1926 Cleveland-Cliffs invested in Cincinnati-area Belfont Steel and Wire in return for a commitment to purchase ore for eight years.[204] Finally, in 1929 Cleveland-Cliffs purchased stock in the Republic Iron and Steel Company and Donner Steel, securing also, in the latter case, a contract to manage a Donner mine on the Mesabi.[205]

By the outbreak of World War I in 1914 Mather's initiatives had succeeded. In many

The Trumbull-Cliffs furnace in Warren, Ohio, under construction, c. 1920. A joint project between Trumbull Steel and Cleveland-Cliffs, the Trumbull-Cliffs furnace was built and operated by Cleveland-Cliffs. When completed in 1921, it was the largest blast furnace in the world. *(Courtesy of Cliffs Natural Resources, Inc.)*

ways, Cleveland-Cliffs had become a diversified, vertically integrated organization, even though still dwarfed by large, vertically integrated, coke-based iron and steel producers like U.S. Steel and Bethlehem Steel. The core of the company's strength still lay in its iron ore mines and its reserves of wholly owned mineral lands on the Marquette Range. Company mines had increased production from around half a million tons of ore a year in the early 1890s to over 2 million tons in the years preceding 1914. The company's other activities supported this core. Its growing fleet of vessels not only carried the ore to iron and steel manufacturing plants in Ohio, Illinois, Indiana, and Pennsylvania, but had begun to carry, as a return load, coal from company mines into the Lake Superior basin for power plants and other operations. The company's extensive landholdings furnished timber for sale, but also provided timber supports for its mines. Cordwood from its landholdings fed its charcoal-fired blast furnaces (which produced 70,000 tons of pig iron in 1913), its wood chemical by-product plants, and the wood-dependent industries near Munising that the company promoted and often partly owned. The same logs and the company's iron ore provided freight for the CCI-controlled LS&I and Munising railroads. The chemical by-product plants, charcoal furnaces, mines, and railroads all provided customers for the company's hydroelectric system. As Mather would later comment: "The reciprocity between these activities of the company [gives it] a unique position among organizations which are connected with the Lake Superior iron-mining industry."[206]

World War I Boom and Postwar Bust

In the late summer of 1914 war broke out between Serbia and Austria-Hungary and quickly spread over all of Europe. Blockades and counter-blockades, the elimination of some European markets for American steel, and general uncertainty initially combined, as in the American Civil War a half century earlier, to depress iron and steel markets.

As the market for ores diminished in late 1914, CCI shut down some of its higher-cost producers—the Salisbury, Austin, Princeton, and Maas mines—and curtailed operations at others. It reduced salaries and wages 10 percent and cut back exploration and development activities. For 1914 ore deliveries and pig iron sales dropped around 30 percent from 1913 levels.

As World War I evolved into a prolonged stalemate in Europe, however, the demand for iron, steel, and chemicals to support British and French war efforts surged, with demand growing even more after America belatedly entered the war in April 1917. The U.S. government quickly classified nearly all the company's primary products—iron ore, charcoal pig, and wood chemicals—as "essential," and all the company's plants "operated to their fullest extent."[207] In 1913, the year before the outbreak of war in Europe, Cleveland-Cliffs had produced about 2 million tons of iron ore and 71,000 tons of charcoal pig iron. In 1918, the year the war ended, it produced almost 3.5 million tons of ore and 81,000 tons of pig iron (down from the wartime peak of 133,000 in 1917).

S. R. Elliott, the company's general superintendent of mining, took leave to enter officer training and was commissioned a major of engineers and sent to France. The company also lost several members of its engineering department. Murray M. Duncan, Elliott's right-hand man, remained, but had other responsibilities. He accepted a government appointment to oversee iron ore production in the entire Lake Superior region, the source of 80 percent of American iron ore.[208] Altogether around 450 Cleveland-Cliffs employees enlisted or were drafted, well over 10 percent of its workforce.[209] Increased demand for production, enlistments, and the draft quickly created a labor shortage in the mining region, driving wages upward. In July 1915 the company's Upper Peninsular managers jointly advised Mather that the wage cut imposed when the war broke out should be rescinded as soon as possible: "We have always to remember," they noted, "that this is a Non-Union district, and unless the companies treat the men fairly and liberally, it will result in their organizing, in which case they will dictate wages and hours of labor."[210]

Although Cleveland-Cliffs focused primarily on increasing iron ore production to support the war effort, it also participated in other government-sponsored initiatives. In the patriotic fervor surrounding American entrance into the war, the newly formed Federal Food Administration appealed to citizens to plant "victory gardens" to permit shipment of more foodstuffs to American troops and allies abroad. In several of its mining districts CCI plowed, disked, harrowed, and fenced large areas and offered garden plots and seed potatoes at cost to employees.[211] When the antiwar IWW (International Workers of the World) sent organizers to the district, attracting "a large following among the miners" and leading to "considerable labor agitation," the company welcomed a detachment of the state constabulary, providing it with two of its unused buildings in Negaunee as a barracks and stable.[212] The army gave the company one specific task: produce 15,000 tons of extraordinarily pure iron ore so its ordnance facilities could fabricate an artillery piece capable of matching Germany's "Paris gun" (sometimes mistakenly called "Big

Bertha"), which was capable of firing shells up to eighty miles. The ore had to contain less than 0.02 percent phosphorus, the element that usually weakened iron. Duncan arranged for CCI's Republic mine to produce the ore. The war, however, ended before the American reply to the German Paris gun could be completed and shipped overseas.[213]

Wartime demand for iron, steel, and wood chemicals, which were used in the production of explosives, gave the company's charcoal iron and wood chemical operations a much-needed boost. The market for charcoal iron revived, and wood chemicals like acetone, an important solvent, found extensive use in explosives manufacture. By the end of 1915 both CCI's Gladstone and Marquette furnaces and their associated by-product plants were "running to their full capacity." The following year the demand for other wood by-product chemicals like acetic acid encouraged further expansion of wood chemical operations. The expansion was delayed when, in January 1918, an explosion demolished the chemical by-product plant at Marquette. Sabotage by an ethnic German employee was suspected but never proved.[214] In any case, between 1917 and 1920 CCI invested over $1.3 million in expanding its two primary charcoal furnace and wood by-products plants.[215]

The expansion was unfortunate. Like many other American companies, Cleveland-Cliffs overexpanded during World War I. When German armies suddenly and unexpectedly collapsed in the fall of 1918, bringing the war to an abrupt end, both the American government and governments overseas cancelled contracts, leaving suppliers saddled with large, unneeded stockpiles of materials and productive overcapacity. These conditions sparked a short-lived depression in 1920–21. But as late as 1924, Mather lamented that many of the company's product areas were still suffering from "excess capacity brought about by the stimulus of the war."[216]

In the depression that followed World War I, Cleveland-Cliffs shut down ten of its mines and operated the remainder, with one exception, on a single four-hour daily shift in an attempt to keep as many of its workers employed as possible. Mather noted that the company had reduced its operations as much as possible consistent with maintaining its properties "and the welfare of our employees who are entirely dependent upon us." He acknowledged that those employed and working half-time were "barely earning a living." Wages, which had risen rapidly in World War I, decreased just as rapidly, with miners suffering three wage cuts in 1921 alone: 14 percent, then 11.35 percent, then 8 percent. The price of goods, which had risen as a result of wartime inflation, did not fall nearly as fast. Wages would return to their August 1918 rates only in 1923.[217] The company's charcoal iron and wood chemical operations were hit hardest. Wages there dropped more precipitously, being cut 15 percent, then 10 percent, and then 35 percent. When these cuts proved insufficient to stop the bleeding, Cleveland-Cliffs closed down both furnaces and their associated by-products plants.[218] In 1921 the vessel department—which had expanded by six vessels during the war—was the only operating department that yielded a profit. All others—ore mines, coal mines, charcoal furnaces, wood chemicals, the Trumbull-Cliffs furnace, timber operations, and railroads—lost money. Total losses for the year exceeded $1.6 million.[219]

The Not-So-Roaring Twenties

Although Cleveland-Cliffs reopened its charcoal furnaces in 1922, the high cost of wood and the scarcity of timber men kept the plants unprofitable.[220] In 1923, acknowledging the inevitable, CCI permanently closed its Gladstone furnace. Mather told stockholders that despite having "fine plants for the recovery of by-products," the outlook was "discouraging."[221] Matters were, indeed, to worsen. On Christmas Day 1925, Austin Farrell, the director of the company's furnace department, died. He had directed the unit for thirty-three years. Although it was struggling, Farrell had at least kept the department close to profitability by tying charcoal iron to wood by-product chemicals to a greater extent than had previously existed in the country.[222] CCI's wood chemical and charcoal iron operations would have one or two good years after Farrell's death, but the end was in sight. Charcoal iron was no longer needed; cheaper coke iron could almost match it in quality and open-hearth steel was superior.

CCI wood chemicals had also begun to struggle. Problems abounded, mostly related to geography and marketing. Cleveland-Cliffs' wood chemicals competed with those of better-located rivals. These rivals had lower railroad freight rates to most points of consumption and were therefore capable of offering quicker delivery. If the Hammond Distilling Company in Indiana, for instance, placed an order with the New York–based Wood Products Company, it could expect delivery in three days; if it ordered from CCI in Michigan's Upper Peninsula, delivery took three to four weeks.[223] Making matters worse, in the mid-1920s several American chemical companies developed processes for making wood chemicals synthetically.[224]

Cleveland-Cliffs also found in the 1920s that operating coal mines was not lucrative. In 1922 a prolonged coal strike decreased profits. In 1923 Mather complained: "There is no money in operating coal mines." He compared the company's coal operations to its struggling charcoal iron operations: "valuable" but "operating unsatisfactorily due to the high costs and overdevelopment resulting from the war."[225] In 1925 CCI's board authorized Mather to sell the Ethel mines for $250,000 or "even somewhat less."[226] Mather was unable to find a buyer at that price, and the operation continued to lose money every year.

Meanwhile, Cleveland-Cliffs suffered the most serious mining disaster in its long history. On the morning of November 2, 1926, following a routine blast, water and debris suddenly flooded into the company's Barnes-Hecker mine. The blast apparently broke open an underground cavity (a vug). The vug had either already filled with water from the swamp located above the mine before the company constructed a drainage system in 1922 or the blast opened a path through the vug to the surface, permitting

Miners in drift of Barnes-Hecker mine, 1926, a month before the most costly mining disaster in the company's history. Several of the miners pictured here would die in the catastrophe. *(Courtesy of Cliffs Natural Resources, Inc.)*

water-saturated overburden and a small nearby lake to enter the mine. In ten to fifteen minutes, the mine filled with mud, water, and debris. Of those underground at the time, one survived; fifty-one perished, with only seven bodies recovered. While outside investigators exonerated the company from blame in the accident, a judgment made easier by the company's previously stellar record in promoting mine safety, the disaster prompted the company to abandon a mine that had just begun to show promise.[227]

In the midst of its postwar troubles, the company had several bright spots. Chief among them, production from CCI mines rose slowly but steadily in the 1920s. In 1929 CCI shipped nearly 4.3 million tons of ore, around 30 percent more than it had shipped annually between 1918 and 1920. Two of its Marquette-area mines—the Maas and Negaunee—expanded to the point that in 1928 CCI had to buy and move a portion of the city of Negaunee since several subdivisions overlay ore deposits.[228]

Cleveland-Cliffs successfully increased its ore output despite increasing labor shortages and rising labor costs. In 1921 Congress passed the country's first restrictive immigration law. As a result, the rate of immigration (per 1,000 population) dropped to one-third of the level it had stood at between 1890 and 1914. On the Lake Superior iron ranges, long dependent on an immigrant labor force, this created problems. The ethnic diversity that had characterized Cleveland-Cliffs' workforce for decades slowly diminished. In the period before World War I noncitizens made up more than 90 percent of the company's workforce; by 1925 a similar proportion were citizens, albeit a large number were foreign born.[229] Labor shortages and the resulting upward pressures on wages, reinforced by economic depression after 1929, forced the company to seek out more laborsaving techniques in every stage of its operations.

To speed up ore extraction, CCI personnel developed the "Athens System," first applied at the 2,500-foot-deep Athens mine in the 1920s. In this system, miners systematically worked at multiple levels of a deposit, extracting ore from lodes in vertical rather than horizontal slices.[230] The system would be widely imitated at other mines. At the same time, the company cut costs in the labor-intensive task of timbering underground passages and mined-out openings (stopes) by using treated lumber in place of untreated. This decision followed experiments conducted in cooperation with the Bureau of Mines showing the average life of untreated timber underground was 3.8 years while timbers treated with zinc chloride lasted 13 years or more.[231] To reduce labor costs further, CCI installed a timber-treating plant at its Athens mine in 1936.[232]

The most labor-intensive jobs in any underground iron mine were mucking (shoveling broken ore into cars underground) and tramming (moving the cars to a hoisting shaft). Mucking and tramming could occupy 35 to 50 percent of total shift time for a work crew handling all aspects of mining (that is, including drilling and blasting as well).[233] These labor-intensive tasks naturally came under scrutiny. As previously noted, in the early 1890s Cleveland-Cliffs had pioneered the introduction of electric locomotives for underground tramming. Mucking proved a bit more difficult. As early as 1917 CCI attempted to use a small underground steam shovel to load ore. This technology proved less effective than a simpler and much less costly device: a mechanically powered scraper, or scoop, pulled along the floor of a mine adit, or tunnel, by rope.[234] In 1898 some western mines had experimented with scrapers. In 1916 and 1917 Ingersoll-Rand had attempted to introduce these scrapers on the Lake Superior iron ranges, but the air-powered devices proved too weak and too fragile for iron mining. Moreover, they operated only in one direction. Scoops had to be returned by hand to where broken ore remained to repeat the process. Labor, moreover, resisted the introduction of the scraper, seeing the technology as one that would benefit companies, but not workers.[235] Not until after World War I did mining-supply companies develop a sufficiently powerful and reliable double-acting, or double-drum, scraper system.[236]

These improved scrapers found ready use in mines using a caving system. These operations usually had a tramming drift, or tunnel, some distance below where the ore body was being worked. This tunnel was equipped with rails and cars to haul ore to the hoisting shaft. Small vertical or steeply inclined shafts (winzes or raises) linked the tramming drift to the area where the deposit was being worked some distance above. Once they blasted ore loose in the work area, miners used the new scrapers, pulled by ropes that wrapped around electrically powered drums, to drag the heavy rock along the floor

A tugger-scraper system on the seventh level of CCI's Morris mine in the 1920s. The electrically powered drum that made up the tugger part of the combination is on the left; the scraper that the tugger's cables pulled back and forth is on the lower right. The scraper would move ore from the point where it was blasted loose to a vertically situated passage (called a raise), which led to a lower level. Ore fell by gravity into ore cars on this level for transportation to a hoisting shaft. *(Courtesy of the Michigan Technological University Archives and Copper Country Historical Collections, Houghton.)*

to one of the passages leading to the tramming drift below. The ore would fall by gravity into storage compartments (mills) built just above the tramming drift. There it would be released into cars for transport to the hoisting shaft. In this way, scrapers eliminated the labor-intensive task of loading ore by hand onto tramcars and reduced the labor involved in moving the ore. Once introduced in improved form, scrapers rapidly spread and became the "chief means of reducing the cost of mining" in the Lake Superior iron region.[237]

In 1925 Cleveland-Cliffs introduced scrapers in several mines. Reports everywhere were enthusiastic. The manager of the Morris-Lloyd mine reported that the scrapers had "worked out wonderfully well." Compared to traditional shovel operations, he noted, they increased efficiency by "almost exactly 90%." At the Barnes-Hecker mine, the tons of ore handled per man per shift almost tripled.[238] In the early 1920s CCI also experimented with a mechanical loader, essentially a tramcar with a power shovel attached at the front that would pick up ore or rock, lift it, and dump it into a car behind. While they could not be used effectively except on main levels, they handled rock at a rate almost double that of hand shoveling in the limited areas they could be used.[239]

In 1929 the company opened the Tilden mine, the first major open-pit mine on the Marquette Range since the late 1870s, when most mines had gone underground. The ores produced were not up to the grade of most of the company's mines. They were siliceous ores, that is, ores with typically less than 50 percent iron and up to 15–20 percent silicon. These ores did not command a high price, but had a niche market. They were used in blast and electric furnaces to produce ferrosilicon, a material used to manufacture chemically resistant steels and transformer and motor cores, or to improve the operation of blast furnaces. Because the Tilden ores could be mined using steam shovels, they could be mined cheaply and occasionally found a buyer.[240]

Not all of the company's cost-cutting experiments in the two decades following World War I found widespread application. One of its more unusual experiments was the use of concrete shaft houses instead of the more traditional wood or newer steel construction. In 1919 CCI needed to replace two old wooden shaft houses at its Cliffs Shaft mine in Ishpeming. Because of the mine's long anticipated life, CCI officials rejected wooden replacements. Cost and delivery delays ruled out steel. CCI engineers decided instead to erect reinforced concrete shaft houses around the existing wood houses while mine operations continued. Because of the prominence of the shaft houses—located on a hill overlooking company offices and the city of Ishpeming—Mather asked that they be given a pleasing appearance. CCI retained an architect who proposed an Egyptian revival design. The resulting shaft houses, resembling Egyptian obelisks, were completed in December 1919.[241] Steel shaft houses were the generally preferred option after 1910, but Ishpeming's unique concrete shaft houses became (and remain) a local landmark and a symbol of Mather's desire to blend utility, economy, and aesthetics.

The Eaton-Mather Partnership

Even as Cleveland-Cliffs consolidated its reputation as a well-managed, technologically progressive mining company in the aftermath of World War I, the iron and steel industry that it served saw a new round of consolidations. A key factor was the rapid replacement of primitive "beehive" coking ovens with by-product coking ovens between 1910 and 1930. (CCI's use of by-product ovens for preparing charcoal anticipated this development by a decade.) The new ovens, while much more expensive than beehive ovens, sharply reduced the price of coke, both by more efficiently converting raw coal to coke

The Cliffs Shaft mine headframes, c. 1895. The wooden headframes had been erected in the 1880s by the Iron Cliffs Company to tap into rich ore deposits identified deep underground through diamond-drill exploration. The building between the two headframes is where a crusher reduced the very hard Cliffs Shaft ore to manageable size. (*Courtesy of Cliffs Natural Resources, Inc.*)

The Cliffs Shaft mine with its new concrete Egyptian revival headframes, looking southeast, October 1920. Lake Bancroft is in the left foreground. The city of Ishpeming lay just to the southeast, between the shaft houses and the hills in the background. (*Courtesy of Cliffs Natural Resources, Inc.*)

One of the concrete Egyptian revival headframes at the Cliffs Shaft mine, c. 1920. The men in the foreground are probably working to beautify the area around the shaft, following directions from Warren Manning. *(From Alexander N. Winchell,* Handbook of Mining in the Lake Superior Region *[Minneapolis: Byron & Learned, 1920], 73.)*

and by producing a range of saleable coal-based chemicals. Smaller independent furnace companies that had survived the consolidations around the turn of the century and continued to supply foundries and smaller steel companies with pig iron could not afford the new ovens and either went out of business or were absorbed by bigger firms. By the end of the 1920s few independent blast-furnace companies remained. Most surviving firms were large vertically integrated iron and steel conglomerates that owned at least some ore supplies.[242]

Cleveland-Cliffs had been a leading ore supplier of these small, independent furnace companies. It now saw its customer base decline further in numbers, leaving the company vulnerable since it had no permanent ties to most consumers of its product. In the mid-1920s Cleveland-Cliffs remained the largest firm devoted primarily to iron ore mining, and it continued to ship, proportionately, roughly the same percentage of ore produced as it had in 1910, that is, a bit over 6 percent of the Lake Superior district's total (and the Lake Superior district supplied almost 90 percent of the nation's iron ore in 1925). But it had slipped from its position of second place among all Lake Superior producers behind the giant U.S. Steel/Oliver operations (44.6 percent of shipments in 1925). Two of the iron ore merchant firms that had integrated backward into mining in response to the consolidation of the industry after 1895 had surpassed it: Pickands Mather (18 percent) and M. A. Hanna and Company (6.7 percent).[243]

Faced with increased vulnerability, Mather and Cleveland-Cliffs scrambled between 1915 and 1930 to ensure that the company had customers for its ores. As noted previously, CCI acquired 51 percent ownership in the Cleveland Furnace Company in 1915, built the largest blast furnace in the world in a partnership with the Trumbull Steel Company in 1921, and invested in Otis Steel and Belfont Steel and Wire in order to guarantee customers for its ores.[244]

Mather also became increasingly involved in merger deals. As a private stockholder, he participated in syndicates that bought stock in various iron and steel concerns to promote mergers.[245] As head of Cleveland-Cliffs, he became even more deeply involved because Cleveland-Cliffs was viewed as an attractive merger partner, not only because it remained the largest independent iron ore–mining firm with the second-largest reserves of ore after U.S. Steel's subsidiary Oliver Iron, but also because the company had a reputation as a well-managed firm.[246] In this context William G. Mather's strategy of ensuring that Cleveland-Cliffs had customers by investing in the firms to which it supplied ore reached its climax.[247]

In the late 1920s Mather and Edward B. Greene, a Cleveland banker soon to be on the board of Cleveland-Cliffs, developed ties to Cleveland entrepreneur Cyrus Eaton. Mather became involved with Eaton as early as 1925, when he met with him to discuss the pending acquisition of Trumbull Steel, a firm closely tied to Cleveland-Cliffs, by Republic Iron and Steel Company, a firm in which Eaton had interests. In 1926 Eaton persuaded Mather to participate in a syndicate formed to purchase Republic Iron and Steel stock, and the two talked later about the possibility of combining the remaining independent ore producers, like Cleveland-Cliffs and Oglebay Norton, with the leading independent steel manufacturers.[248]

Eaton had broad vision. He saw the Republic Iron and Steel Company as the logical vehicle for constructing a large midwestern steel company that would consolidate the host of small and medium steel firms in the Midwest to create a company that would dominate the upper midwestern steel market and, on a national scale, rival the two giants of the industry, U.S. Steel and Bethlehem Steel.[249] In 1926, as a first step in creating this midwestern steel giant, Eaton organized an investment company: Continental Shares. In the late 1920s, through Continental Shares, he secured substantial stockholdings in a number of medium-sized steel companies, including Republic Iron and Steel, Inland Steel, Youngstown Sheet and Tube, and Wheeling Steel.

More important, Eaton secured controlling interest in Republic in 1927. Under Eaton's direction, Republic between 1927 and 1929 acquired a number of iron and steel firms, including Trumbull Steel and the Trumbull-Cliffs Furnace Company. Because these acquisitions gave Eaton control over two ore-consuming properties with strong ties to CCI, Mather invited Eaton in May 1927 to accompany him and several new CCI directors on an inspection of Cleveland-Cliffs properties in Michigan's Upper Peninsula. They stayed at the Mather Cottage during their week in Ishpeming.[250] In turn, because Cleveland-Cliffs had acquired substantial stockholdings in Republic as a result of the acquisition of the Trumbull operations by Republic, Eaton had Mather and Greene elected to Republic's board of directors. When Eaton pushed his program further by merging Republic Iron and Steel with the Central Alloy Steel Corporation, Bourne-Fuller, and the Donner Steel Company in April 1930 to create the Republic Steel Corporation, CCI's involvement in Eaton's plans and its holdings in Republic increased further, for CCI had also owned stock in and supplied ore to both Central Alloy Steel and Donner Steel.

Because of his concern about the diminishing number of ore customers resulting from the new spurt of mergers, Mather hoped to ensure that Cleveland-Cliffs had close links to whatever emerged. Thus he worked with Eaton in the late 1920s to bring Eaton's Midwestern Steel Company to life, calling Eaton "the most untiring and persistent man" that he knew.[251] Eaton, in turn, believed that his prospective new company needed a reliable ore supply and guaranteed ore reserves, and he saw these as something Cleveland-Cliffs could provide. A marriage, or at least a close alliance, between Eaton's proposed midwestern steel giant and CCI's ore supplies seemed natural to both men.

Mather and Eaton met in March 1929 in Cleveland to try to work out the means of linking CCI's ore supplies with the array of ore consumers that Eaton had consolidated or planned to consolidate under Republic's umbrella. They continued conversations after Mather departed for his annual vacation to Pasadena, California, culminating in a four-day meeting in Pasadena in April with most of CCI's board of directors, including William G. Mather's half brother, Sam, on CCI's board and still a senior member of the rival firm Pickands Mather. They reached a tentative agreement on April 18.[252]

The agreement was not all that Eaton wanted, for Mather was determined to preserve CCI's identity, even while participating in Eaton's plans for a giant new steel company. They agreed to create a holding company, called the Cliffs Corporation, as a preliminary to the formation of Eaton's Midwestern Steel Company. Cliffs Corporation would hold all of Cleveland-Cliffs stock and the $40 million of stock Eaton had accumulated from four of the steel companies (Youngstown, Inland, Republic, and Wheeling) that he hoped to merge to create his Midwestern Steel Company. Half of the common shares of the new holding company would go to the Eaton interests. CCI stockholders exchanged their shares of CCI common stock for the other 50 percent of the common shares of Cliffs Corporation on a one-for-one basis. In addition, however, Cleveland-Cliffs shareholders received one and a quarter share of Cliffs Corporation preferred stock for each share of Cleveland-Cliffs' common stock, since Cleveland-Cliffs' contribution to Cliffs Corporation was valued at $90 million, whereas the Eaton block's contribution had a market

Cyrus Eaton (1883–1979), 1957. Canadian-born, Eaton served his apprenticeship in business with John D. Rockefeller before settling permanently in Cleveland, where he became involved in banking and investment. In the late 1920s, Eaton's investment company Continental Shares acquired controlling interest in a number of firms supplying the rapidly expanding automobile industry, including Firestone and Goodyear. Eaton then sought to form through merger and acquisition a large integrated midwestern steel firm capable of supplying the auto industry with its steel needs and rivaling steel industry giants U.S. Steel and Bethlehem Steel. *(Courtesy of Cleveland Press Collection, Special Collections, Cleveland State University.)*

value of only $40 million.[253] The agreement that created Cliffs Corporation, moreover, contained a number of provisions designed to "continue the unquestioned control of the . . . management of Cleveland-Cliffs" over the company. For instance, Mather insisted that the shares of both Cleveland-Cliffs common stock turned over to the Cliffs Corporation in trust and the Cliffs Corporation common stock issued to Cleveland-Cliffs stockholders be voted in a block by CCI's board of directors. In addition, Mather secured a provision that, in case of a default in dividends, gave each share of Cliffs Corporation preferred stock (distributed to CCI shareholders) and Cliffs Corporation common stock (distributed equally to both CCI shareholders and Eaton's interests) one vote each.[254] These steps ensured Cleveland-Cliffs' autonomy in the new organization, and, in fact, effectively gave CCI's board a veto over actions by Cliffs Corporation. Mather described the proposed arrangement to Cleveland-Cliffs stockholders as an alliance, rather than a merger, and assured them that the alliance would give CCI an inside track on ore sales to those companies in which Cliffs Corporation held large voting stock and diversify stockholder income, since Cliffs Corporation share holders would receive dividends from both Cleveland-Cliffs operations and the iron, steel, and finished steel companies' stock that Eaton's group had contributed.[255]

Just as the Great Depression broke across its bows, CCI entered the last throes of its drive to ensure customers for its ores through investment in iron and steel firms. In March 1930 for $35.5 million Cleveland-Cliffs acquired the McKinney Steel Holding Company. This purchase gave Cleveland-Cliffs around 55 percent of the common (voting) stock and 62.5 percent of the total capital stock of the Corrigan, McKinney Steel Company, which operated blast furnaces and steelworks in Cleveland and mines on the Menominee, Gogebic, and Mesabi ranges. Mather planned to eventually sell the blast furnaces and steel mills of Corrigan, McKinney to the projected Midwestern Steel Company on which he was working with Eaton. As part of the process, however, Cleveland-Cliffs hoped to retain Corrigan, McKinney's mine holdings, giving it a strengthened presence in the Lake Superior ore district outside of the Marquette Range.[256] The Corrigan, McKinney acquisition, however, required CCI to borrow $25 million on six-month notes from a consortium of eight banks, in addition to taking $5 million from its own treasury.[257] This was a huge amount by contemporary standards for a firm the size of Cleveland-Cliffs. Mather assumed a merger of Corrigan, McKinney with Eaton's interests would follow rather quickly. Unfortunately, outside events were overtaking the American iron and steel industry that would make both this purchase and the company's involvement in the Cliffs Corporation utter disasters.

4

DEPRESSION, WAR, AND DEPLETION, 1930–1950

On Black Tuesday, October 29, 1929, after a prolonged period of over specula-tion, stock prices on the New York Stock Exchange began to drop rapidly. Attempts to halt the decline failed. By the time stock prices hit bottom in 1932, the market had lost 89 percent of its value. Financial panic, deflation, and bank failures ensued. As confidence in the economy waned, businesses closed, unemployment climbed to almost 25 percent, and those who still had jobs saved rather than spent, fear-ing for the future. Consumer demand evaporated and plummeting sales, especially in the growth automobile industry, sent ripples through the American economy. Orders for iron and steel dried up. Orders for iron ore soon followed suit.[1]

Impact of the Depression: The Mining Regions

Like many others, Cleveland-Cliffs' management initially assumed the economy was entering a short-lived recession, like those of 1907 and 1921. In this belief, Mather con-summated the purchase of the McKinney Steel Holding Company in early 1930. Cliffs continued to mine ore at near-normal rates, producing a bit over 3 million tons in 1929 and 1930, above the company average of 2.4 million tons for the previous four years. Actual ore shipments, however, dropped rapidly: from 4.3 million tons in 1929 to 2.5 million in 1930.

By 1932 the American economy was in shambles, and the iron ore industry reflected that condition. That year *all* the companies mining ore in the Lake Superior iron district *combined* shipped only around 3 million tons, down from the 50-million-ton annual average of the 1920s and significantly less than what Cleveland-Cliffs alone had shipped in 1929. CCI's share of this sharply diminished pie was a mere 300,000 tons, less than one-tenth of its 1929 ore shipments.[2]

By 1931 CCI management had begun to sense that this economic downturn might not be short lived and began to curtail operations. That year Stu Elliott, general manager of the company's ore operations, shut down some mines completely, operating the remaining properties on a two-day-a-week basis "largely for the benefit of our employees and the communities where they are located."[3] The company needed no new ore; stockpiles overflowed with unsold ore. Even at this reduced schedule, the mines produced more ore than Cleveland-Cliffs could sell. Production totaled 400,000 tons in 1932; only 300,000 tons sold. In both 1932 and 1933, as the Depression dragged on, Elliott closed all of the company's mines in the spring, reopening them in November on a two-day-a-week basis "solely . . . to do our part to help out in the very bad unemployment situation" by providing work in winter when families needed funds the most.[4] The superintendent of the Lloyd mine noted in 1933 that his mine was kept open more as "a welfare relief project," with efficiency being sacrificed "to give as many men employment as possible."[5] The failure of banks holding CCI funds and lack of income complicated matters further; in some instances the company was unable to meet even its now sharply diminished payroll.[6]

Cost cutting soon went beyond shutting down mines and reducing workweeks. On February 1, 1932, CCI closed the small research laboratory it had established on the Mesabi Range in 1929 to investigate ways of concentrating low-grade iron ores.[7] Equipment purchases, exploration, and development work all but stopped.[8] In 1935 one of the company's mining superintendents noted that diamond-drill work in his mine had been shut down for five years.[9] Cleveland-Cliffs had no need to uncover new reserves; it could not even sell what it had on hand. The company payroll fell from $6.6 million in 1930 to $1.9 million in 1932.[10]

Because the communities in which CCI operated were generally one-industry towns, the company's problems were quickly community problems. Stuart Elliott, for instance, observed in early 1935 that "practically every employee" had gone into debt, and practically every merchant in the mining towns of Ishpeming and Negaunee carried a "very large amount of credit on their books."[11] Those miners who still had at least partial employment clung to their jobs for dear life. As one mining superintendent noted, labor turnover in his mine was zero, with a waiting list of men for "every kind of job."[12] Even those miners with jobs suffered. Their reduced working schedules—often only two days a week—provided them with earnings just sufficient to buy only plain food and some fuel, but not enough for clothing and other necessities.[13] In 1931 a small group of socialists in Negaunee paraded, demanding relief, claiming several thousand in Negaunee were on the verge of starvation and criticizing Cleveland-Cliffs for not doing more.[14] In 1933

Stuart R. Elliott (b. 1874), c. 1929, general manager of Cleveland-Cliffs' mining operations in the Great Lakes area from 1927 to 1943. Elliott attempted to moderate the impact of mine closings on the communities dependent upon the company during the Great Depression in a variety of ways. *(Courtesy of Cliffs Natural Resources, Inc.)*

banks in Negaunee and Ishpeming failed, tying up the life savings of many of the company's employees and adding to the misery and stress.[15]

Cleveland-Cliffs did what it could. When it sold the Holmes mine to the Oliver Mining Company in 1930, Oliver laid off all former CCI men save one. Despite not needing to increase his workforce, Elliott took most of them on, since many were long-term company employees. In justifying his position to Cleveland headquarters, he wrote: "It was our duty to take care of these old employees and our reputation for fair dealing has not suffered by the policy."[16] Besides trying to take care of long-time employees and keeping its mines open at least part of the year, CCI in Gwinn offered men owing rent the opportunity to work and take credit against their accounts by repairing houses, water mains, and fences and cleaning streets and alleys to restore the town site and nearby locations to "a very neat appearance."[17] The manager of the Negaunee mine provided some repair work to long-term employees with large families or significant hospital bills.[18]

With Mather's approval company superintendents also duplicated what they had done in World War I: they promoted community gardening by supplying plots of land to miners, and then plowing, fertilizing, and supplying seed in an attempt to keep some food on miners' tables. When drought hit in 1932, the company ran water lines to the garden plots or put water barrels in the center of the garden areas to salvage the crops. When the region was hit by a blight of grasshoppers, the company purchased and distributed poisons. On his own authority Elliott laid out free wood lots on company property, provided supervision for those unfamiliar with cutting, and then helped them get the wood home to stave off the region's bitter winters. Elliott admitted that the costs for this program and the gardening program were "considerable," but considered the money well expended. "Nothing the Company has ever done," he commented, "has furthered its good name more than these activities." In 1932 the company supported 1,673 gardens and issued over 3,000 wood permits.[19] The company's land agent, John Bush, reported that "people everywhere" commented that CCI was "helping people through this depression more substantially than any other company."[20]

Operational Retrenching

In 1932 the company's indebtedness hit a record $33.65 million; operating losses had begun to accumulate.[21] Even while CCI sought to provide some relief to those dependent on it in Michigan and Minnesota, it also began to cut back on its comprehensive paternalistic programs. The pension system, in force since 1909, took the first hit. In 1932 CCI reduced pension payments by 50 percent to many on its rolls and suspended all new additions.[22] The company quit subsidizing Ishpeming's YMCA, which closed its doors in 1935.[23] The company-subsidized hospital in Ishpeming, long a symbol of CCI's commitment to the community, also came on the chopping block. From 1918, when it had led the efforts that erected the hospital, Cleveland-Cliffs had managed the facilities and covered around 70 percent of its expenses. In the late 1930s it began to look for ways—without success—to have the community assume operation of the facility.[24] Company housing likewise suffered. CCI in 1935 still owned around 500 houses in its Michigan locations. During the early Depression it spent "very little" on their maintenance.[25] By 1934 the company houses in Ishpeming had become "exceedingly shabby," in the words of one mine superintendent.[26] CCI now began to increase efforts to get out of the housing business by selling its houses.[27] By 1948 only 15 of 226 miners employed at the Cliffs Shaft mine who rented homes still rented company-owned homes.[28] The company also abandoned its annual prizes for best-kept premises and gardens, and dispensed with the services of its landscape architect Warren Manning, who had annually reviewed company mining sites for beautification purposes.[29]

The Depression also made Mather's policy of diversification unsustainable. In the crisis of the 1930s, like the crisis of the 1870s, rather than try to hold on to everything and risk losing all, CCI returned its focus to mining iron ore. Cleveland-Cliffs' divestiture of nonmining operations began relatively early in the Depression. One of the first victims was the Ethel coal properties, a money loser throughout the late 1920s. By 1930 its losses "had become serious," and CCI chose to pay a $150,000 penalty and abandon its lease fifteen years early rather than continue to operate the mines at a loss.[30] Over the next decade and a half, CCI steadily shifted from being a producer of coal to being a company whose coal business was almost entirely reselling purchased coal.

Another of the early victims on the chopping block was CCI's long-suffering charcoal iron furnaces. CCI closed down its furnace and by-products plant in Marquette in 1931 as the Depression deepened. When Marquette and Negaunee banks offered to loan CCI funds to reopen the Marquette plant to alleviate unemployment, Mather accepted the offer.[31] But the market for charcoal iron had evaporated. CCI reopened the by-products portion of the plant as a charcoal and wood chemicals operation, but the charcoal-fired blast furnaces remained closed. They would never again produce charcoal iron. Moreover, Cliffs soon sought to get out from under the wood chemicals operations as well, negotiating with General Chemical, American Cyanamid, and DuPont in 1933, hoping to find a buyer.[32] When this initiative failed, Cleveland-Cliffs next sought a partnership arrangement. In 1935 Dow Chemical Company and CCI organized the Cliffs-Dow Chemical Company. Dow owned 60 percent of the new company and managed it; Cliffs owned 40 percent. The venture allowed the chemical portion of the charcoal iron works at Marquette to continue in operation. CCI remained with the venture largely because Dow was unwilling to purchase the entire operation and because the plant used CCI timberland for its wood and the LS&I Railroad for wood transport, providing some income to these operations and continued work for some company employees.[33] The new company quickly dismantled the charcoal blast furnace and overhauled and updated the wood chemicals portion of the plant. Cliffs-Dow, however, suffered from the Depression slump in demand for chemicals of all sorts, losing money for the remainder of the decade.[34] An attempt by CCI to persuade Dow in 1936 to buy out its share failed.[35]

Other divestitures came in wood products. In 1917 the Munising Woodenware Company had absorbed the Munising Veneer Company, with CCI as majority stockholder. When both Munising Woodenware and the independent Piqua Handle Company faltered during the Depression, CCI promoted their consolation in 1935 to form the Piqua-Munising Wood Products Company. Cliffs owned 72 percent of its preferred and 52 percent of its common stock, as well as the first mortgage bonds.[36] Even though the consolidated company was unable to pay for the logs it used, CCI supported it for a period since there was "practically no other outlet" for the timber department's logs. In 1940, however, CCI quit subsidizing the operation and let Piqua-Munising slip into bankruptcy.[37] In 1944 CCI would also dispose of its stock in the Munising Paper Company.[38] Cliffs had operated its own lumbering crews on its large landholdings from the 1890s. In 1938, after some years of losses and labor problems culminating in a major strike of regional lumber workers, CCI dismantled its timber department, turning logging operations over to independent jobbers.[39]

Other company properties also went on the block. In 1932, as the demand for its ores hit rock bottom, CCI leased, but did not sell, the Morris mine to Inland Steel on the grounds that the company did not need the output of the mine and the lease would provide some income from royalties, as well as from the electric power and railroad transportation services that Inland would purchase from CCI-affiliated companies like Cliffs Power and Light and the LS&I Railroad.[40] In 1933 Cleveland-Cliffs surrendered its lease on the Greenway mine on the Mesabi and the Prickett lease on the Menominee.[41] In 1936 it disposed of the lands of the American Iron Mining Company—2,800 acres west of Lake Michigamme. Company geologists indicated the properties had very little possibility of having any mineral value.[42]

CCI also considered selling Cliffs Power and Light, with its six hydroelectric dams, to a Milwaukee utility company. Unable to reach satisfactory terms, CCI chose, for the time being, to retain its electric generation system, especially because Cliffs Power had provided at least some revenue during the Depression years.[43]

One thing CCI did *not* sell in the Depression was any land suspected of having minerals. This decision was in keeping with company tradition. In 1926, for instance, in a discussion of sale of forest land to the Ford Motor Company, Mather had observed that the company's policy was to retain all mineral rights when it sold properties.[44] A year later, in discussing land sales with Philo P. Chase and C. H. Hecker, Mather had insisted that deeding mineral land or potential mineral land "was directly contrary to the policy of the company."[45]

While most of its operations suffered, some of the company's diverse units helped it limp through the Depression and discouraged further divestitures. The company had a railroad-tie mill that was occasionally profitable, as was Cliffs Power and Light.[46] So was the marine department. Even in 1932, when ore shipments were 7 percent of their pre-1930 level, company vessels made a small profit carrying odd cargoes.[47] Toward the end of the Depression, the company even increased its fleet, owning fifteen vessels and managing the fleets of three other companies with seven vessels by 1941.

The Struggle with Indebtedness

The greatest casualty of the Great Depression was not the company's diversified interests in charcoal iron, wood chemicals, coal, or wood products, but the company's ambitious attempt to expand significantly its operations on other iron ranges and ensure its markets by becoming *the* primary ore supplier to Eaton's proposed Midwestern Steel Company and a major stockholder in the independent steel companies that were to be merged in that enterprise. Already in 1930 company auditors warned of "the serious decline in the market value of the securities we own in other companies."[48] Making matters worse, the preferred stock created as part of the Cliffs Corporation arrangement required 5 percent dividends annually. This amounted to around $2.5 million, a sum "considerably out of line in relation to its [the company's] earning power," even before the onset of the Depression.[49]

The more serious problem was the Corrigan, McKinney Steel Company, the firm in which CCI had purchased a controlling interest early in 1930 by taking out a $25 million short-term loan. Already in 1931 CCI had problems making payments on that loan. Negotiations for a bond issue to convert it into a lower-interest, longer-term debt with manageable payments failed, a victim of the nation's economic woes. The best Cleveland-Cliffs could do was one-year notes, which were renewed "with great difficulty."[50] In 1932 CCI lost nearly $2.7 million, much of it due to interest charges on the Corrigan, McKinney loan. Making matters worse, while CCI had controlling interest of Corrigan,

William G. Mather, c. 1920s. When the Great Depression hit in 1929 Mather was already in his seventies. His efforts to keep the company afloat caused his health to deteriorate. Even after yielding CCI's helm to Edward B. Greene in 1933 (at around age seventy-five), Mather continued to be involved in the company as chairman of the board for more than ten additional years. *(Courtesy of the Western Reserve Historical Society, Cleveland.)*

McKinney's common stock, it did not have controlling interest of its preferred stock, which under company bylaws replaced the common stock as the sole voting stock if preferred stock dividends were not paid. To keep control of Corrigan, McKinney, which began to lose money in the Depression, CCI had to pay Corrigan, McKinney preferred stockholders dividends out of already tattered CCI pockets.

In late 1931 a bankers' committee took over Cleveland-Cliffs' finances because the company was unable to make payments on the money borrowed to purchase Corrigan, McKinney. The committee objected to CCI advancing any more cash to cover dividends on Corrigan, McKinney preferred stock.[51] Trying to raise funds however possible, the company not only made the cuts previously noted in its mining operations and began to sell off what nonmining properties it could, but it also slashed the wages of officers and department heads in its Cleveland offices, as well as those of its mining superintendents. In 1933 two of the Cleveland banks that held Cleveland-Cliffs notes failed, complicating the already difficult problem of securing funding to keep the company afloat.

William G. Mather, already seventy-two when the Great Depression hit, struggled desperately to keep the company for which he had worked since 1878 afloat, even, at times, pledging part of his personal fortune as collateral for corporate loans. He spent most of his time negotiating quarterly and then annual renewals of the special bank loans that kept the company solvent while seeking with little success to find longer-term, more stable financing.[52] By June 1933 financial issues "had become so pressing and important" and conditions so "extremely difficult" that he requested additional assistance, something which, apparently, the banks participating in the loan agreement urged on him, especially in light of his age. Mather, now almost seventy-six, suggested that Edward B. Greene, chairman of the executive committee of Cleveland Trust, be named executive

Edward B. Greene (1878–1957), formerly of Cleveland Trust, a leading
Cleveland bank, succeeded William
Mather as CEO of Cleveland-Cliffs
in 1933 and guided it out of the
financial difficulties created by the
company's purchase of Corrigan,
McKinney and by its affiliation with
Cyrus Eaton's plans for a midwestern
steel conglomerate. *(Courtesy of Cliffs
Natural Resources, Inc.)*

vice president.[53] Two months later, perhaps under pressure from the bankers' committee, Mather asked, instead, that Greene replace him as president. Mather stepped back to the more ceremonial position of chairman of the board. Greene took over responsibility for the daily detail work of the company. After the board approved this arrangement, Mather wrote in his diary: "This is great relief to me."[54]

Edward B. Greene, Mather's successor, had intimate knowledge of Cleveland-Cliffs and its financial arrangements. A 1900 graduate of Yale, Greene had joined the Cleveland Trust Company, one of the banks with which Cleveland-Cliffs often worked, on gradu-ation. He soon became a member and eventually chair of that bank's executive com-mittee. His reputation was such that Ohio's governor appointed him to the Emergency State Banking Committee during the banking crises of 1932. Greene had been on the Cleveland-Cliffs board since 1928 and was married to the great-granddaughter of Jeptha Homer Wade.[55] In the midst of a dangerous financial crisis, he was a rational choice to succeed William G. Mather. On taking the reins, Greene warned the board that perma-nent financing was essential and that it might "be necessary to dispose of some of the company's properties in order to satisfy its creditors."[56]

The Corrigan, McKinney Steel Company was chief among those properties. In 1934, unable to continue providing dividend payments, CCI quit subsidizing Corrigan, McKinney preferred stock. Corrigan, McKinney preferred shareholders responded by taking control of the company and electing a new board of directors that included *no* representatives from CCI despite the fact that CCI still controlled the majority of the company's common stock and around 20 percent of its preferred.[57] CCI had gone deeply into debt to secure control of Corrigan, McKinney; in 1935 it still had the debt, but it had lost control. Moreover, the whole purpose of purchasing the company had been lost when the Depression had undermined the creation of Eaton and Mather's proposed Midwestern Steel Company.

Faced with sacrificing its mines to keep Corrigan, McKinney or sacrificing the steel company to keep its mines, Greene reaffirmed Mather's policy of holding onto iron ore

The Corrigan, McKinney steel plant, 1943, after Cleveland-Cliffs sold its financial interests to Republic Steel. Cleveland-Cliffs had secured control of Corrigan, McKinney in 1930. Its attempts to maintain control of this mill and associated properties during the Great Depression crippled the company's finances and led to the sale of Corrigan, McKinney to Republic in 1935. *(Photograph courtesy of Clayton Knipper, Cleveland Press Collection, Special Collections, Michael Schwartz Library, Cleveland State University.)*

reserves at the cost of everything else and sacrificed the steel company. In 1934 Republic Steel offered to purchase CCI's holdings in Corrigan, McKinney. CCI's board and stockholders quickly approved. Although the Department of Justice filed suit to prevent the transaction, the courts ruled in Cleveland-Cliffs' favor in early 1935.[58] The sale went ahead. Greene, Mather, and CCI's board breathed a collective sigh of relief. The transaction relieved CCI of the burden of trying to prop up both itself and Corrigan, McKinney and provided CCI with substantial holdings in bonds and both preferred and common stock in Republic Steel, some of which it sold to reduce further its outstanding debt. The company's losses from the transaction, nonetheless, totaled more than $20 million. The dream of becoming the primary supplier of ore to a giant steel company and of expanding its mining holdings significantly on other iron ranges by absorbing Corrigan, McKinney's mines was gone.[59]

With his background in finance, new Cleveland-Cliffs president Edward Greene was better equipped than Mather to keep the company afloat in the morass of financial problems into which it had sunk. Characterizing CCI's financial situation as "especially dangerous" because of the large amount of short-term bank debt that was "extremely difficult to carry" in the Depression, Greene, through 1933, 1934, and 1935, continued Mather's work of finding short-term loans to keep the company solvent.[60] Once Corrigan, McKinney had been sold, however, Greene managed to secure long-term refinancing, working with a consortium of banks in Cleveland and New York. The December 1935 refinancing enabled CCI to reduce interest charges and dissolve the bank creditors committee that had monitored the company's operations for over four years.[61] Between 1935 and 1937 Greene reduced the company's outstanding debt from $25 million to $14.5 million, and for the first time in years the company owed no money to banks.[62]

CCI was not yet out of the woods. Cliffs Corporation, technically the parent company of CCI and holder of its common stock, was also in trouble. Cyrus Eaton, Mather's partner in creating Cliffs Corporation, had been forced out due to personal financial difficulties early in the Depression. The market value of the iron and steel company securities he had contributed as his share of the enterprise had dropped from $40 million in 1929 to less than $5 million by the end of 1931.[63] The price of its voting trust certificates fell from a high of $157.50 in 1929 to a low of $2 in 1933.[64] In addition, Cliffs Corporation had not been able to pay Cleveland-Cliffs preferred stockholders from the onset of the Depression. By the end of 1934 accrued and unpaid dividends and arrears on the sinking fund on CCI preferred stock already totaled $11 million. By the end of 1940 the figure was $22.7 million.[65]

On top of these issues, Greene had to deal with problems that neither Samuel L. nor William G. Mather had had to face. From the time the Lake Superior iron ore region opened in the mid-1840s, Cleveland-Cliffs and other iron-mining companies had been able to deal with their employees largely free from government interference. Depression-era legislation ended this. The National Industrial Recovery Act of 1933, the National Labor Relations Act of 1935 (Wagner Act), and other New Deal legislation mandated increased wages, government regulation of working conditions, and collective bargaining. In 1936, for instance, Cleveland-Cliffs had to rearrange quarters on many of its vessels to accommodate the increased crew size required by government regulation.[66] This was just the beginning of much more complicated relations with both government and labor. Trying to moderate the impact of the new labor legislation, in 1933 Cleveland-Cliffs created its "Plan of Employees' Representation" as an alternative to external union organization.[67] Under this system, joint committees made up of management and labor representatives discussed issues brought to them by either party and made recommendations, which the company usually implemented. CCI used its Employees' Representation Committee to beat back early attempts by the Mine, Mill and Smelter Workers Union to organize the Marquette Range after that union's successes on the Gogebic and Menominee Ranges. Mine supervisors felt the

Employees' Representation Committee system worked well and that employees were satisfied.[68] In 1937, however, the U.S. Supreme Court dealt a blow to Cliffs and other Lake Superior mining companies using an assortment of employees' representation systems to counter the organizing activities of the Mine, Mill and Smelter Workers and the United Steelworkers of the CIO. The Court ruled that employee representation plans, because of heavy company involvement, did not really give labor collective bargaining power.[69]

On most of the other Lake Superior iron ranges, the United Steelworkers quickly stepped in and organized miners. Cleveland-Cliffs encouraged those who had worked with it on the Employees' Representation Committees to form a nonaffiliated local union rather than affiliate with the United Steelworkers, which represented not only iron ore miners but the employees of many of the nation's steel mills. While developments portended ill for the future, Cleveland-Cliffs' employees' representation committees had apparently worked well enough that Marquette Range miners chose to go the locally organized union route, forming the Marquette Range Industrial Union, instead of affiliating with the United Steelworkers.[70]

In 1937 it appeared that the long Depression was at last over. The company's sales of iron ore that year were the best in its history, its net profits second only to those of 1929. When steel companies raised wages in response to the apparent return of prosperity, CCI did the same, simultaneously expanding its mining employees' workweek from twenty to forty hours.[71] But in 1938 the economy took another nosedive. The company's ore shipments declined 68.7 percent. For the district as a whole, it was, apart from 1932, the worst year for ore shipments since 1900. CCI joined other operators on the Lake Superior iron ranges in reducing salaries, once again, for both management and labor and reduced shifts to three days a week.[72]

World War II

The outbreak of World War II in Europe in September 1939 rescued the company. By 1940 hostilities had created sufficient traffic in exports to ease the United States out of the Depression. In 1939 and again in 1940, Greene refinanced the company's debt, reducing interest payments each time. By 1940, for the first time in over a decade, the company had no mortgage on its property. Cleveland-Cliffs' funded debt, which at its highest point had reached nearly $26.3 million, had been reduced, partly from earnings and partly from the sale of company securities in steel firms, to $11 million.[73] In 1941 ore shipments from the company's mines hit an all-time high, over 7 million tons, and both the company's underground mines on the Marquette Range and its open-pit operations on the Mesabi operated at close to full capacity.

Reflecting the improved outlook, Cleveland-Cliffs in 1940 negotiated another of its traditional joint ventures. To develop a new mining property north of Ishpeming,

The Mather A mine, 1950s. The elevated trestles on the right were used for stockpiling ore. A joint venture between Cleveland-Cliffs and the Bethlehem Steel Company, with CCI as manager of operations, the Mather A tapped into deposits of ore—some over 2,000 feet down—discovered during exploratory drilling in the late 1930s. Sinking began in January 1941; production began in 1944. The Mather B shaft (the small tower faintly visible in the upper left center), 9,000 feet distant, was completed in 1950 and eventually linked underground to Mather A. The shaft of the Cambria-Jackson mine, also owned by Cleveland-Cliffs, can be seen at the right closer to Teal Lake. *(Courtesy of Cliffs Natural Resources, Inc.)*

Cleveland-Cliffs and Bethlehem Steel created the Negaunee Mine Company, owned equally, but with CCI acting as leaser of the land and mine operator. The depth of the deposit, requiring a shaft sunk to 2,900 feet, made development costs high, estimated at $3 million, and necessitated the partnership. Because of the extensive development work required, the mine, named the Mather in honor of William G. Mather, did not begin production until 1944.[74]

On December 7, 1941, the Japanese attacked Pearl Harbor, dragging the United States directly into World War II. Five days later Greene reported to the board of directors on precautions being taken to prevent sabotage at company facilities.[75] CCI even-

The Cleveland-Cliffs steamship *William Gwinn Mather* going through the locks at Sault Ste. Marie on March 23, 1942, then the earliest seasonal opening of the locks in their history, and part of the wartime effort to speed up the flow of iron ore. A year earlier a fleet of thirteen CCI vessels, with the *William G. Mather* as flagship, had pushed through Lake Superior ice floes to reach the ore docks of Duluth in record time. *(Courtesy of Photo Collections, Cleveland Public Library.)*

tually fingerprinted all its Upper Peninsula and Minnesota employees, issued identification cards, and posted guards at the entrances to its properties.[76] In early 1943 the Coast Guard detailed armed guards to patrol the company's ore docks in Marquette.[77] In April 1942, as an act of patriotism, CCI granted its Cleveland salaried employees who had been drafted or volunteered ninety days' pay from the time they left for government service.[78] The company suffered the loss of some key personnel. The company's treasurer, James L. Look, accepted a commission in the Army Air Corps, and A. C. Brown, one of the company's vice presidents, took leave to serve as deputy chief of the iron and steel branch of the raw materials division of the War Production Board.[79]

America's entrance into World War II ensured that the company's mines and its fleet of ore carriers would operate at full capacity, for during the war the Lake Superior district produced over 85 percent of the iron ore that American factories transformed into ships, bombs, aircraft parts, tanks, jeeps, and artillery. In November 1942 corporate minutes noted: "[P]roduction records at our mines are being broken repeatedly."[80]

In this environment, Cleveland-Cliffs expanded mining operations wherever possible. In late 1941 it reopened the Princeton mine, closed since 1921, and expanded Mesaba-Cliffs operations.[81] It also began processing low-grade ore in the Champion mine's long-abandoned stockpiles.[82] In a joint venture with the North Range Mining Company, CCI created the Missouri-Cliffs Company to investigate iron ore deposits in

CCI's Canisteo mine, near Coleraine, Minnesota, on the Mesabi Range in the 1930s. Cleveland-Cliffs first acquired an interest in this open-pit operation in 1929, acquiring full control in 1945 as part of its efforts during World War II and immediately after to increase its reserves of iron ore. *(Courtesy of Cliffs Natural Resources, Inc.)*

Missouri. By 1943 Missouri-Cliffs operated a mine in Missouri, and in 1944 it opened a second property on the Mesabi Range.[83] In 1943 CCI acquired the Cambria-Hartford mine lease on the Marquette Range from the Teal Lake Iron Mining Company and Republic Steel.[84] In early 1945 it bought up the outstanding shares of the Canisteo Mining Company on the Mesabi Range.[85]

Unfortunately, expanded wartime demand did not ensure profits in mining operations. Anxious to avoid the war-profiteering charges that had surfaced after World War I, the U.S. government, through the Office of Price Administration, froze iron ore prices at 1940 levels shortly after America's entrance into World War II. In the meantime, all other production costs rose—especially labor. In March 1943 the War Labor Board ordered a wage increase for all iron ore mine employees, retroactive to July 1942. This hit CCI hard, since most of its production came from underground mines, and labor costs represented a far higher proportion of the total cost of producing a ton of iron ore in underground mines than in open-pit mines.[86] Thus, while the price of ore stood 10 percent lower in 1943 than in 1939, artificially held at that level by government fiat, the cost of producing ore had risen 20 percent.[87]

Underground iron ore producers desperately attempted to get the Office of Price Administration to end the freeze on ore prices or raise the price allowed. In hearings in 1942, Greene pointed out that iron ore was being treated differently than other raw materials. Price levels for zinc stood 75 percent above 1939 levels and coal 24 percent higher; iron ore prices were 10 percent lower.[88] In mid-1943 Greene reported that the company was "losing money on production of all ore except Cliffs Shafts Lump" and noted that at some mines the company could not even reimburse expenses.[89] The Office of Price Administration finally granted a small price increase in 1945, but at roughly the same time the War Labor Board ordered another wage hike. The net result was that while the Office of Price Administration between 1940 and 1945 raised the mandatory price of iron ore by 20¢ a ton, the War Labor Board ordered wage increases amounting to 62¢ per ton on ore mined from underground mines. In 1945 Greene noted that Cleveland-Cliffs would have made no profit at all in the war years if it had not been for earnings from the company's nonmining operations (shipping, railroading, timber) and its stockholdings in steel companies.[90]

Besides price freezes and imposed wage hikes, a host of other factors contributed to the company's inability to profit from its mining operations in World War II. Early on the Selective Service System declared iron ore mining a critical occupation, deferring miners from the draft. In 1944, however, in desperate need of military manpower, the government eliminated deferments on all males under age twenty-six, including miners. Besides losing skilled miners to the draft, the region simultaneously lost able-bodied men to defense plants outside the region that paid higher wages. In 1944 many CCI mines—despite the press for "more ore to win the war"—dropped from three to two shifts per day simply for lack of labor.[91] Making matters worse, the government set up a station in Marquette in late 1944 to recruit men for the Bremerton Navy Yard on the Pacific Coast, costing the company another 132 men.[92] As the company's younger miners were drafted or headed south or west for higher-paying jobs, those left behind to work the mines were older, less experienced, less fit, and hence less efficient. Productivity suffered.

In the press for increased production, the mines suffered as well. In 1942 the superintendent of the Negaunee mine cautioned that the rapid rate of mining to meet wartime demands did not allow sufficient time for overburden in the sublevel caving system to settle and compact, increasing hazards. He also reported more breakdowns of equipment due to the heavy operating schedule. The director of the company's safety department warned that mine supervisors were pursuing fewer disciplinary actions related to safety "because of the dire need for production."[93] Charles Stakel, who had succeeded Stu Elliott as general manager of the ore department in 1943, commented on January 1, 1944, that "every mine and facility" in the company's operations suffered from labor shortages, absenteeism, poor housekeeping, and dropping morale as heavy operating schedules entered their third year.[94] By 1944 the company's engineering department had lost so many men that it had to dispense with weekly inspections of soft-ore mines, where the dangers were greatest, and one mine manager noted that "cleanliness and repairs are being badly neglected."[95] Stakel commented to Greene in the cover letter accompanying his 1944 annual report that the time he and his associates spent dealing with "labor matters, grievances, wage adjustments, rationing and proprieties, etc.," combined with the "multitudinous and conflicting orders" often issued by various agencies of the federal government, meant they had little time to watch costs and study conditions at the mines.[96]

In August 1945 World War II finally came to an end. Cleveland-Cliffs had contributed substantially to the war effort through the ore it produced and its affiliated operations in shipping, chemicals, and energy production. Despite limited wartime profits, Cleveland-Cliffs had continued to reduce its debt load. From a Depression-era high of about $33 million, the debt had sunk to only around $5 million by 1945.[97] But ominous clouds loomed on the horizon.

While other companies in the iron ore and steel industries had grown during the period from 1930 to 1945, Cleveland-Cliffs had not. If anything, it had lost ground to other ore producers in the Lake Superior district. In the words of one observer, the interest and dividend payments the company had been obliged to make on its preferred stock (and on the purchase of Corrigan, McKinney) had "bled CCI Co. white" and helped account for its loss of position in the ore trade to rivals like the M. A. Hanna Company.[98] The organizational structure of the company was confused since it was, technically, a wholly owned subsidiary of the Cliffs Corporation, the organization created in 1929 by Mather and Eaton as a precursor to the aborted Midwestern Steel Company. Making matters worse, no sooner had peace arrived than the company faced the possibility of its first significant mining strike in fifty years. In October 1945 the United Steelworkers (CIO), which had finally organized the company's mines during the war, demanded increased wages for steelworkers and called for a strike vote pending ongoing labor negotiations between its parent organization and steel companies.[99] Its membership approved. For CCI, with its preponderance of labor-intensive, deep-underground mines, rising labor costs were potentially crippling.

Equally ominous, the abnormally high consumption of iron ore during the war had seriously depleted the company's reserves of high-grade, direct-shipping ores.[100] In the words of Franklin Pardee, head of Michigan's geological survey, the iron mines of Michigan had responded to the war effort by "depleting and skimming the cream from their ore bodies."[101] Despite sustained wartime demand, CCI's ore output in fact declined, from 4.2 million tons in 1942 to 3.4 million tons in 1945, even though the company had reopened several old mines and opened one new one. Making matters worse, by 1944 the theoretical profit margin in underground mines, due to rising labor costs, had already reached the point of no return, only $0.003 per ton, according to one expert.[102] Crippled by Depression-era losses, hampered by an obsolete operating arrangement with the Cliffs Corporation, cursed with declining ore reserves ever more costly to reach, confronted by an organized and more militant labor force demanding higher wages, and led by an aging upper management in ill health, Cleveland-Cliffs faced a very uncertain postwar future.

Merger with Cliffs Corporation

Early in the postwar period, CCI officials attacked the problem of the company's relationship with the Cliffs Corporation. Although the Depression had killed Eaton's dream of creating a midwestern steel company to rival U.S. Steel and thus the raison d'être of Cliffs Corporation, Cliffs Corporation remained. It held all Cleveland-Cliffs common stock. By 1947 Cliffs Corporation owed CCI preferred stockholders $26.16 per share in back dividends and was $14.5 million behind in sinking fund payments designed to retire that stock. Attempts to pay the annual dividends due on the preferred stock (almost $2.5 million) proved, in the words of H. Stuart Harrison, later president of Cleveland-Cliffs, "a handicap . . . and . . . considerably out of line in relation to its [Cliffs Corporation or CCI's] earning power."[103] Computations of anticipated earnings indicated Cliffs Corporation would require over twenty years to settle the back claims of preferred CCI stockholders. Making matters worse, until preferred shareholders had received all their back dividends, common stockholders could not receive any dividends.

The only thing that kept the whole arrangement from being catastrophic was that CCI officers held large blocks of both Cliffs Corporation and CCI preferred stock, and they also served as officers in Cliffs Corporation.

Nonetheless, the problem of the relationship between the two entities and the drag of dividends on preferred stock clearly needed cleaning up. CCI and Cliffs Corporation officials floated some forty recapitalization and twenty merger plans between 1937 and 1946, but the issue festered. This was in part because CCI was struggling with a myriad of more serious crises, in part because holders of CCI preferred stock—who now had a role in running Cliffs Corporation due to the failure to pay dividends—and the holders of Cliffs Corporation common stock (half of whom were part of Eaton's old group) could not agree on how to terminate Cliffs Corporation.[104]

When World War II ended, Mather and Greene tried again, proposing to liquidate the Cliffs Corporation and proportionally distribute its assets. This proposal, however, led to a major confrontation with some Cliffs Corporation stockholders who wished to extract a higher return from their holdings and who used the advanced age and declining health of Mather and Greene as a weapon. By the end of World War II Mather was over eighty, and both Mather and Greene, as well as several other members of CCI's upper management team, were apparently in ill health. Those opposing the liquidation plan called for "immediate and drastic action to clean house" at CCI by removing Mather, Greene, and other upper management before any liquidation of Cliffs Corporation.[105]

Cyrus Eaton, no longer the key figure in Cliffs Corporation but still an interested party, mediated the dispute. At Eaton's urging, Greene and Mather abandoned the liquidation plans and agreed instead to merge the two companies.[106] Cleveland-Cliffs came out of the merger in possession of significant blocks of shares in four major steel companies (Inland, Youngstown, Republic, and Wheeling). In fact, the steel stocks comprised around 27 percent of the company's total assets.[107] Cleveland-Cliffs stock, in the meantime, became much more broadly distributed since owners of Cliffs Corporation common stock—many of whom had never before owned shares in CCI—as part of the merger agreement received 2.25 shares of new Cleveland-Cliffs common stock for each share of Cliffs Corporation stock they held.[108]

Greene used the merger to pass the baton to Alexander C. Brown. Brown, president of the Industrial Brownhoist Corporation, had been invited to become first vice president of CCI and executive assistant to Greene in 1934. Brown now became president of the merged company; Greene became chairman of the board; and William G. Mather, now almost ninety, became honorary chairman.

The Labor Problem

The merger solved an organizational problem that had troubled the company for a decade. Other problems were not so easily solved. In 1942 and 1943 the United Steelworkers (USW) of the CIO successfully organized CCI's mines in both Minnesota and Michigan. Seeing the handwriting on the wall, General manager Stuart R. Elliott wrote Greene in early 1943 that it was inevitable that the CIO would eventually represent the employees at all CCI mines. The company, he added, "had nothing special to offer its employees and it was in no position to combat this movement." He concluded: "We are in bed with these fellows, so to speak, and I have made up my mind, and have impressed my views on all of the Superintendents . . . that we have to make it our business to get along with these fellows. I am sure this can be done and it is to the Company's advantage to cooperate."[109] Indeed, in the closing years of World War II, relations between the company and the United Steelworkers were mixed, but generally good.[110] This situation was to change.

In late 1945, after the end of the war with Japan, the United Steelworkers union announced to all American steel companies and iron-mining interests whose employees it represented that unless workers were granted a wage increase of 25¢ an hour to compensate for wartime cost-of-living increases and the loss of overtime pay now that peacetime conditions had returned, it intended to strike. The steel companies pointed to government-imposed price limits placed on their products and argued that unless these were removed, they could not afford to grant the raise. With a strike looming, President Truman in January 1946 suggested a compromise of 18.5¢. The union accepted; the steel industry did not, arguing that the government-imposed cap on steel prices had to be raised first. On January 21 steelworkers walked out. Iron miners—including those in

Pickets at the Mather mine during the 1946 strike. A back-to-work movement initiated by the company for operations at its Mather mine led to rock throwing and tire slashing. *(Courtesy of the Marquette County Historical Museum, Marquette.)*

CCI mines—joined them on February 8. Three-party talks and the government's agreement to raise steel price caps led to an agreement between the USW and steel companies on the 18.5¢ raise. Work resumed at steel mills on February 17.

The agreement with the steel companies, however, did not lift the cap on ore prices. The largest iron-mining firm, U.S. Steel's subsidiary Oliver Iron Mining Company, which had mainly open-pit mines in Minnesota, nonetheless consented to the 18.5¢ raise relatively quickly. In Michigan and parts of Minnesota where labor-intensive underground mines predominated, however, the strike continued. In spite of efforts by Cyrus Eaton and others to persuade them to settle, companies with significant underground operations—like Cleveland-Cliffs—maintained that they could not raise wages that much and remain profitable with existing government price controls on the price of ore and the higher labor costs of their enterprises.[111]

Much of the goodwill accumulated by CCI in the previous half century dissipated in the long, bitter strike of 1946, especially since, as one of the largest underground mining companies, CCI assumed leadership of the struggle to defeat the union's demands.[112] Particularly unsettling was CCI's promotion, in the middle of the strike, of a back-to-work movement based on a 10¢ per hour raise with more to follow if and when the Office of

Price Administration consented to increase the price of iron ore. Around 25 percent of the workforce responded on March 22, with an especially heavy turnout at the Mather mine. This attempt to get workers across picket lines, however, led to large-scale demonstrations, importation of union organizers from outside the area to reinforce picketers ("thugs," according to some company witnesses), and moderate violence—throwing rocks, tipping cars, slashing tires, breaking windows, picketing homes of miners who attempted to return to work. The company responded by seeking injunctions to limit picketing and eventually secured warrants for the arrest of union officials for violation of picketing injunctions and for inciting violence. The United Steelworkers responded by filing an unfair labor practice complaint with the National Labor Relations Board because company mine bosses and superintendents approached miners individually about returning to work instead of bargaining collectively.[113] Union officials claimed some of their people were badly beaten and accused the company of spreading a keg of roofing nails over the lot where picketers parked, causing at least 100 flat tires; company officials accused the union of importing goons from out of state to intimidate employees who wished to return to work.[114] In less than a week the back-to-work movement played out, the company blaming its failure on lack of protection offered workers by the sheriff's department and state police.

The strike ended on May 22, after 104 days, with the company conceding the 18.5¢ demanded by the USW. While the Office of Price Administration's agreement to a retroactive increase in the price of iron ore in June alleviated the company's concerns about the economics of the wage hike, the strike's long-term effects were serious. CCI's back-to-work movement and the USW's intimidation of those inclined to return to work—many World War II veterans—split families and friends and led to enduring hard feelings between local management and labor.[115] Ernie Ronn, a union official, looking back at the strike some forty years after, observed that "some of the scars" from the 1946 strike still existed and would remain "as long as those people who went through those picket lines live."[116] A company official, looking back over the same time horizon, commented that the strike was "our first-hand experience with the big league, getting into hardball instead of just playing softball."[117] The inexperience had shown.

A tradition of good labor relations was not the only casualty of the strike. After the strike, company officials closed the Princeton mine due to deterioration of underground workings caused by picketing strikers' refusal to allow full maintenance crews in and the higher price of labor in what was already a marginal operation.[118] CCI also accelerated the dismantling of the paternalistic system it had initiated in the preceding half century. In 1946, immediately after the strike, it closed the clubhouses it had long funded for employees at North Lake and Gwinn, reduced its support for the local hospital and its contributions to a number of local organizations, and terminated its post-Depression house and garden prizes.[119] More cuts would come. By 1950 CCI had ended its support of the Ishpeming hospital, turning it over to private interests. A few years later it terminated the popular visiting-nurses program.[120]

Tugger-scraper in operation at the Mather mine, early 1950s. Note the use of steel sets in place of the traditional timber supports to hold up the roof of the drift, or mine tunnel. The tugger is upper center; the scraper that it powers is in the lower center. The raise, or small vertical shaft to which the scraper pulled its content for deposit in a mine car at a lower level, is in between, covered by rails to prevent oversized rock from entering the raise and clogging it. *(Courtesy of Cliffs Natural Resources, Inc.)*

For the remainder of the 1940s labor relations within the mines were strained but, for the most part, generally good.[121] However, because USW representation linked company miners to the steel industry as a whole, whenever steelworkers went out on strike, CCI miners, in solidarity, followed. The company endured strikes—mostly tied into national steel strikes—in 1949, 1951, 1952, 1956, and 1959, before a period of labor peace once more emerged. Most led to salary or benefit increases and steadily put greater economic pressure on the labor-intensive underground mining operations on which the company's prosperity depended. While the leading iron-mining concern in the United States—U.S. Steel's Oliver Iron Mining Company—and other CCI rivals like Pickands Mather and Hanna operated more open-pit mines, the vast bulk of CCI operations in the 1930s and 1940s were underground mines. This made labor costs a more serious problem to Cleveland-Cliffs. By 1948 labor made up 36 percent of the cost of mining a ton of ore underground, compared to only 7 percent in an open pit.[122] Average labor costs at Cleveland-Cliffs mines rose 83.5 percent between 1939 and 1947, while the price of iron ore rose only 10.6 percent.[123]

The company's mine managers in the postwar period fought a rearguard action to keep underground operations economical and competitive. They continued aggressively to introduce new technologies. For example, one of the major costs of operating underground was the expense of timber supports whose life, even with preservation treatment, was limited. In 1945 CCI began experimenting with structural steel to support permanent openings, finding costs competitive with timber due to easier installation and longer life.[124] In 1955 it installed a new, more economical hoisting system at a new shaft at the Cliffs Shaft mine, replacing the two concrete shaft houses erected in 1919.

Significant for future developments, in 1947 the company also began working with Linde Air Products on jet piercing in one of its open-pit operations. Jet piercing, a new

Aerial view of Cliffs Shaft mine, c. 1955, looking north. The new Koepe shaft house, which replaced the 1919 concrete shaft houses, rises between them. Some of Ishpeming's residential housing is visible on the right. (*Courtesy of the Marquette County Historical Museum, Marquette.*)

drilling technology, promised to rapidly speed up penetration of very hard rock by using an oxygen and fuel oil flame as hot as 4,500° F to heat the rock being drilled. Water cooling kept the burner and drill bit from being consumed. The intense heat weakened and spalled the material being drilled, and the steam from the cooling water transported the cut stone to the surface. Tried at Cliffs' then-small Tilden open-pit mine with its hard jaspilite rock, the experiments suggested that, if the price of oxygen dropped significantly, the new drill was potentially revolutionary. It accelerated drilling through hard rock by a factor of around eight.[125]

Ore-Depletion Concerns

Potentially more serious than the new challenges of dealing with organized labor and the troubling economics of underground mining was the threat of ore exhaustion faced by virtually all operators in the postwar Lake Superior iron ore district. In May 1942, shortly after the United States entered World War II, Edward W. Davis, director of the Mines Experiment Station of the University of Minnesota, warned in a report to the materials division of the War Production Board that at war production rates the known ore reserves of the Lake Superior district would be exhausted by 1950. Without government subsidy, he cautioned, ore shortages could come even earlier. "It is shocking to realize," he declared, "that in a comparatively few years, the great steel industry dependent upon Lake shipments will find itself short of the necessary ore to meet emergency steel requirements."[126] Davis's concerns over the future of the Lake Superior iron ore industry were echoed by E. L. Derby Jr., the head of CCI's geology department. He warned company officials that the war was placing "an abnormally heavy demand" for ore on the district and that its high-grades ores were being depleted "at a dangerous rate."[127]

Concerns about the depletion of Lake Superior iron ore were not new. As early as the 1880s, American industrialization had accelerated consumption of iron to a point where many people feared the iron ores of Lake Superior would soon run out. This prompted various entrepreneurs to erect plants for beneficiating (enriching) lean waste ores near the Jackson, Republic, and Michigamme mines.[128] Sensing opportunity, Thomas Edison briefly operated an experimental ore-concentration plant near the Humboldt mine, west of Ishpeming.[129] None of these operations succeeded commercially. Edison's venture ended in December 1890 when his Humboldt concentrating mill burned down.

The demise of Edison's operation coincided with the discovery of the Mesabi Range in the early 1890s. With immense quantities of high-grade ores suddenly available, fears of depletion subsided until the 1920s and 1930s, when concern began to reemerge. By then the average iron content of ores being shipped to steel mills had fallen from above 60 percent to well under that old benchmark. Much of the ore left was less marketable medium-grade ore requiring additional treatment to raise its iron content by removing as much silica as possible. The accelerated use of remaining high-grade, direct-shipping ore reserves during World War II meant Davis's and Derby's alarms about imminent depletion had to be acted upon quickly.

After World War II the search for a solution to the problem of ore depletion took three paths: location of new reserves of domestic iron ore; overseas exploration for new, undeveloped deposits of direct-shipping ore (60 percent or higher iron content); and research and development of industrial processes to extract the iron from lean taconite and jasper ores. At first it appeared that the second path—development of overseas deposits—would be the salvation of the American steel industry. American mining companies and mining subsidiaries of American steel companies located and purchased controlling interests in high-grade direct-shipping ores in Venezuela (U.S. Steel, Bethlehem), Liberia (Republic), and Canada (M. A. Hanna), as well as in Brazil and Australia. Iron ore imports rose from around 4 percent of domestic consumption in 1946 to nearly 10 percent by 1950 to almost 40 percent by 1959.[130]

Cleveland-Cliffs was one of the American firms looking for ore supplies abroad. In 1948 Cleveland-Cliffs joined six other companies to create the United Dominion Mining Company in Quebec, Canada.[131] In Venezuela in 1952 and 1953 the company exposed at El Trueno, south of the Orinoco River, what was, according to company geologist Burt Boyum, "the largest single ore body Cliffs had ever proved, up to that time."[132] But after spending $2 million, the company decided to let its option expire, citing the difficulty of transporting ore from a remote interior location, lack of a local market, and inability to find a joint-venture partner to share expenses.[133]

Most of the company's explorations, however, took place in the United States. The number of land offers investigated by the company's geology department in Michigan grew from fourteen in 1946 to forty-five in 1948 and ninety-four in 1951.[134] Exploration expenditures rose from $212,000 in 1946 to a peak of $2.3 million in 1953. The company's geology department in the same period grew from eight to twenty-seven employees. According to Burt Boyum, the company's assistant geologist in the period, CCI developed the most advanced geological exploration department of all the Lake Superior mining companies. This department tried every practical geophysical technique in the book in its search for additional ore and was the first to apply oil field exploration techniques to iron ore.[135]

In the old mining districts of Lake Superior, Cleveland-Cliffs ran as many as twelve diamond-drill rigs annually on exploration projects, in addition to constant diamond-drill work within existing mines. Generally, these efforts failed to yield new untapped veins. Indeed, by 1947 both the Negaunee and Maas mines on the Marquette Range had almost exhausted their ore and would soon close.[136] However, in 1951 C. W. Allen, the general manager of the ore department, was able to report that the company's ore reserves were trending upward. He pointed to the discovery of a large ore body under the Mather B mine (opened in 1950) and the identification of "enormous reserves" of low-grade iron in the area of the Tilden mine south of Ishpeming and Negaunee. He also acknowledged, however, that nothing had been discovered that would prolong the life of other properties.[137]

Supplementing its attempts to locate new domestic reserves of direct-shipping ores by exploration, the company also acquired reserves through purchase. The primary acquisitions between 1945 and 1950 came on the Mesabi Range in Minnesota, where in March 1947 Cleveland-Cliffs acquired three mines operated by the Wisconsin Steel Company, a subsidiary of International Harvester Company: the Hawkins, Sergeant, and Agnew mines, the latter two underground operations. International Harvester's operations were, in a sense, victims of the 1946 USW strike, for, unlike CCI, International Harvester did not settle with the United Steelworkers; at the time of CCI's purchase the mines had been idled for more than a year.[138] CCI also purchased the lease to the Canisteo mine on the Mesabi in 1945–46, bought an increased interest in the Mesaba-Cliffs Mine from Republic Steel in 1948, and became the chief marketing agent for Cyrus Eaton's Steep Rock Iron Mines, Ltd., a Canadian-based mining venture that produced direct-shipping ores.[139] In the 1950s Cliffs would mount other, largely unsuccessful exploration activities in Missouri, Montana, Wyoming, and elsewhere.[140]

Because of the company's failure to uncover significant, exploitable, new deposits at home or abroad, it was the third option—the concentration of the low-grade ores of the Marquette Range—that assured the survival of Cleveland-Cliffs. Although CCI never expended large sums on research and development prior to the 1950s, it was not a complete novice at the process. As noted previously, as early as 1929 the company had erected a small lab in Minnesota to experiment on enriching Mesabi ores. The effort ended in the Great Depression. But in 1944 CCI erected and equipped a new test laboratory at the Holman mine in Minnesota to investigate various types of beneficiation.[141] By the late 1940s Cliffs had become a leader in research on heavy-density media (sometimes called sink-and-float separation) as a means of ore concentration. In heavy-media separation operators use a mixture of a finely ground dense material suspended in water to create a fluid (slurry) with a higher density than water to better separate low-density waste materials from higher-density useful ores. CCI also developed new and improved techniques to separate silica from iron ore by other means, placing into operation at its Hill-Trumbull mill in 1948 the first commercial-size Humphreys spiral plant ever built for beneficiation of ores.[142] This device used both centrifugal force and specific gravity to enrich ore crushed to sizes smaller than those treated by more conventional means.

It was generally recognized, however, that the ores processed by these crude gravity methods of beneficiation would not solve the depletion problem. While useful in enriching some medium-grade ores, the processes were not readily applicable to the large deposits of lean jasper ore the company owned in Michigan. To beneficiate these ores required the development of much more complicated and expensive technologies. The first steps in developing these technologies were initiated outside the company in Minnesota, where Edward W. Davis began in the 1910s to work out means for extracting iron from his state's vast reserves of taconite, a very hard, greenish gray rock that contained flecks of magnetic iron oxide (magnetite). Like Michigan's jasper, Minnesota's taconite was a lean ore, containing less than 40 percent iron, compared to high-grade or direct-shipping ore, which contained at least over 50 percent and sometimes up to 60–69 percent iron.

Davis, a professor at the University of Minnesota, began his work on processing taconite about 1913. Prior to World War I, he helped set up the Mesabi Iron Company, located near the town of Babbitt. Mesabi Iron bought up thousands of acres of taconite deposits and erected a plant in the hope of commercially extracting and enriching taconite to the point where it could compete with direct-shipping ores. The venture was premature. Drilling the hard rock was too expensive; so was processing. High-grade ores were still relatively abundant and cheap. The company failed in 1924. Davis and others, however, continued their search for methods of commercially treating taconite, making slow but steady progress.[143]

In 1939, a decade and a half after the collapse of Mesabi Iron, another group of entrepreneurs, believing Davis and his colleagues had made sufficient progress, took up the torch. They formed the Reserve Mining Company, adopting that name because taconite had been considered a "reserve" to be developed at some point in the future. Cleveland-Cliffs, modestly invested in Minnesota iron-mining properties by this time, took a 10 percent interest in the new venture. Its partners in Reserve Mining were several iron ore and steel companies: American Rolling Mill Company, or Armco (33 1/3 percent), Wheeling Steel (33 1/3 percent), and the Montreal Mining Company (23 1/3 percent). Oglebay, Norton and Company, a mining management and ore-transport and marketing firm, served as the operating agent for Reserve Mining. At this time Cliffs owned a large block of Oglebay Norton stock, and Crispin Oglebay, the project's champion, served on the Cliffs board, so CCI's investment was logical.[144] The Mesabi Iron Company remained in corporate existence, retaining mineral rights on its extensive taconite properties. Reserve simply leased mining rights on these properties from Mesabi Iron and took over the company's abandoned taconite mining and processing facilities.[145]

At first, success of the Reserve project appeared dubious because Minnesota's high taxes on mining properties made the steel industry reluctant to invest in a capital-intensive venture like taconite. Minnesota's ad valorem law taxed all iron properties, regardless of the methods used to extract the ore, at 50 percent of their assessed value from the time iron was discovered until the iron was mined out. Since they were taxed at the highest rate, high-grade ores were mined as fast as possible until they were depleted. Minnesota mining communities retained 90 percent of the taxes collected, using them to pay for their schools and community improvements. Naturally, they resisted changes in the tax law that would deprive them of this windfall. "You were either a friend or an enemy of the people," Davis recalled, "depending upon whether you were for or against the 'steel trust' in any discussion."[146] Steel companies were equally adamant in their refusal to invest in taconite. Davis recalled attending a meeting in Cleveland with steel industry representatives in the 1930s where he explained the progress they were making at the pilot plant in the fine grinding of taconite and wet concentration of magnetite. He invited Charles M. White, vice president in charge of operations for the Republic Steel Company, to come up to Minnesota to see for himself. White replied he had no interest "in that God-damned hard stuff or anything else out there in Minnesota until you get over the idea of taxing everything to death."[147]

In 1940 U.S Steel's subsidiary Oliver Mining announced that it would sell its excess iron ore on the open market. Previously Oliver's entire output had been absorbed by the parent company. This development seems to have galvanized the iron and steel industry to action. Oliver Mining owned or controlled about 70 percent of the ore reserves on the Mesabi. The high quality of the iron ore in its reserves allowed Oliver to mine this ore relatively cheaply in open-pit mines. The less high-quality ores of other mining concerns on the Mesabi often required labor-intensive beneficiation. These mining companies had remained profitable because they had not had to compete directly against the Oliver Iron Mining Company. But with the entrance of Oliver into the open ore market, many feared that the price of ore might fall 40 percent or more. Cliffs' geologist E. L. Derby was one. He worried that the new policy would cause the "strangulation of independent companies."[148]

Repercussions went beyond putting the independent iron companies out of business. Davis pointed out in a widely circulated 1940 article titled "Change in U.S. Steel Iron Ore Marketing Policy Endangers Range Communities and State" that by drastically reducing the price of ore, U.S. Steel would, in effect, also decide how much tax it would pay, since the existing tax rate was partly based on prevailing ore prices. Taxes on iron mining throughout the Lake Superior region would fall as ore prices fell.[149] These arguments convinced northern Minnesotans that their long-term economic future depended on the development of means to enrich taconite. With an alternative source of high-grade ore available, U.S. Steel could not control the supply and thus the price of ore.

Reform of the tax law was the key to the development of low-grade ore. Davis and others won grassroots support for a bill replacing the ad valorum tax on low-grade ores with a more modest tax on the manufactured product (namely, enriched, processed low-grade ore) at the time of shipping. The tax reform act passed in 1941. As taconite champion James Morrill, the president of the University of Minnesota, wrote, "Industrial interest in taconite dates almost from the hour that the new taconite law was signed."[150] Within months of the passage of the new tax law, Oglebay, Norton and Company pushed forward with the development of the Reserve Mining Company. The company hired Henry K. Martin in 1942, a graduate of the School of Mines at the University of Minnesota and an expert in the grinding and classification of all types of ores, to oversee the company's research efforts. To Davis, Martin's hiring was the "first sign we at the Mines Experiment Station had that the firm had any immediate interest in taconite."[151]

In 1940 Pickands, Mather and Company organized a second major ore beneficiation project on the Mesabi. The Erie Mining Company was owned by a partnership that included Bethlehem Steel (45 percent), Youngstown Sheet and Tube (35 percent), Interlake Iron (10 percent), and Steel Company of Canada (Stelco) (10 percent).[152] At this time Pickands Mather was headed by Elton Hoyt II, its senior managing partner and prominent iron industry spokesman. While serving on the War Production Board's Iron and Steel Industry Advisory Committee, Hoyt had become a champion of taconite processing. Erie leased some of the state-owned taconite reserves on the Mesabi Range in 1941, including lands held in trust by the University of Minnesota.

Hoyt convinced Pickands Mather's board to fund a taconite research laboratory at Hibbing directed by Fred D. DeVaney, a graduate of the University of Minnesota. DeVaney developed a method called froth flotation, a chemical process that separated the pulverized iron ore particles in crushed taconite from unwanted silica. Bubbles rafted the iron-bearing particles to the surface of the mixture, where they could be skimmed off.[153] Erie Mining later dropped flotation for simpler and less expensive methods of magnetic separation developed by E. W. Davis at the Minnesota School of Mines and already successful at Reserve. However, flotation processes related to DeVaney's would later find a role in the beneficiation of Michigan jasper ores.

Through the 1940s, both Reserve and Erie tackled the problems associated with each step of the process of manufacturing a commercially viable, enriched lean-ore product.

First, taconite had to be extracted economically. Then it had to be ground to the consistency of a fine talcum powder to free the tiny particles of iron ore contained in the hard rock. The crushed taconite was next concentrated magnetically. About two-thirds was waste, or gangue, a nonmagnetic stony material with only a small amount of iron. The remaining third was a concentrate with about 64 percent iron. During the subsequent step—agglomeration, or pelletization—the iron concentrate was rolled in huge drums where it was mechanically shaped into small balls. These so-called green balls were then fired in a kiln. At first the balls often broke apart, causing troubles in transportation and furnace handling. This seemed an intractable problem until an engineer at Bethlehem Steel's plant in Lebanon, Pennsylvania, revealed that adding bentonite, a claylike binding agent, not only produced a tougher ball but also allowed the pellets to be more rapidly fired at high temperatures. This was the breakthrough that made commercial development possible.[154]

In 1946, with World War II at an end and the negative impact of wartime demands on the remaining reserves of high-grade ore all too visible, the Reserve project gained momentum. Oglebay Norton selected a site at Beaver Bay (later renamed Silver Bay) for Reserve Mining's taconite-processing plant. After public hearings in 1947, Reserve Mining obtained permits to take 130,000 gallons of water per hour from Lake Superior, which were returned to the lake along with the crushed waste materials, called tailings, within a zone that stretched three miles along the coastline and three miles out into the lake. At the time it was widely believed that the tailings posed no damage to the water quality of the lake.[155] The company built a new rail line, hydroelectric plant, harbor, and two new towns for the families of the 2,500 men employed by the new operation.

The scale and high costs of Reserve and Erie were far beyond any previous twentieth-century iron-mining venture. They required huge open pits and very large earthmoving equipment, in addition to very large-scale processing equipment and reservoirs spread over many acres of land to secure the economies of scale needed to bring the cost of beneficiated and agglomerated ores down to those of natural ores. Erie, with an annual production capacity of 7.5 million tons per year and cost of $300 million, was a much larger project than Reserve, at 3.75 million tons and cost of $190 million. To provide a measure of the relative scale of these projects, in 1944 the thirteen mines of Cleveland-Cliffs (the third-largest mining company in the Lake Superior region) produced only 6.5 million tons total, less than the capacity of just the Erie mine.[156]

In 1948 Erie set up a large-scale pilot operation for concentrating taconite near Aurora, Minnesota, designing and installing a series of gigantic machines for continuous ore processing. To provide staff for the operation, a new community, Hoyt Lakes, was founded for workers' families and named in honor of Elton Hoyt, who died in 1955. Erie Mining also had its own port on Lake Superior called Taconite Harbor. When shipment of taconite pellets began in 1954, their size and shape facilitated shipping and handling. Accustomed to harder-to-handle irregularly shaped chunks of ore, dockworkers called them blueberries because the dark pellets, seen against the rusty sides of a railroad hopper, looked blue.[157]

Three members of the "old guard" discuss a booklet prepared for the Cliffs centennial, 1950. *Left to right:* Carl Brewer, chief mining engineer; Walter F. Gries, superintendent–welfare department; and Charles J. Stakel, general manager–Michigan mining. All three would retire in the early 1950s as the company made the transition to open-pit mining and large-scale pellet operations. *(Courtesy of Cliffs Natural Resources, Inc.)*

Cleveland-Cliffs closely followed the Reserve and Erie pellet projects. In addition to its initial 10 percent stake in the Reserve Mining venture, in 1943 the company increased its investment in Reserve while providing modest support, with nearly a dozen other iron-mining companies, to the Battelle Memorial Institute to carry out contract research on the beneficiation of lean ores.[158] In July 1946 the board renewed this contract.[159] In 1947 it entered into an "Iron Ore Pelletizing Agreement" with the other companies invested in the Reserve Mining Company, as Reserve began to implement plans for an open-pit mine, a commercial beneficiation plant, and a dock and harbor.[160] By this time Cleveland-Cliffs was planning its own research program to tap its Michigan reserves of jasper. The magnetic concentration processes developed by Reserve and Erie for Minnesota taconite with its magnetite ores could not be applied directly to these reserves because jasper contained nonmagnetic hematite ores.

By the early 1950s the company had recovered from the near-fatal disaster of its involvement in Eaton's plans for a midwestern steel giant and the related creation of Cliffs Corporation and purchase of Corrigan, McKinney. However, it had failed to uncover significant new deposits of direct-shipping ores in either the Lake Superior basin or in the United States as a whole. Nor had its efforts to find iron deposits abroad met with anything more than limited success. The high labor costs of underground mining and projections that the ore in its mines was just about exhausted made it imperative for the company to find a way to develop its reserves of low-grade jasper ore on the Marquette Range. However, whether a commercially viable method for separating and concentrating nonmagnetic Michigan iron ore could be found was still an open question. The company would move to remedy this situation, closing all but one of its mines on the Mesabi and using its capital and mining expertise to transform itself over the coming decades into a large-scale, technology-based mining company.

5

PELLETS AND PARTNERSHIPS, 1950–1974

In June 1950 the people of the mining communities of Gwinn, Negaunee, and Ishpeming, as well as those working in the company's six mines on the Mesabi, gathered to celebrate the centennial of the Cleveland-Cliffs Company. Executives and members of the board of directors arrived from Cleveland aboard two historic ore boats, the *William G. Mather* and *Jasper H. Sheadle.* Neither William Mather nor E. B. Greene could attend the festivities because of failing health but they must have been there in spirit. President Alexander Brown paid tribute to the Mather family in his introduction of dinner speaker Harlan Hatcher, author of *A Century of Iron and Men,* a popular history of Cleveland-Cliffs written for the centennial. "How amazing it is to realize," Brown said, "that throughout the one hundred years of this Company's history now being commemorated, a father and his son—Mr. Samuel Livingston Mather and Mr. William G. Mather—have, during their two lives, witnessed every page being written in this hundred-year history and have themselves personally engaged in a large portion of those historical events. During the lives of those two men, the Company has shipped nearly 190 million tons of iron ore."[1]

The next day the company celebrated the opening of Shaft B of the Mather mine. Brown donned overalls and descended in the cage to the mine's sixth level, accompanied by mine superintendent C. W. (Pete) Allen, among others. The "holing through" to Shaft A allowed Brown and his entourage to walk along the drift to the point underground where Shaft B in Negaunee joined Shaft A in Ishpeming. Brown pounded a gold spike into one of the rails to mark the spot. The Mather, the company's largest underground

President Alexander Brown points to a sign marking the point where the tunnel from Mather B (located in Negaunee) joined the tunnel from Mather A (located in Ishpeming) at the Cliffs centennial celebration, June 1950. The total length of the tunnel was around 9,000 feet. *(Courtesy of Cliffs Natural Resources, Inc.)*

Aerial photograph of Mather B in Negaunee, c. 1950. Mather B, opened in 1950, was the company's last underground mine. The houses behind the mine buildings were later moved to make room for mine expansion. *(Courtesy of the John L. Horton Collection, Special Collections, Michael Schwartz Library, Cleveland State University.)*

mine, had begun production in 1944, and in 1948 production had exceeded 1 million tons of direct-shipping iron ore. The joining of the two shafts would make the "mighty Mather" among the world's largest underground mines.[2] All the company's sixteen mines closed at 4:00 p.m. to allow workers to take part in a "family day" at the Ishpeming winter sports arena, where the Negaunee, Ishpeming, and Gwinn bands performed and guests enjoyed free pop, coffee, and ice cream. The celebration culminated with fireworks attended by 17,000 people.

Few who attended the Cliffs centennial would have disputed the *Marquette Mining Journal*'s prediction of "a rosy future for the Marquette Range and the Lake Superior mining region."[3] At this time Cleveland-Cliffs operated mines on the Marquette, the Menominee, and the Mesabi. Of the eight mines on the Marquette Range, seven were underground mines—Cliffs Shaft, the Maas, the Lloyd, the Athens, the Cambria-Jackson, the Negaunee, and the Mather. The Tilden mine was the company's only open-pit mine in Michigan. The company also owned Spies-Virgil, a nearly exhausted mine on the Menominee Range. Cleveland-Cliffs operated five open-pit mines and two underground mines on the Mesabi Range, which produced both direct-shipping ore and intermediate-grade ore requiring beneficiation. Ore from mines in Minnesota accounted for almost half of the company's total ore tonnage. The company's gross tonnage for 1950 was 10,222,000, with approximately 64 percent from underground mines.[4]

For the next three years, the Korean War assured the company of record ore sales. So great was the need for additional vessel tonnage during the Korean War that Cliffs contracted with the Cleveland-based American Ship Building Company for a state-of-the-art freighter to be called the *Edward B. Greene*. Because the vessel could not be ready before 1952, Harrold L. Gobeille, vice president–marine, purchased a mothballed Victory-class cargo ship from the U.S. Maritime Commission. It took just ninety days for the Bethlehem Ship Building Company in Baltimore to lengthen its hull by 165 feet and modify its superstructure. Christened *Cliffs Victory*, the vessel was sent down the East Coast, around the tip of Florida to New Orleans, up the Mississippi to Chicago. The company garnered enormous publicity as Chicagoans watched the *Victory* (which was slightly longer than the Chicago harbor lock) ease into Lake Michigan.[5]

Despite record sales during the Korean War, the early 1950s were a time of uncertainty. Inflation and the growing militancy of labor were immediate concerns, but even more worrisome for the officers of the company was whether it could succeed in its plans to enrich and pelletize its lean Michigan ores. Would this new product enable the company to compete with high-quality imported ore?

Cleveland-Cliffs' management paid close attention to the progress of the Reserve Mining Company's pelletizing efforts not only because of the company's financial interest in the project, but also because pellets from enriched low-grade ore represented the future of the Lake Superior iron industry. As the scale of the project increased and legal problems surfaced over the terms of Reserve's lease of the mineral rights, Cleveland-

Christening of the *Edward B. Greene,* Toledo, 1952. *Left to right:* Alexander C. Brown, president, 1947–54, chairman 1954–57; Mrs. Dean Perry, daughter of E. B. Greene; and A. C. Ackerman, president of the American Shipbuilding Company. *(Courtesy of the Cleveland Press Collection, Special Collections, Michael Schwartz Library, Cleveland State University.)*

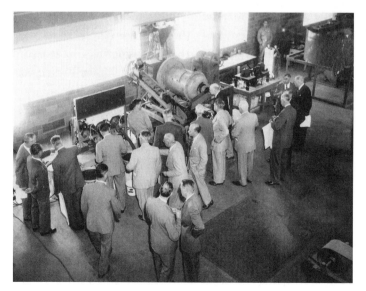

Members of the board of directors visited the Cleveland-Cliffs Research Laboratory during the Cliffs centennial, 1950, where they inspected pelletizing apparatus similar to that developed by E. W. Davis of the Mines Experiment Station at the University of Minnesota. *(Courtesy of Cliffs Natural Resources, Inc.)*

Cliffs, Montreal Mining, and Wheeling Steel sold the shares they owned in the Reserve mine to the Republic and Armco steel companies in September 1950. Cliffs received cash equal to its original investment and was entitled to receive a royalty of not less than 10¢ per ton on 10 percent of the finished product.[6] Presumably to raise capital for its escalating investment in the Reserve mine, Republic Steel sold 100,920 shares of the stock it held in Cleveland-Cliffs in October 1950.[7] Cliffs continued to participate in the less ambitious Joint Pelletizing Enterprise in Ashland, Kentucky, which may have served as a benchmark against which to measure its own pelletizing projects in Michigan.[8]

Federal legislation facilitated research and development of the iron pellet. Because of the anticipated demand for raw materials during the Korean War, Congress passed the Defense Production Act in 1950. The federal tax code now allowed mining companies to rapidly depreciate capital investment in new mining ventures like the one Cleveland-Cliffs envisioned for enriching low-grade jasper ore. Instead of twenty-year write-offs, mining companies could depreciate approved projects over five years. Depreciation also served as an important incentive for steel companies to invest in mining partnerships—the preferred method of defraying the large capital costs of processing low-grade ore.[9]

The cold war pushed the union of pellets and partners. By the early 1950s it was clear that regardless of the discovery of iron reserves in South America and other parts of the world, national security required a domestic supply of iron ore. During the Korean War, President Truman asked William S. Paley to chair a commission to assess the country's materials policy in the cold war context. The 1952 report, called *Resources for Freedom,* predicted increasing pressure on domestic iron ore supplies as the steel industry expanded. The Paley report recommended that steps be taken to secure a continuing supply of iron ore in the event of another war, including the use of domestic reserves of low-grade iron ore. The report also recommended completion of the St. Lawrence Seaway to facilitate shipment of Canadian iron ore.[10]

Partnerships were not new to Cliffs. Indeed, they were as old as the company. Both Samuel L. and William G. Mather had used partnerships to raise capital for new ventures or when outright purchase of mining properties was not possible. Cliffs had partnered with Lackawanna Steel in the development of the Negaunee mine in 1903, and again in 1940 with Bethlehem Steel to develop the Mather mine, retaining at least equal ownership stakes while maintaining control over operations. Although CCI owned in fee seven of the eight mines it operated on the Marquette Range, the seven mines the company managed on the Mesabi were all partnerships.[11] Cleveland's location a short distance from the midwestern steel mills had proved advantageous in forming and maintaining past partnerships. However, the new postwar partnerships, or "joint subsidiaries," Cliffs would form with various steel companies were nearly all minority partnerships, often owned by three or more competing firms. Since no one company owned a majority, they operated below the radar of the Securities and Exchange Commission, creating a seamless relationship between producers and consumers of iron.[12]

The new mining partnerships, necessary to make pelletizing operations commercially feasible, required financial resources on a scale far beyond the cash reserves and credit of most steel companies, let alone the usually lower-capitalized mining companies. These partnerships created a mutual dependency. The mining companies operated the mines. Each steel partner received its pro rata share of the mine's output at a cost determined by a long-term sales contract. The partner was also required to pay its pro rata share of the mine's operating expenses. Unlike the industry model developed for the Reserve and Erie mining companies in Minnesota, where Oglebay Norton and Pickands Mather did *not* invest in the partnership, Cliffs purchased shares that entitled the company to sell its complement of pellets at the Lake Erie base price—a price determined by the first shipment to arrive at the port of Cleveland each spring.

Moreover, unlike most of its competitors in the mining industry, Cliffs usually owned the land where its mines and future pellet plants were located—a Mather legacy that had stressed that mining properties should never be sold. This entitled Cleveland-Cliffs to receive royalties in addition to management fees from its steel industry partners. Both were indexed to rise with inflation. When demand for steel was high, partnerships benefited both steelmakers and mining companies. When demand for steel fell, the partnerships benefited the mining companies because the steel companies were locked into their partnership commitments and sales contracts. These differences, decades later, proved important to Cleveland-Cliffs' survival and reemergence as an independent mining company after the turn of the twenty-first century. By the end of 1973 partners' interests were 78 percent of the total 20.7 million tons CCI shipped.[13]

With national policy supporting its development, the iron pellet infused new life into the economy of the Lake Superior region, creating new jobs and keeping the economies of northern Minnesota and the Upper Peninsula of Michigan alive.[14] During the Eisenhower administration George Humphrey, former head of M. A. Hanna, served as secretary of the treasury. This appointment increased the region's political and economic clout. The steel industry's preference for pellets ultimately permitted the Lake Superior region to retain its historic dominance of the iron ore market in the United States, though the region's proportional contribution declined from 81 percent in 1950 to 74 percent in 1965.[15]

The lower labor costs of open-pit mining were another powerful incentive to develop a commercially viable product from low-grade ore. After a two-month industry-wide strike ended in 1952 with a wage increase, manager of Michigan mines C. W. Allen reported to Walter Sterling that a comparison of costs between open-pit and underground mining "served again to illustrate the flexibility of the open pits and their ability to absorb increased labor costs through greater productivity, thus highlighting the intended opening of the Humboldt and Republic pits to improve the economic balance of the producing properties in Michigan."[16]

Pellets from Jasper

Cleveland-Cliffs' first step in its transition to mining and processing low-grade ore was the hiring of Grover Holt as chief engineer in 1942. A native of North Dakota, Holt had worked for several mining companies on the Mesabi, where he was highly regarded as an expert on beneficiation. He was responsible for the formation of the mineral beneficiation division of the American Institute of Mining, Metallurgical, and Petroleum Engineers (AIME) in the 1940s. When Cleveland-Cliffs promoted him to general manager of all ore-mining operations in 1953, he was at the height of his career. After his election as president of the 30,000-member AIME in 1957, he remained at Cleveland-Cliffs as a special advisor to the president.

At this time Hugh Leach took over management of the company's Minnesota mines and James Westwater of its Michigan mines. Westwater had joined the company in 1940. Born in Aberdeen, Scotland, he was raised in Michigan and graduated from the Michigan School of Mining and Technology. Innovations during the tenure of these two talented managers included a new method of undercutting from inside the shaft, first used at the Mather mine in 1951. In 1955 at the Cliffs Shaft mine the company installed the first Koepe hoist built in the Western Hemisphere to replace the concrete shaft houses erected in 1919. The Koepe hoist, a technology developed in Germany, relied on a form of elevator technology using a system of steel ropes and pulleys to move loads up and down the mine shaft instead of the traditional cable drum. The system's greater use of counterweights sharply reduced net hoisting loads and made it more economical than traditional hoists.[17]

Leach, a graduate of the Minnesota School of Mines, had his office in Hibbing. Although most of the Mesabi ore produced during this period required some form of beneficiation, Leach recalled the 1950s as boom years when they could not fill orders fast enough. "We all thought 'oh boy, great, it will never stop,' but it sure stopped. . . . there was the beginning of the realization that these mines or ranges were going to play out and we better start looking around for properties and projects elsewhere."[18]

Research, it was hoped, would provide the key to unlocking the iron oxide in Michigan jasper. Manufacturing pellets from Michigan jasper, however, presented a set of technical challenges different from those of Minnesota taconite. Although jasper had a higher iron content (from 33 to 40 percent, compared to 30 to 35 percent for taconite), jasper was harder to crush. The iron ore contained in most jasper was nonmagnetic hematite and required the development of a flotation process for separation from its stony matrix. Flotation had been used since the nineteenth century to extract valuable minerals, such as copper, lead, zinc, gold, and silver, from waste rock. Typically, the ore-bearing rock was first pulverized to a very fine powder. The powder was then immersed in a water bath containing chemical reagents to facilitate separation. The reagents were generally oils that adhered to metallic particles, but not silica, and floated them to the top of the bath.

Cleveland-Cliffs Research Laboratory in Ishpeming, completed in 1949. In this structure the company began large-scale research for methods of concentrating and pelletizing the lean jasper ores found on its Michigan properties. *(Courtesy of Cliffs Natural Resources, Inc.)*

The minerals were then skimmed off and dried. The enriched mineral was then sent to be smelted in large-scale processing plants. The waste went to tailings ponds, where the moisture from the flotation bath slowly evaporated, leaving a finely powdered residue.[19] Although flotation methods had been used successfully in the nonferrous industry, it was not certain whether commercial flotation methods could be developed for separating iron oxide from jasper.

Louis Erck, chief metallurgist for Cleveland-Cliffs, headed a research group in Ishpeming that would tangle with the problems of processing jasper. Erck had a background in the beneficiation of Cliffs' intermediate-grade ore. He had developed several basic patents in the heavy-media method of concentration and headed Cleveland-Cliffs' test laboratory in Minnesota prior to his transfer to Ishpeming.

The group worked out of the boiler room of the Cliffs Shaft mine until a new laboratory building was completed in 1949. The new laboratory building cost a half million dollars; it was supplemented by a $750,000 pilot plant erected the following year.[20] By 1951 the lab employed six engineers and eleven technicians. The Ishpeming facility contained an assortment of devices for working on lean ore beneficiation projects: spiral classifiers, flotation cells, a Humphreys spiral, magnetic separating devices, a Wilfley table, and a vibrating screen, as well as a wide variety of testing equipment.

At the new laboratory Erck, Calvin Bjorne, Ned Johnson, his assistants, and a half dozen other staff members began work concentrating lean, nonmagnetic ores. They focused on two methods that at first seemed promising: flotation methods using heavy media, in which the company had developed some expertise already in its Minnesota operations, and artificial magnetization to convert nonmagnetic ores to magnetic so the magnetic means of separation already developed by E. W. Davis and used by Reserve and Erie could be applied.[21] Eventually they hit dead ends on these tracks and would have to turn to other methods. Edwin (Ned) Johnson, a graduate of Michigan Technological University in Houghton, Michigan, who began his career at Cliffs' research laboratory in 1948, recalled being surrounded by a group of enthusiastic, intelligent people who were determined to succeed in developing the new process for concentrating iron ore. "Whether we liked it or not," Johnson said, "we were pushed into the low-grade ores to survive."[22]

On the corporate level, Walter A. Sterling spearheaded Cliffs' transition from underground mining to large-scale open-pit pelletizing operations. Sterling was a second-generation Cliffs man, born and bred in Ishpeming. He had worked summers in high school as a compass man in Cliffs' land and lumber department. After graduation from the University of Michigan in 1916 with a degree in civil engineering, he took a job at a power plant in Toledo, Ohio. During World War I he served in the Army Corps of Engineers, and at war's end returned to Michigan to work in mining. He joined Cleveland-Cliffs in 1929 as supervisor of the Canisteo mine in Coleraine, Minnesota, and was promoted to manager of the company's Minnesota operations in 1940. Sterling became vice president of operations in 1950, president in 1953, and chief executive officer in 1955. According to H. Stuart Harrison, who succeeded Sterling in 1961, he was "the right man in the right place at the right time."[23]

On the political front, Cleveland-Cliffs lobbied the State of Michigan to modify its tax laws to encourage efforts to exploit low-grade ore deposits. From 1911 on, Michigan mines had been taxed on the basis of their "value," which included not only operational equipment, buildings, and ore stockpiles, but also known ore reserves. The state geologist's office, rather than local authorities, assessed mining properties. Generally, Michigan mining companies had found the system tolerable, since it took local politics out of the assessment process. However, the system also discouraged exploration, since the discovery of new reserves, even if not needed for some years, would immediately be taxed. A change in Michigan tax law from an ad valorem tax to a specific tax on low-grade ore at the time of shipping, passed in 1951 and expanded in 1959, encouraged the transition to pellets.[24]

Cleveland-Cliffs' venture into the uncharted territory of commercial development of low-grade Michigan ores began in September 1951 when it opened the Ohio mine two miles west of Michigamme in Baraga County, Michigan. It was a discouraging initiation. Drilling into the hard limonite-type ore with a churn drill proved slow and arduous. The mill for crushing the hard ore also proved inadequate. It broke down frequently, driving

Walter A. Sterling (1891–1980) spearheaded the company's transition to pellets. Sterling became president in 1953 and chief executive officer in 1955, serving until 1961. *(Courtesy of Cliffs Natural Resources, Inc.)*

up costs. A heavy media plant set up nearby produced an iron concentrate, but the product was of such low quality it had to be stockpiled.[25] Nevertheless, the engineering and research teams learned from the aggravations and setbacks; the company's commitment to the new initiative did not waver.

Cleveland-Cliffs anticipated that the new technology-intensive processes would require more electric power than was available from the six hydroelectric and one steam turboelectric power plants of Cliffs Power and Light Company. Thus, at the end of 1953 Cliffs initiated another of its joint ventures: a new corporation, the Upper Peninsula Generating Company, owned jointly with the Upper Peninsula Power Company of Houghton, Michigan. The new company built a power plant on the shore of Lake Superior near Marquette, with an initial capacity of 22,000 kilowatts. As the company's power needs grew, this plant was progressively enlarged over the next two decades and became the focus of environmental concerns in the 1970s. In January 1954 Cliffs liquidated Cliffs Power and Light Company and got out of the public utility business. Cliffs also sold the company's coal business to the Island Creek Coal Company in September 1954, presumably to raise additional capital for its new ventures in lean-ore development.[26]

In 1951, just as it got its small-scale Ohio mine beneficiation plant into production, Cliffs formed the Humboldt Mining Company, a joint venture with the Ford Motor Company. Cliffs' contribution to this initiative was financed through dividends and $3.5 million realized from the sale of stocks in five steel companies. The portfolio at that time represented 27 percent of the company's assets.[27] In announcing the joint venture, Ford vice president Irving A. Duffy expressed the "intense interest" of the Ford Motor Company in the project. He told the *Daily Metal Reporter,* "[T]he magic of Humboldt is the converting of deposits heretofore considered to be nothing but 'country rock' into huge iron ore reserves."[28] The Humboldt property, purchased in 1949, had been the site of an old underground mine that produced high-grade lump ore for several companies between 1865 and 1920. It had also been the site of Thomas Edison's failed experiment with beneficiation in the 1890s.

CCI built a concentration plant at Humboldt that replicated processes already tried on a small scale at the Ohio mine. To avoid the many pitfalls encountered at the Ohio mine, researchers at the research laboratory in Ishpeming set up a pilot plant to test each phase of processing prior to full-scale operation.[29] Pilot plant testing proved indispensible for all future mine development.

Humboldt, which began producing concentrate in 1954, was the company's first mine to use the oil flotation process on a commercial scale. Oil flotation (a type of froth flotation) had been patented by Francis and Alexander Elmore in England in 1898 and first used commercially at Broken Hill, Australia, in 1905.[30] In oil flotation water and oil-based chemical reagents were added to pulverized ore. The iron ore particles, coated with an oily film, attached themselves to bubbles in the mixture that spontaneously rose to the surface, where they could be skimmed off, leaving the silica and other waste at the bottom. The author of an article in *Business Week* pointed out that one of the advantages of the new technology was that it was a "push button and automatic affair," requiring fewer, more highly trained workers. "The one vital point where mechanical gadgets have to step aside for the human touch is when the bubbles holding the iron are skimmed off the flotation cells," the author pointed out. "It takes an experienced metallurgist to tell by the size of the bubble and the thickness of the froth, whether he's recovering as much of the iron as possible, whether he has just the right mixture of jasper, reagent, and water."[31]

The iron concentrate (about 60 to 63 percent iron) was stabilized by mixing it with standard ore before shipment to Detroit. At the Ford Motor Company's Rouge Plant the concentrate was fused to produce sinter, which was then added to the blast-furnace charge. An article in *Skillings' Mining Review* announcing the opening of the concentrating plant in 1954 remarked that the project had attracted "wide attention and will be watched with great interest as the pioneer in the utilization of this type of iron formation."[32] The "hot float" process the company developed to separate the iron ore from the pulverized rock reduced silica content to about 4 percent. "That, of course," recalled mining engineer Robert M. DeGabriele, "made a very superior product in the eyes of the users because it reduced their requirements for [lime]stone and of course coke, and made for better yield in the furnace."[33]

Another innovation used at Humboldt involved a new drilling technique called jet piercing, invented by the Linde Air Products Company. Jet piercing speeded up penetration of very hard rock. The hole was then loaded with blasting powder. After the blast, electric shovels loaded the ore into railroad cars or trucks for transport to a processing

Mining equipment in the Humboldt mine pit, early 1960s. Note the jet piercing drill (*back right*), a critical new technology that burned holes into the hard Michigan jasper in preparation for blasting. Humboldt, which opened in 1954, was among Cliffs' earliest operations on the Marquette Range to concentrate low-grade iron ore. *(Courtesy of Cliffs Natural Resources, Inc.)*

area where it was crushed and concentrated. The new drilling technique replaced the churn drill, which had proved inadequate for drilling through the extremely hard cherty hematite of the Humboldt formation.[34] Jet piercing contributed greatly to making the recovery of low-grade iron economically feasible and was adopted throughout the iron-mining industry.

Marketing Empire and Republic

As the Humboldt Mining project got off the ground, the company's geology department discovered a vast reserve of low-grade magnetite on land the company owned in the area of the Empire mine. This was important because up to this time, as chief geologist E. L. Derby reported to Grover Holt, exploratory drilling had seemed to indicate "questionable economic reserves as of present"—probably a reference to the company's reserves in the Tilden area that would require an expensive, and as yet unproven, flotation process for separation. What made the Empire discovery exciting was not only the size of the ore body but also the fact that it was made up of magnetite ores. These could be separated magnetically to yield a high-grade product. Magnetic separation was already being used successfully on the Mesabi Range at the Reserve and Erie mines. The discovery prompted Derby to predict a "bright future" for the company.[35]

To develop this new ore body, the company issued a "Prospectus on the Republic and Empire Mines Development" in 1954. This was sent to Inland Steel, Jones & Laughlin, Wheeling, Pittsburgh Steel, and International Harvester. Sterling thought steel partnerships would assure Cliffs of "definite long term ore sales contracts." These partnerships would provide not only guaranteed customers for the company's ore but also the financial backing needed to develop and install new beneficiation technologies and to pelletize the resulting enriched product, something the company had not done at the Ohio mine or Humboldt projects. If successful, the iron pellet would be superior in iron content *and* structure to the ores produced in Canada and Venezuela. Sterling was so convinced that the company's future depended on pellets that he wanted to go ahead with development of the new operation regardless of whether the company secured partners.[36]

Nevertheless, the company's decision to make pellets from jasper was controversial. When the Empire-Republic proposal to proceed with the development of pellet plants at each mine came up for the vote of the board on October 20, 1954, eight of the thirteen directors voted in favor, four against, with Cyrus Eaton abstaining.[37] Eaton commented in a private memo that the Wade, Greene, and Mather families "would be better served by conserving cash and paying debt rather than by making expenditures of the type proposed for the Republic Mine."[38]

In 1956, to finance the Republic mine portion of the project, CCI formed the Marquette Iron Mining Company, a joint venture owned by Cleveland-Cliffs (47.5 percent), Inland (20 percent), Jones & Laughlin (15 percent), Wheeling Steel (10 percent), and International Harvester (7.5 percent), and managed by Cleveland-Cliffs. Cliffs leased the Republic property to Marquette Mining for ninety-nine years. Cliffs' partnership interest entitled the company to receive almost half of the pellets produced by the mine and to sell them on the open market. In explaining his company's decision to become a Cliffs' partner, P. D. Block Jr., Inland's senior vice president, said: "There are no better deposits of low-grade ore in the Lake Superior district and this project is particularly attractive to our company because of its location close to Chicago."[39]

The Republic mine and its associated pelletizing plant opened in 1956. Cliffs had acquired the Republic iron mine in 1914 and operated it for twenty-three years as an underground mine before it was shut down when its high-grade deposits played out. The new open-pit operation tapped the previously useless lean-ore deposits in the location and replicated operations at Humboldt. Large trucks hauled crude ore from the Republic mine pit to an adjacent plant where it was pulverized, concentrated through oil flotation, and filtered to remove most of its moisture. The concentrate was then sent to the Eagle Mills Pellet Plant, located at Eagle Mills between Ishpeming and Marquette on the main line of the Lake Superior and Ishpeming Railroad. An article in *Skillings' Mining Review* noted that among the innovations at Eagle Mills was an agglomerating disc called the "flying saucer" that facilitated the balling of the concen-

Republic mine and pellet plant (*upper right*), c. 1960s, at Republic, Michigan. *(Courtesy of Cliffs Natural Resources, Inc.)*

trate into marble-sized pellets and the first use of an up-draft traveling grate where the pellets were baked.[40]

As with any new technology, problems surfaced. The design of the traveling grate, for example, with air traveling upward through the pellets, made it difficult to control the heat and temperatures in the bed. The company dismissed the McDowell Company of Cleveland, the plant's contractors, and Cliffs' managers Harry Swanson and Hjalmer Anderson took over. Swanson and Anderson succeeded in producing an acceptable product, though costs were high.[41] The company's achievement was nevertheless hailed by *Mining World.* "These pellets are more than just pellets," the trade journal reported. "They are the first commercial pellets ever made from Michigan specular hematite (jasper) ore—and, what's more, the first ever made from hematite flotation concentrate in the United States."[42]

By 1956 discoveries of ore fields in other parts of the world and the successful development of iron pellets from low-grade ore in the Lake Superior region had eased fears of a worldwide scarcity of iron ore. Walter Sterling urged support for a domestic price structure and rapid amortization of new mines to encourage capital investment.[43] Stu Harrison, vice president–finance, also emphasized the coming importance of large capital expenditures for concentrating taconites and jaspers in the *Analysts Journal.*[44] In 1960 the

The Republic mine with pellet plant (*upper left, top of pit*), 1972. When the Republic mine neared the end of its life, a major portion of the Republic tailings basin system was developed into the Republic Wetlands Preserve mitigation project in 2001. (*Courtesy of Cliffs Natural Resources, Inc.*)

company decided to pelletize its Humboldt concentrate and set up a pelletizing operation at the Humboldt mine.[45] The Arthur G. McKee Company designed and constructed the plant, which featured the first use by the company of the Allis-Chalmers grate-kiln method of firing the pellets.[46] Allis-Chalmers perfected this technology with each successive pellet plant, installing what was then the world's largest grate-kiln system in the Tilden Pellet Plant in 1974.

The Contrarian: Cyrus Eaton

Although the majority of Cliffs' directors supported the development of the company's reserves of jasper, one member of the board consistently and emphatically voiced his disapproval. Although his effort to create a midwestern iron and steel giant had failed, Cyrus Eaton had recovered much of his fortune during World War II. He purchased a majority stake in the Steep Rock Iron Mines, located at Steep Rock Lake, 142 miles

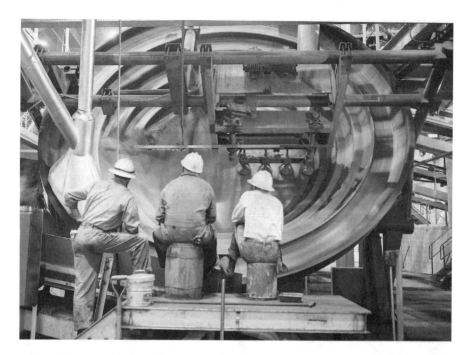

Employees at Eagle Mills monitor the "flying saucer agglomerator," 1956. This technology for rolling the concentrated iron slurry into balls was later replaced with more efficient balling drums. Eagle Mills closed in 1970. *Left to right:* Bob Hemmila, Ken Uren, and Bob LaJoie. *(Courtesy of Cliffs Natural Resources, Inc.)*

west of Port Arthur, Ontario. Well into the 1950s Steep Rock supplied Cleveland-Cliffs with a high-grade standard ore through the Premium Ore Company, Steep Rock's U.S. agent. Steep Rock was among the largest Canadian mines—originally thought capable of producing 10 million tons a year of high-grade lump ore prized as feed for open-hearth steel. Cleveland-Cliffs' management and Eaton were often at odds over the high prices that Steep Rock charged for its ore and the Canadian mine's inability to consistently supply ore in the contracted amounts. Compared to the Cleveland-Cliffs mining operations, Steep Rock was poorly run and suffered from high labor turnover.[47] Nevertheless, Cleveland-Cliffs' standard ores could not match the quality of Steep Rock ore. Ironically, Steep Rock's failure to make a timely shift to pellets may have led to the decline of the mine's profitability after 1956.[48]

Eaton's seat on Cliffs' board derived from his control of a large block of the company's shares of stock. He owned a controlling interest in the Portsmouth Steel Company and through Portsmouth had purchased 85,000 shares of Cleveland-Cliffs from the Alleghany Corporation in July 1953.[49] This allowed him to appoint Edmond Lincoln of Wilmington, Delaware, a former economist for the DuPont Company, to Cliffs' board. Lincoln and Eaton opposed the pellet initiative and accused the board of sanctioning

The grate-kiln system for firing pellets at the Republic Pellet Plant, erected in 1963, was manufactured for Cleveland-Cliffs by Allis-Chalmers. Balls made of concentrated iron ore were dried, heated, and hardened as they moved along on the traveling grate in the background. They were tumbled and fully hardened at 2,400 degrees Fahrenheit in a rotary kiln (*foreground*). (*Courtesy of Cliffs Natural Resources, Inc.*)

wildcat explorations in Canada and South America when ore from the Steep Rock Iron Mines was available, "a property already developed and from which the risk element had already been removed."[50] Although a merger between Steep Rock and Cliffs was considered in 1956, it was rejected by the board in favor of a long-term contract.[51]

By 1955 Eaton personally owned 7,514 shares of Cliffs' common. Through his ownership of the Portsmouth Steel Company and its subsidiary Detroit Steel, he also controlled another 365,272 shares of common stock and 100 shares of preferred, or 13.6 percent of Cliffs' voting stock. In an effort to win control of the board, he appealed to George Garretson Wade, Cliffs' largest stockholder, who apparently rebuffed him. Then he turned to George Gund, president of Cleveland Trust, who voted the stock of the

Wade, Mather, and Greene families held in trust by the bank. "The company's future prospects have been undermined," Eaton wrote in a confidential letter to Gund, among other things "by the disintegration and dissipation of its once valuable power properties and the giving up of its long-established coal business at a time when other far-seeing organizations are acquiring coal reserves."[52] Eaton failed to sway Gund to vote against management.

Standing in the way of control of Cleveland-Cliffs, Eaton discovered, were 183,000 shares of Cleveland-Cliffs common held by Schmidt and Company, a New York investment firm. Eaton retained Abe Fortas, member of the prestigious Washington law firm of Arnold, Fortas and Porter, to put pressure on Cliffs' management to reveal the owners of this block of stock. Not above using politics to get what he wanted, he also contacted Senator Paul Douglas (D-Illinois) after it was revealed that the 183,000 shares in question were owned by Bethlehem Steel's pension trust fund. Douglas called for an antitrust investigation by the Justice Department. The *Wall Street Journal* reported that the senator "was especially concerned that the alleged ownership of a big block of Cleveland-Cliffs stock might give Bethlehem some kind of control over its competitor, Youngstown Sheet and Tube Company, since Cleveland-Cliffs in turn owns 176,500 shares in Youngstown."[53]

The antitrust issue had been under investigation since 1950, the *New York Times* noted in a discussion of interlocking directorships. A 1951 report submitted to the House Monopoly Subcommittee by the Federal Trade Commission cited relationships between the four Cleveland ore companies (Cliffs, Pickands Mather, Oglebay Norton, and M. A. Hanna) and Republic Steel, Youngstown Sheet and Tube, Inland Steel, Wheeling Steel, Jones & Laughlin, and the National Steel Corporation. It described the Cleveland Trust Company and the National City Bank as "meeting places" for this group of iron and steel company officers. The *Times* noted that "the interlock consisted of both stock ownership and common directors."[54]

In 1955 Eaton thought that he might be able to use the antitrust issue as leverage with Bethlehem Steel. He offered to purchase the 183,000 shares directly from the steel company to relieve it of the appearance of a conflict of interest. Allowing him to gain control of Cleveland-Cliffs, Eaton told R. E. McMath, Bethlehem's chairman of finance, would be a service to the stockholders of Cleveland-Cliffs, given "the age and infirmities of W. G. Mather, followed by the physical and mental decay of E. B. Greene and now emphasized by the age and lack of capacity of A. C. Brown." McMath informed Eaton that Bethlehem had no interest in disturbing its cordial relations with Cleveland-Cliffs at a time when the steel company was seeking to take over Youngstown Sheet and Tube. The sale of its stock, McMath said, would "arouse great protest from some of their friends in Cleveland who were essential in putting over the Youngstown merger."[55] The merger, denied by the Justice Department in 1958, resulted in Bethlehem's decision to build a plant at Burns Harbor, Indiana. The last greenfield (new) plant built by an integrated steel producer, the plant proved extremely profitable for Bethlehem.[56]

Eaton had underestimated the cohesiveness of Cliffs' board, composed of a self-effacing group of representatives of old Cleveland families known for their generosity to the city's cultural and civic institutions. Over several generations Case Western Reserve University, the Cleveland Museum of Art, Trinity Cathedral, and the Western Reserve Historical Society had benefited from the philanthropy of these families, which reached back into the nineteenth century. Boston-based Philip R. Mather, son of Sam Mather, represented the Mather family, serving on the board of directors from 1942 to 1970. George Garretson Wade served from 1926 to 1957, Jeptha H. Wade from 1957 to 1997. James D. Ireland II, the son of Elizabeth Ireland Mather, held a seat on the board of directors between 1951 and 1985. The board's tenacity and commitment to the city's iron-mining traditions and legacy held Eaton at bay.[57]

Turning Point

By 1959 Eaton's influence was on the wane, the company's portfolio of steel stocks was increasing in value and, most important, the company's investment in pellet technology was beginning to bring modest gains. Even a strike by the United Steelworkers in 1959—serious though it was—could not halt the forward momentum of the mining industry's transition to pellets.

The 1959 strike began in July and interrupted mining for 116 days during the height of the shipping season. Even before the strike, Cliffs had been forced to cut back production from its underground mines because of competition from higher-grade imported ore. The Mather mine, with a capacity of about 3 million gross tons annually, was cut back to an output of 1.7 million gross tons. Production at the Maas, Bunker Hill, and Athens mines was cut to 650,000 gross tons. During the strike Sterling warned that almost one-third of America's 1958 supply of iron ore was already coming from foreign sources—"enough to keep all Cliffs' people on the job for an average of about four years!"[58]

Although the industry-wide strike was over wages, one of the roadblocks to settlement by the steelworkers was a clause in the new contract that gave management the right to improve plant efficiency. When President Eisenhower invoked the Taft-Hartley Act on October 7, steel and mine workers did not go back to work. After the case had been in the courts a month, the Supreme Court upheld the constitutionality of the Taft-Hartley Act. Although workers returned to the job, the strike, Walter Sterling told a reporter, had cost the company 50 percent of its lake shipping season. The company had counted on moving "a sizable tonnage of ore" before the lakes froze. This had proved impossible because of three weeks of record-breaking bitter cold in November.[59] When the strike was finally settled, the iron and steel industry agreed to a 40¢ raise in hourly wages, making steelworkers among the highest paid industrial workers in the country. A writer for the trade journal *Iron Age* called the settlement a capitulation to the unions.[60]

The strike marked a turning point in the balance of trade. Imported steel rose from 2.9 percent of domestic consumption to 6.1 percent and imported steel exceeded exported steel for the first time. The strike had an impact on ore producers because it coincided with the opening of the St. Lawrence Seaway. The seaway facilitated shipping of ore from Labrador, Canada. Ore from South America and Africa was also delivered to eastern ports, where it was stockpiled until the strike was settled. This gave foreign companies a chance to forge new relationships with American steelmakers.[61] Even though steel imports fell slightly the next year, foreign steel and iron ore had gained a foothold that could not be dislodged.[62] As historian Paul Tiffany commented in a book on the decline of the steel industry, "By the time Eisenhower left office in 1961, the American steel industry was in far worse condition than it had been in 1953, and one of the most significant causes of that deterioration was the growing volume of cheaper imported steel into domestic markets."[63]

Despite the 1959 strike, Cleveland-Cliffs managed to almost equal the previous year's earnings. Ironically, Cliffs' agreement with Steep Rock to purchase ore for resale in the United States proved its salvation during the strike. Although ore sales from its domestic mines fell 14 percent, ore purchased from Canada and a small amount from other foreign sources rose 77 percent. Walter Sterling called the increase in working capital to a year-end total of $65 million the year's financial highlight. But what gave him even more reason for optimism was the strong demand for the company's concentrates and pellets from its Michigan jasper properties.[64]

A 1960 Wall Street investors' report, which coincided with the company's listing on the New York Stock Exchange, confirmed Sterling's positive assessment of the company's future prospects. It accorded high praise for Cliffs' efforts to pelletize low-grade ore and predicted that production would grow from 800,000 tons in 1959 to 7 million by 1969. The report noted Cleveland-Cliffs' ownership of both high-grade reserves and about twenty square miles of low-grade jasper. In addition to its mining interests, Cliffs operated the fifth-largest ore-carrying fleet on the Great Lakes, owned 75 percent of the Lake Superior and Ishpeming Railroad, its own power-generating facilities as well as 50 percent of the Upper Peninsula Generating Company, and 390,000 acres of timber. Its portfolio included shares of the five steel companies, originally acquired in the 1930s, now with a market value in excess of $116,096,000. According to the report, about 47 percent of the company's iron ore production in 1959 was still direct-shipping ore from underground mines, 44 percent from beneficiated intermediate ores, and 9 percent from low-grade jasper.[65] The rising demand for Cleveland-Cliffs' stock reflected Wall Street's positive view of the future potential of the iron pellet.

After two decades, the mining industry's commitment to pellets was beginning to bear fruit. In July 1960 the *Wall Street Journal* carried a story, headlined "Iron Ore Upheaval," stating that pellets melted faster than raw ore in blast furnaces and contained less moisture so that they could be transported in cold weather without freezing.

"When a plant with five blast furnaces switches to pellets, the effect is the same as adding a sixth blast furnace," it was reported. The shift to pellets saved $55 million—the cost of building a new blast furnace.[66] Thomas M. Rohan, the Cleveland regional director of *Iron Age,* was even more enthusiastic. "Producers of pellets are in a seller's market with customers scrambling to get the output," he wrote. "Total shipment of pellets from U.S. and Canadian plants has gone up at an astonishing rate—in the face of a general loss of markets for direct-shipping ores to foreign sources." He pointed out that while there was a glut of direct-shipping ore (now called standard or natural ore), there were not enough pellets to satisfy demand. And Cleveland-Cliffs, the only mining company that sold pellets on the open market, had the capacity to supply pellets to steel firms "who don't want to put up, or haven't got the money to join the taconite club."[67]

Rohan's article made public the results of a test that established the superiority of pellets when used as 100 percent of the ore burden. This test, carried out at Armco's Bellefonte blast furnace at Middletown, Ohio, proved that pellets used alone in the blast furnace produced almost twice the normal yield of pig iron.[68] Because of their uniform size and shape, they melted faster than standard ore. Pellets made operation of the blast furnace more efficient because the charge required less coke and limestone. Indeed, the net savings on energy and labor offset the higher cost of pellets, and their lighter weight meant a 15 to 20 percent saving in freight costs.

To Stanley Sundeen, manager of research and ore development for Cleveland-Cliffs, the combined effect of pellets and high-grade imported ore had made standard ore "unsalable at a profit." The survival of the state's mining industry would depend on replacing the tonnage lost to underground mining with pellets.[69] Changes in Michigan tax law from an ad valorem tax to a specific tax on low-grade ore at the time of shipping, passed in 1951 and 1959, encouraged this transition.[70]

While domestic pellets could beat all but the highest-grade foreign ores, John S. Wilbur, vice president–ore sales and marine, sounded a cautionary note. One attraction of the ores from Chile, Peru, Brazil, and Liberia was the low freight rates for ocean transportation made possible by lower labor costs. "An ordinary deckhand on a U.S. Freighter gets more money than a captain on a foreign ship," he told members of the Lake Carriers' Association in 1960.[71] Ocean shippers could carry ore more cheaply than lake vessels because they were not limited to a maximum of 25,000 tons and could operate year-round instead of only seven months of the year. Lower lake railroads charged exorbitant rates for short hauls from the ports to the mills, making rates for rail transport of foreign ore inland to mills in the Pittsburgh-Youngstown area more competitive than before.[72] This was not the first time that Wilbur would challenge the industry's apparent complacency.

In an effort to trim costs, in 1964 marine superintendent John L. Horton oversaw a joint research project with the U.S. Maritime Commission to develop automatic boiler

Cliffs Victory on one of its many voyages, seen against the Cleveland skyline, c. 1957 or later. The company purchased the World War II cargo vessel in 1950 for service during the Korean War. At that time the hull was lengthened 165 feet. Another 96 feet were added to the hull in 1957, making it briefly the longest and fastest vessel on the Great Lakes. *(Courtesy of the John L. Horton Collection, Special Collections, Michael Schwartz Library, Cleveland State University.)*

controls, manufactured by the Bailey Meter Company of Cleveland.[73] These automatic controls, installed for the first time on the *William G. Mather,* not only saved fuel but also lowered labor costs because they eliminated the need for watch-standing firemen. The same year, the marine department installed the first dual propeller bow thruster system on the *Mather.* The system, developed by the American Ship Building Company, enabled ore boats to navigate the Cuyahoga River without the added costs of tugboats. Between 1964 and 1971 the vessels *Cliffs Victory, Sterling, Champlain,* and *Pontiac* were outfitted with the new technology.[74]

H. Stuart Harrison (b. 1909), who became president in 1961, managed the company's portfolio of steel stocks, which he gradually sold to raise funds to expand the Empire and Tilden pelletizing operations in Michigan and to develop and expand the Robe River mine and plant in Australia. He stepped down as president in 1974, remaining as chairman until 1977. *(Courtesy of Cliffs Natural Resources, Inc.)*

Expansion and Diversification under Stu Harrison

H. Stuart Harrison, who became chief executive officer in 1961, would continue the expansion of the company's pellet capacity and diversification into related mineral fields. Harrison had graduated with honors from Yale with a degree in finance and economics in 1932. His father, an executive at Corrigan, McKinney Steel Company, had encouraged his son to make his career in the steel industry. After graduation Harrison had briefly worked for Corrigan, McKinney but he was more interested in finance. He moved to New York City to work for the trust department of the Hanover Bank, where he was in charge of portfolios related to steel and other heavy industry. In 1937 E. B. Greene convinced him to return to Cleveland. In 1945 Harrison became treasurer of Cleveland-Cliffs, rising to vice president–finance in 1953. Harrison's expertise in steel stocks was an important asset to the company, since the value of Cliffs' portfolio of steel stocks (acquired in the 1947 consolidation with the Cliffs Corporation) continued to rise. In 1955 34 percent of the net earnings of the company came from its steel stocks and other investments.[75]

Harrison's first challenge as CEO came in 1962 when the Oliver Iron Mining Company cut the price of its surplus standard ore by 80¢ per ton. This price reduction forced Cliffs to reduce its standard ore prices, despite rising labor and material costs. In his letter to Cliffs' employees, Harrison described the price reduction as a "real jolt for it means that serious problems face us in the operation of our underground mines." He explained that Oliver could cut prices and still make money because its mines in Minnesota were all open pit, in which labor costs were lower. Underground ores were still Cleveland-Cliffs' bread and butter, and profit margins were being squeezed not only by higher wages but also by the increased cost of beneficiating standard ore to make it competitive with pellets and foreign ore.[76] Young Sam Scovil, then a member of the sales department, reassured the members of Cliffs' board of directors in a talk on ore prices that increased pellet production would reduce dependence on standard ores.[77] This increased production would come from the Empire mine, Cliffs' third concentrating and pelletizing operation.

The Empire Mine: Magnetic Jasper

The Empire mine, located northwest of the town of Palmer on the eastern end of the Marquette iron range, proved one of the company's most efficient and lucrative mining operations. From 1907 to 1928 the Empire mine had produced direct-shipping ore for the Oglebay Norton, M. A. Hanna, and Clement K. Quinn companies. After Cleveland-Cliffs acquired the lease in 1962, the company organized the Empire Mining Company, owned by Inland Steel (32 percent), McLouth (25 percent), International Harvester (15 percent), and Cliffs (28 percent). In contrast to the specular hematite ore bodies of the Humboldt and Republic mines, Empire's ore was magnetite, making it feasible to use the magnetic separation process developed on the Minnesota ore ranges to liberate the iron from the waste rock. This was a less expensive process than froth flotation, employed at Humboldt and Republic. Another advantage of magnetic separation was the exothermic reaction produced during the process, which reduced energy costs.

The chemical and physical composition of Empire's magnetite allowed Cliffs to use fully autogenous grinding mills instead of the conventional rod and ball mills used at the Humboldt and Republic mines. Autogenous grinders break up ore by tumbling it against itself. This innovation, first applied to iron ore by Cliffs' mining engineers, reduced the cost of labor and materials. Steel balls and rods in traditional grinding mills quickly wore down and frequently had to be replaced. Grinding in autogenous primary mills reduced the Empire's ore to the consistency of beach sand. Secondary grinding took place in smaller pebble mills, also designed to operate autogenously. The pebble mills used pebbles of ore about two inches in diameter as the grinding media and reduced the ore to the consistency of very fine powder.

The addition of water at this point in the process produced a slurry that was sent through magnetic separators called cobbers, which began to separate the magnetic iron ore from the nonmagnetic silica to produce a concentrate. To remove additional impurities from the concentrate, the company's engineers put it through a reverse flotation stage after magnetic separation. That is, the concentrate passed through flotation chambers containing a reagent that floated the waste to the top to be skimmed off. The iron concentrate was drawn off the bottom of the chambers. This was the reverse of most flotation

Primary grinding mills at Empire II, completed 1966. Autogenous grinding, in which the ore tumbles against itself, reduced the ore to the consistency of a fine powder. The plant was the first to use fully autogenous grinding. In 1963 it was capable of producing 1.6 million tons of pellets per year. It was expanded in 1966, raising its output to 3.5 million tons; in 1975 to 5.3 million tons; and in 1980 to 8 million tons. *(Courtesy of Cliffs Natural Resources, Inc.)*

Empire balling drum and vibrating screen, 1970s. After the binding agent bentonite was added, the iron concentrate was rotated in large balling drums. Green (unfired) balls were screened so that only those of a uniform size were carried along a belt to the grate-kiln. *(Courtesy of the John L. Horton Collection, Special Collections, Michael Schwartz Library, Cleveland State University.)*

processes, in which the useful mineral was floated to the top for skimming. The enriched concentrate was then pumped to disk filters, where the water was filtered out. Balling drums shaped this concentrate, now about 65 percent iron, into small pellets that dried and hardened as they traveled along a heated grate through two different heating zones into a rotary kiln, where they were baked into hard pellets.

The Bechtel Corporation of San Francisco handled the Empire mine's engineering,

Superintendent Bernhardt Petersen in the control room of the Empire Pellet Plant, 1973. After grinding the ore to a fine powder, Empire's huge magnets separated the iron particles from the waste material in the slurry. Magnetite requires less energy than hematite to concentrate the iron because of the exothermic reaction that is produced during processing. *(Courtesy of Cliffs Natural Resources, Inc.)*

procurement, and equipment installation with the exception of the agglomerating, or pelletizing, plant. Allis-Chalmers of Milwaukee again won the primary contract for the grate-kiln system. The project required the construction of a dam on Schweitzer Creek, three miles south of the concentrator, to facilitate the supply of water for the plant. In early 1963, in announcing the impending opening of the Empire mine, Harry C. Swanson, manager of the company's Michigan mines, credited the company and its steel company partners with "a lot of guts and a lot of foresight and knowledgeable people who contributed a lot of time and endless effort to prove to everyone that low-grade mining in hematite ores to produce a high-grade pellet was possible."[78] It had taken exactly one decade from discovery of the magnetite reserve to begin production.

Within a few weeks of the opening of the Empire mine in December 1963, it was operating 15 percent above its rated capacity, allowing the company to stockpile 400,000 tons of pellets over the winter. Upon the opening of the shipping season in April 1964, the first carload of pellets carried an evergreen tree—a tradition marking the first shipment of ore from a new mine. From Escanaba the pellets were transported down Lake Michigan for delivery to the Inland Steel plant at Indiana Harbor.[79] When the Empire mine completed its first year of operation, it had exceeded its rated capacity of 1.2 million tons of pellets. Two years later the company added a second line to the Empire plant, doubling production.

The Pioneer Pellet Plant: Processing Underground Ore

The declining demand for direct-shipping, or standard, ore—along with the rising cost of labor—accelerated the closing of underground mines across the Lake Superior ore district. Between 1960 and 1961 alone the number of active underground iron mines in Minnesota, Michigan, and Wisconsin dropped from thirty to nineteen.[80] By 1964 only nine underground mines remained in the Upper Peninsula of Michigan.[81] The last mine on the Gogebic Range would close in January 1966. Cleveland-Cliffs closed down the company's Bunker Hill mine in December 1963, even though its reserves had not yet been exhausted. A far more difficult decision involved the Mather mine, which in its heyday had employed more than 1,000 miners. The Mather mine was leased by the Negaunee Mining Company from Cleveland-Cliffs, with shares owned by McLouth Steel (43.5 percent), Bethlehem (34.5 percent), Republic Steel (10.5 percent), and Cliffs (11.5 percent), with Cliffs acting as manager.

To keep Mather ore competitive with higher-grade imports and agglomerates, the company set up an ore improvement plant in 1957. The ore from Shaft B was hauled on large underground railcars to a new crushing and loading station located inside the mine. After the ore was brought to the surface, it was transported by rail to the ore improvement plant at Eagle Mills, where additional operations reduced moisture and silica content.[82] However, even with this form of beneficiation, the direct-shipping ore from the Mather mine—a high-grade soft hematite—was less marketable than pellets. Although the mine produced a record 2,631,000 tons in 1957, by 1963 output had declined to 736,000 tons. It was necessary to lay off several hundred miners when the company closed Shaft A in December 1960. Jim Westwater, then vice president–mining, recalled that during a meeting in the early 1960s Bethlehem officials informed him that even though they owned a large percentage of the mine, they were no longer willing to accept Mather's soft and sticky ore. Westwater got them to agree that if the company were able to pelletize this ore, they would take the pellets.[83]

To save the Mather B mine, in 1965 the companies that owned the Negaunee Mining Company developed a plan to pelletize half of the Mather mine's output at a new plant, located just beyond the ore improvement plant at Eagle Mills. The Pioneer Pellet Plant was a $15 million joint venture with steel company partners McLouth Steel (50 percent), Bethlehem (20 percent), and Republic (15 percent), with Cliffs retaining the remaining 15 percent. Pioneer was the first and only plant in the Lake Superior region to

The Eagle Mills ore improvement plant (*right*) and the Pioneer Pellet Plant (*left*), June 1972. Pioneer (opened in 1963) was the first commercial plant in North America to pelletize natural iron ore from an underground mine. The adjacent ore improvement plant (opened in 1957) dried, screened, and ground three types of crude ore produced by the Mather B mine. The ore with the highest iron content was sent to the Pioneer plant for processing into pellets. The remainder was sold as direct-shipping ore. (*Courtesy of Cliffs Natural Resources, Inc.*)

produce pellets from an underground mine. "This project counteracts an alarming trend which in the last few years has seen many of the Michigan underground mines closed in the face of competition from high-grade imported ores and domestic pellets made from beneficiated low-grade ores," an article in the *American Metal Market* reported.[84] The company planned to expand output of the mine from 700,000 tons to 2.4 million tons annually, saving 350 jobs and providing employment for an additional 400 miners.

The new pellet plant was economically feasible because a change in Michigan's tax laws allowed pellets from underground mines the same tax treatment as pellets produced from open-pit, low-grade jasper mines. Working together, the company's operating and research departments, the Michigan Mining Association, the United Steelworkers of America, and Dominic Jacobetti (D-Negaunee) drafted the legislation. The Michigan

Underground mining superintendent Eino Koski, Mather B, 1973. By pelletizing some of the ore produced by Mather B in the Pioneer Pellet Plant, Cleveland-Cliffs succeeded in keeping the mine open, but operating costs were much higher than those of open-pit mining. Mather B closed in 1979. *(Courtesy of Cliffs Natural Resources, Inc.)*

legislature unanimously passed the Specific Tax Bill, and it was signed into law by Governor George Romney May 8, 1962.[85]

Cliffs again awarded the primary contract for the plant to the Allis-Chalmers Manufacturing Company, with the Arthur G. McKee Company responsible for the outside facilities, including conveyors, ore storage, pellet stockpile, and fuel oil system. At the pellet plant the ore was finely ground in a dry rod mill, pelletized, and then placed on a traveling grate, which conveyed it to a rotary kiln for firing. Because of the earthiness of the underground ore, no binder was required.[86] However, the pellets produced from underground ore had a higher silica content and lower iron content. Finished pellets averaged about 60 percent iron, compared to 64 percent in pellets made from low-grade jasper; they also lacked the strength of pellets produced from Cliffs' open-pit mines. The Pioneer Pellet Plant operated around the clock, 365 days a year, with pellets stockpiled through the winter until the shipping season began. In addition to producing 1.2 million tons of pellets per year, the company shipped 1.2 million tons of sinter feed and coarse ore, sold as two distinct products named Cliffs Group Fines and Cliffs Group Coarse.[87]

By 1968 90 percent of Cliffs' domestic ore shipments, about 9.8 million tons, came from just five pellets plants—Humboldt, Republic, Empire, Pioneer, and Eagle Mills—with Empire by far the largest producer.[88] The expansion of pellet production in its mines allowed the company to rehire men who had been laid off, some of whom had been unemployed since the 1959 strike. The list had exceeded 1,100 in 1960. Even more satisfying was the hiring of 302 new men, bringing total employment in Michigan to 2,100. At this time Cleveland-Cliffs and the United Steelworkers Union joined forces to take advantage of state and federal manpower-training programs to encourage miners to take voluntary layoffs in order to train for jobs as electricians, welders, and mechanics. The new open-pit mining and pelletizing operations needed skilled labor to keep their huge trucks and specialized machinery humming. A 1964 article in the *American Metal Market* celebrated the first new hire in the mines of the Marquette Range since the late 1950s.[89] Cliffs' capital expenditures of $75 million, principally to expand its pellet operations, were paying off in the form of higher revenues, better pay, and rising employment in Michigan's Upper Peninsula.[90]

In a 1965 address to the American Mining Congress, President Stuart Harrison marveled at the transformation of mining that had occurred since World War II. The return to open-pit mining, the use of trucks and conveyor belts instead of locomotives for hauling ore, the increasing size of equipment, and the introduction of heavy-media separation in the early 1940s were merely incremental changes. What had happened with the successful development of pellets was nothing short of a revolution. Pellets had saved the mining economies of the Lake Superior region by making domestic ores competitive with high-grade foreign ore. Harrison predicted that by 1975 three-fourths of the ore production in the United States would be in the form of pellets, a growth from 30 million in 1965 to 71 million tons by 1975.[91] In a talk to the New York Society of Security Analysts in 1966, Thomas E. McGinty, assistant treasurer of Cleveland-Cliffs, emphasized that pellets had proved a boon to the steel industry as well. Millions of dollars that might have gone into building additional blast-furnace capacity could be used for other capital expenditures, such as adopting the basic oxygen furnace (BOF), a more efficient steelmaking process.[92]

As the industry made the transition to pellets, the center of the American steel industry continued to shift from Pittsburgh to the Chicago-Gary area. Although the Marquette Range had been the nearest major iron ore range to Pittsburgh, shipping to the Chicago area was even more advantageous. Inland Steel, for example, was willing to increase its partnership interest in the Empire mine because pellets could be shipped from the port at Escanaba down Lake Michigan instead of using Marquette or Duluth. This allowed the company to haul twice as many tons in a season. Although Lake Superior ore was no longer competitive with imported ore in the coastal regions of the country, Michigan and Minnesota still enjoyed the advantage of lower transportation costs to midwestern mills.

Also, most foreign ore was not made into pellets, now the preferred feed for American blast furnaces. Harrison told stockholders in 1965 that "pellet sales were limited only by our capacity to produce them and we are taking steps to expand this as rapidly as possible."[93]

A lawsuit brought by Wheeling Steel against Cleveland-Cliffs, the Hanna Mining Company, and its mining subsidiary, the M. A. Hanna Company, in September 1966 highlighted the dependency of the steel companies on their suppliers. It alleged that the mining companies were a monopoly engaged in fixing the prices of iron ore in the Great Lakes iron market.[94] Wheeling was Cliffs' partner in the Marquette Iron Mining Company, which operated pellet plants at the Republic mine and Eagle Mills. The suit charged that Cleveland-Cliffs and Hanna, each of which had directors on Wheeling's board, exerted undue influence on Wheeling's decisions related to partnerships and long-term ore sales contracts. The steel company's partnerships prevented it from getting bargain-priced iron ore when demand fell. Cleveland-Cliffs and Hanna, it was alleged, influenced the steel industry through the large blocks of stock in steel companies they continued to own and called the shots when Cleveland bankers voted the stock they held in trust. No case law had yet developed to handle the antitrust implications of partnerships, or "joint subsidiaries," owned by three or more competing firms—no one of which held a controlling interest. As historian Daniel Fusfield has pointed out, these firms were not required to report their ownership interests to the Securities and Exchange Commission. The partnerships thereby sustained a web of common interests out of reach of antitrust litigation.[95]

Press coverage of the Wheeling suit mentioned the large block of Cleveland-Cliffs stock held by Cleveland Trust (and voted by Cleveland-Cliffs director George Gund, who was president of Cleveland Trust). An article in the *Cleveland Press* quoted Cyrus Eaton on the pernicious influence of interlocking directorships on the American iron and steel industry.[96] An investigation by the Justice Department into the ore companies' alleged monopoly took more than a decade. It was dropped in May 1980—a point when steel imports represented a far greater threat to the diminishing fortunes of the domestic steel industry than restraint of trade.[97]

Foreign Acquisitions, Diversification, and Detroit Steel

In the early 1960s, to lessen its dependence on the domestic steel industry, Cleveland-Cliffs began to consider foreign acquisitions and diversification into related natural resource industries. "Cliffs is now entirely dependent on the steel industry both in its mines and investments," the board minutes stated. "Such a policy of total dependence on a non-growth industry does not appear wise."[98] It seemed prudent to lighten its portfolio of steel stocks. In 1967 Harrison told the New York Society of Security Analysts that the company was "moving more and more into a diversified mining organization with a backbone of iron ore but filling out in other directions." He credited partnerships with liberating the company from the volatility of the merchant iron ore market. Profits from royalties and management fees might not be as large as selling ore on the open market, but they were more secure. And the location of the company's mines was proving a strategic advantage in selling iron ore to Chicago steelmakers. Nevertheless, it made good business sense to diversify into natural resources–related mining, such as uranium, copper, nickel, and shale oil.[99]

Cliffs formed a partnership in 1964 with the Standard Oil Company of Ohio and The Oil Shale Corporation (TOSCO) to form the Colony Development Company to explore the potential of recovery of oil from 7,500 acres of deposits of shale in Colorado. The three companies jointly financed the building of a pilot plant, with the Colony Development Company acting as agent. Cliffs also obtained options on reserves located in eastern Utah and purchased additional properties in Colorado.[100]

In 1967 Cliffs Copper Corporation, a wholly owned subsidiary, and G. M. Wallace and Company became partners in a venture to develop copper mines in Nevada and Utah. Cliffs signed an agreement with Union Pacific Railroad Company and W. S. Moore Company of Duluth to mine and concentrate copper in southwestern Colorado. The company also invested in the exploration of uranium deposits in six areas of Wyoming, Colorado, and Utah.[101] Cliffs created a western division based in Rifle, Colorado, to manage these acquisitions. In 1968 Cliffs also participated in exploratory drilling of forty-two nickel claims in western Australia.

At the same time, Cleveland-Cliffs reengaged in geological exploration for iron in North America. Hugh Leach, then vice president–research and iron ore development, recalled that the company looked at properties in Montana, New Mexico, Missouri, and Idaho. "Unfortunately, I guess I'd say that all of them really proved to be not worth developing," he said, "and I guess our judgment has proven correct because I don't think anybody else has developed them to any substantial degree."[102]

The company's Canadian ventures showed more promise. In 1964 the company began planning a new $57 million mine and concentrating and pellet plant on Lake Temagami, Ontario. The property, purchased and mapped by Cliffs in 1959, became the Sherman mine, a joint venture with Dominion Foundries and Steel, Ltd. (Dofasco), which owned 90 percent. The mine was named for Frank A. Sherman, a former chairman of Dofasco and a pioneer in the Canadian steel industry. It had a production capacity of 1 million tons of high-grade pellets per year.

Robert DeGabriele, vice president–project engineering, supervised the construction of the Sherman mine. In an interview, he recalled the effort it took to convince Dofasco's vice president of engineering to adopt autogenous grinding mills for the plant. After studies by DeGabriele's team showed that autogenous grinding would contribute significantly to the keeping the mine economically viable, it prevailed. "Considering the recovery [of iron oxide] at that property and the remoteness of everything else, I think it's been a very, very successful plant," DeGabriele reminisced.[103] Production began in 1967 with the first shipment transported by rail 337 miles to Dofasco's mills in Hamilton, Ontario, in March 1968.[104] The Adams mine, located at Kirkland Lake, Ontario, started up in 1971. It was also owned by Dofasco and managed by Cleveland-Cliffs.

In 1965 Cliffs announced a plan to develop a mine and pellet operations in the Robe River District of Australia, about seventy miles south of Cape Preston on the Indian Ocean. Cliffs and its partners, which included Mitsui and Company, Ltd. of Japan and Garrick Agnew, Ltd. of Perth, Australia, obtained funding to build a mine capable of producing 10 million tons of ore, a town, a railroad, a pellet plant, and a loading port.[105] Again DeGabriele was sent abroad with a team to manage construction. Cliffs contracted with the Bechtel Corporation to design and build most of the infrastructure. The Australian government even demanded that the company provide landscaping for the new town, as well as churches and jails.[106] In April 1969 Cliffs' subsidiary, Cliffs Western Australian Mining Company, Pty., Ltd., signed a contract with seven Japanese steel companies to supply them with 87 million tons of pellets over a period of twenty-one years.

At the end of the year, buoyed by the performance of the Robe River plant and the prospects for Tilden, Cliffs announced a five-year plan for doubling the company's pellet-making capacity from 10 million tons to 20 million, with half the expansion in Michigan and the other half in Australia. "We are expanding," CEO Harrison told *American Metal Market*, "because we feel that now as the standard ores run out, pellets are basically the best way to reduce iron in a blast furnace."[107]

By the late 1960s U.S. Steel and M. A. Hanna had joined the "pellet club," which already included Pickands Mather (Erie), Oglebay Norton (Reserve), and Cleveland-Cliffs. Unlike the other steel and iron companies of the Lake Superior region, U.S. Steel's tremendous reserves of direct-shipping iron ore and its success in securing control of iron reserves in other parts of the world had allowed it to wait and see whether the efforts of the other companies would succeed. A well-equipped laboratory in Duluth and a pilot plant in Mountain Iron succeeded in producing the company's first taconite concentrate in 1953. But as late as 1955—the year Reserve and Erie began shipping commercial pellets—Oliver was still experimenting with various agglomeration techniques to produce sinter, nodules, and pellets.[108] Oliver's pellet capacity remained relatively small until its Minntac mine at Mountain Iron came onstream in 1968.[109] M. A. Hanna, another fierce competitor in the iron-mining industry, also waited until the 1960s to develop its wholly owned Groveland mine on the Menominee Range, with a pellet capacity of 2 million tons annually.

Development of the iron pellet caused consolidation in the iron industry. In the early 1950s there were about twenty-five to thirty companies shipping from the Lake Superior region. Oliver Mining (with about 45 percent of the total) was ranked first, with M. A. Hanna and Pickands Mather second and third in ore production (with about 16 and 15 percent, respectively). Cleveland-Cliffs was fourth, with a mere 8 percent. Cleveland-

Cliffs had a commanding lead on the Marquette Range, accounting for 82 percent of ore output. Michigan competitors included the Inland Steel Company, North Range Mining Company, and the M. A. Hanna Company. Cliffs' nearest competitor on the Marquette and Menominee ranges was the North Range Mining Company, with less than 1 percent. North Range's Blueberry and Champion mines on the Marquette and Warner and Book mines on the Menominee were all underground properties.[110] In the 1960s this and other small mining companies, unwilling or unable to make the transition to pellets, went out of business. Only five Lake Superior mining companies would remain: Oliver, M. A. Hanna, Pickands Mather, Oglebay Norton, and Cleveland-Cliffs.

If Cleveland-Cliffs had successfully negotiated the transition to pellets by the late 1960s, management was still troubled by the ownership of a large block of stock by Cyrus Eaton. When Eaton sold his interest in Detroit Steel in 1968, this raised the possibility that an outsider might acquire Detroit Steel in order to take over Cleveland-Cliffs. The fall of Detroit Steel's stock from $22 a share to $8 in 1970 increased the company's vulnerability. If Detroit Steel were to offer to purchase Cliffs stock at an attractive price, Cliffs stockholders might not be able to resist the temptation to sell, allowing control of Cleveland-Cliffs to pass to Detroit Steel. "The company was in dire trouble," CEO Harrison recalled. After the Cyclops Steel Company purchased a minority interest in Detroit Steel, Cliffs took the one-eyed bull by the horns. Cleveland-Cliffs purchased a majority interest in Detroit Steel, despite the risk that Cliffs might get stuck with an aging steel mill and $50 million in liabilities.[111] Fortunately, Cliffs' management was able to convince Cyclops to swap the outstanding shares of Cliffs stock for the physical assets of Detroit Steel.[112]

Resolution of the Detroit Steel problem came as an enormous relief to Cliffs' management, which had operated the company since the early 1950s under the threat of takeover by the Eaton interests. It marked the beginning of a new era when the investment community took notice of Cleveland-Cliffs. "The stock really began to move in a large measure, not because of the Detroit acquisition, but because the pellet revolution then was well underway," Tom McGinty, vice president–planning and development, recalled. "The fact that Cliffs' royalties and management fees were hedged by inflation and we were in the midst of a huge inflation in the United States, all combined to make Cliffs a very attractive investment. From that point on, Cliffs' stock rose rather dramatically. . . . Mr. Eaton was gone, and all of the problems associated with that were gone, and so we had a whole new outlook on things."[113]

As pellet operations expanded, the company was able to make new hires, including its first women miners in the spring of 1973. The inclusion of women in the labor force happened slowly and without apparent resistance by male workers. Cliffs' first female workers, Linda Bancroft, Jeannette Dunquist, and Donna Garceau, were assigned to the Humboldt Pellet Plant, where the company installed a new women's dry for changing their clothes. Young and physically strong, they were expected to do both the light and

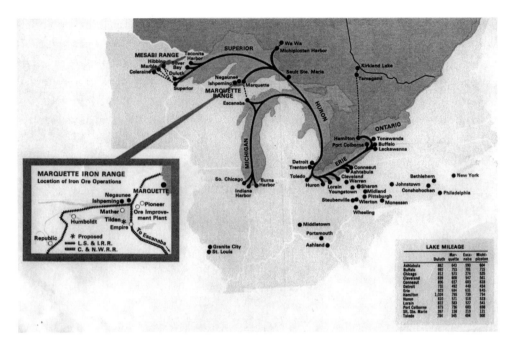

Map with vessel routes and location of Cleveland-Cliffs iron ore operations in North America, 1973. The location of the Humboldt, Republic, Mather, Empire, and Tilden mines, as well as the pelletizing plants, are shown on the insert. Cliffs offloaded its iron pellets at ports on the lower Great Lakes for the trip by rail to the steel mills. *(Courtesy of Cliffs Natural Resources, Inc.)*

the heavy jobs, "no more or less than we would give a male laborer," said an operations engineer interviewed for an article in the *Cliffs News*. The women liked the variety, but the best part of their jobs was the pay, medical insurance, and other benefits. However, since there was no paid leave for new mothers, the women workers usually retired once they started families.[114] The positive experience of these women was in marked contrast to the sexual harassment of women at the Eveleth Taconite Company in Minnesota, who brought a class action against Oglebay Norton in 1988.[115] After this experiment proved successful at Humboldt, women were regularly considered for places in the workforce.

In his 1974 speech to the Newcomen Society shortly before he stepped down as president, Harrison celebrated the company's "plunge into pellets." Introducing this new technology "entailed risks equal to any undertaken by the company in its previous pioneering history."[116] But it had paid off. Partnerships had enabled the company to generate the huge capital requirements of the Empire and the Robe River mine and pellet plant in Australia. In the more than two decades since the company had celebrated its centennial, Cleveland-Cliffs had evolved from an independent iron ore producer to a manager of technology-intensive operations in partnership with various steel companies. With economists predicting an unprecedented demand for iron ore, in March 1973 Harrison wrote in his annual letter to shareholders that the company was engaged in a $600 million capital-expenditure program that would increase Cliffs' North American production capacity by 55 percent. He looked forward to the opening of the Tilden mine, when he expected the company's total iron ore capacity to be 41 million tons.[117] Without a crystal ball, it was impossible to predict that the very partnerships that cushioned the company from the boom-and-bust cycles of the steel industry would prove the company's Achilles' heel.

6

GREAT EXPECTATIONS AND UNEXPECTED CHALLENGES, 1974–2000

B y far the company's most ambitious pelletizing venture, the Tilden mine, came onstream in 1974 at a time of intense demand for iron ore. "We are in a worldwide shortage of iron ore," President Stu Harrison wrote in March 1974. "While there will undoubtedly be periods of surplus in the future, the long-term trend is definitely one of scarcity."[1] For the first time in the history of Cleveland-Cliffs' fleet, shortages of iron ore made it necessary to operate all of the company's Great Lakes ore carriers through December and into January.[2] That year the company completed an expansion of Empire mine, raising the mine's annual pellet capacity from 4 million to 5.3 million tons. Cleveland-Cliffs produced a record of 15 million tons of iron ore, with 4.1 million tons produced by its Robe River plant in Australia. With demand for steel in the United States predicted to grow 2.5 percent a year, the company anticipated an annual pellet production of 16 million tons from the Tilden and Empire mines alone.[3]

Expansion of pellet capacity by Cliffs came at a time when industry experts were predicting an increase in the worldwide demand for steel. Steel industry experts like Father William T. Hogan urged American steelmakers to respond by increasing production.[4] Although the steel industry recognized the threat of imported steel, in the early 1970s devaluation of the dollar made domestic steel more competitive. Optimism was also fueled by the Experimental Negotiating Agreement with steelworkers, which promised to end the vicious cycle of strikes and rising prices of American steel.[5]

What was inconceivable was the unraveling of the American steel industry in the early 1980s. The integrated steelmakers were forced to cut production as less expensive, higher-quality imported steel flooded the domestic steel market. The distress of the steel industry directly affected Cleveland-Cliffs. Two important partners, the Wisconsin and McLouth steel companies, declared bankruptcy. With its pellet operations shut down, Cleveland-Cliffs declared a loss in 1982—the first since the Great Depression. Seriously hobbled by a steep decline in its stock price and low cash reserves, the company nevertheless went forward with the acquisition of Pickands Mather. This acquisition, made at the nadir of both companies' fortunes, would provide Cleveland-Cliffs with valuable mines in Minnesota, Canada, and Australia and contribute significantly to the company's recovery in the 1990s.

The Tilden Mine Project

The Tilden mine, with a projected annual capacity of 4 million tons of pellets, was the centerpiece of Cleveland-Cliffs' plans for coping with the expected demand for iron ore in the 1970s. Acquired by Samuel Tilden and his associates in 1865, it had been owned by Cleveland-Cliffs since 1891, when Iron Cliffs and the Cleveland Iron Mining Company had merged. The Tilden property was located three miles south of Ishpeming, adjacent to the Empire mine. In 1929 the company had begun open-pit mining there. The Tilden mine had produced about 6.5 million tons of direct-shipping high-silica crude ore averaging 38 percent iron before the mine was shut down in 1968.[6]

The Tilden reserve contained an estimated 1 billion tons of crude ore. The iron in the Tilden reserve consisted predominately of hematite, with smaller amounts of goethite and magnetite. Because the company owned it "in fee," it could anticipate generous royalties from pellets produced from Tilden ore. However, developing a system for recovering the iron involved considerable risk. The first hurdle was grinding the ore. Tilden crude ore was very hard and required very fine grinding. Grinding mills wore out quickly.

The greatest challenge, however, was that hematite could not be separated magnetically—a proven process already in use at the Empire mine. Researchers first tried a process called magnetic oxide conversion (MOC). This involved roasting the concentrate to convert hematite to artificial magnetite. Although this proved technically feasible, the company abandoned MOC because of the high energy requirements associated with the development of the process.[7]

When the Ishpeming laboratory found it impossible to work out a satisfactory flotation method for separating Tilden ore, the company turned to the U.S. Bureau of Mines for assistance. In 1961 the bureau had developed a flotation method called selective flocculation at its Twin Cities laboratory. In 1966 Cleveland-Cliffs and the Bureau of Mines signed a cooperative agreement whereby the bureau would test its new flotation method.

Because the process had not been patented, it was necessary to keep careful records so that when the time came, the Bureau of Mines would receive the patent and the company the license for its use. Cliffs built an integrated pilot facility at Eagle Mills for evaluating Tilden's different ore types and testing flow sheet configurations. This plant

The Tilden mine's flotation cells, 1970s. Tilden's unique flotation process for separating the iron after grinding to a fine powder was developed in cooperation with the U.S. Bureau of Mines. *(Courtesy of the John L. Horton Collection, Special Collections, Michael Schwartz Library, Cleveland State University.)*

was capable of processing ten tons of crude ore per hour. Emert Lindroos, chief metallurgist and assistant manager for research and development, recalled long hours in the pilot plant and running daily tests in the laboratory.[8] It was a time of high stress.

Eventually, the company adopted a two-phase flotation process called "selective flocculation and siliceous dispersion." Unlike the flotation method developed by the Bureau of Mines, in which the desired iron oxide floated to the top of a bath and was skimmed off, the company adapted the process to use "reverse flotation," whereby the unwanted materials were floated to the top and skimmed off. In the first stage of the process, selective flocculation, cooked cornstarch selectively attached to the iron oxide particles, causing them to stick to each other, forming "floccules" that sank to the bottom of the tank. At the same time the unwanted silica was floated off by a large quantity of water that overflowed the top of the thickener tank.

In the "siliceous dispersion" phase, another reverse flotation, starch was again added to the slurry along with an amine flotation collector. In a series of 500-cubic-foot cells, equipment agitated the slurry at the same time air was drawn into the slurry by a spinning rotor. The amine attached itself to the remaining silica particles, causing them to adhere to air bubbles. The bubbles "floated" away from the iron-rich slurry. In this stage the starch addition acted as a "depressant." It selectively coated the iron oxide particles, preventing the amine from attaching to and floating the iron oxide away as well. The iron oxide was then recovered in the underflow. Part of the challenge of processing the Tilden ore was the complex chemical interaction between the water and the metallurgy at different stages of processing. Parameters such as temperature, suspended solids, pH, calcium hardness, amine, dissolved silica, and turbidity needed to be constantly monitored.[9]

James Villar, who became manager of R&D in 1968, had the job of overseeing pilot studies and final implementation of the technology. Villar had earned a degree in geology from Michigan State. After serving in the U.S. Marines during the Korean War, he had returned to the Upper Peninsula to work in Cliffs' geology department, where he had served as project supervisor for geological exploration in Chile and Peru. He had received his Ph.D. from Michigan State in 1958 and joined the Cliffs laboratory in Ishpeming in 1962. In addition to managing the laboratory, he kept Cleveland up to date on progress. In an interview by Burt Boyum in 1984, Villar recalled the reaction of President Stu Harrison when he told him they were using tapioca flour from Siam as the flocculent: "I think he envisioned these Chinese junks in Marquette Harbor would bring our stuff." Villar assured Harrison that now that they had found the key to concentrating the Tilden ore, they would find cheaper domestic substitutes.[10] This turned out to be cornstarch.

The flotation process was so sensitive that the chemistry of the different waters of the Carp, Greenwood, and Schweitzer rivers could affect performance. "We never really understood the complexity and importance of the water system until we got into full operation," Villar said. The role of water temperature was another surprise. Once they had succeeded in producing an iron concentrate, they confronted the challenge of filtering out the water.[11] To obtain a dryer filtercake, it was necessary to install huge steam dryers. This added to the system's complexity and cost. The subsequent processing step of agglomeration was no different than the process at the Empire plant. The iron filtercake was mixed with Wisconsin bentonite, rolled into green balls, and fired in one of two Allis-Chalmers rotary kilns. The end product was a pellet of 65 percent iron.[12]

In December 1971, convinced that the process would work, Cleveland headquarters turned to financing the Tilden mine. It was to be the costliest venture in company history. Though unsure whether the Algoma Steel Company would be willing to join forces with Cleveland-Cliffs on such a large project, Stu Harrison sent Sam Scovil, known for his charisma and can-do attitude, to ask Algoma Steel to become a partner in Tilden. Algoma was a large Canadian integrated steel company located on the St. Mary's River at Sault Ste. Marie. A graduate of Yale, Scovil had worked in sales under John Wilbur since 1950. He became vice president of sales in 1963 and senior vice president in 1970.

During World War II he had served as an army captain. Upon his discharge he had found a job as a blast-furnace operator at Republic Steel. This steel background gave him credibility with customers.[13]

Scovil knew that the company's future was riding on his ability to bring Algoma to the table. Minority partnerships like the ones Cleveland-Cliffs had used to finance the Republic and Empire pelletizing operations were the preferred method of financing new mines throughout the iron and steel industry. What made finding partners for the Tilden mine difficult was the high cost of the project in 1970s inflated dollars. After several meetings, to Scovil's surprise, the Algoma people informed him that the company would take a 30 percent ownership stake (later increased to 47 percent) in the mine. Ecstatic, he hurried back to Cleveland. At the company's holiday party on the evening of his return, he waited until just before the festivities broke up to announce the news. "I just wanted to tell Mr. Harrison, our boss," he said in an impromptu speech, "that we have now got a 30 percent partner for the Tilden project, so I think it's going to be a go."[14] Once Algoma had committed, other North American steel companies agreed to join the venture.

The agreement, announced March 30, 1972, brought about the formation of the Tilden Mining Company. Financing was extremely complicated, consisting of two partnerships: the Tilden Iron Ore Company (64 percent) and the J&L-Cliffs Ore Partnership (36 percent). The Tilden Ore Company's partners were Algoma (47 percent), Steel Company of Canada (15.5 percent), Wheeling-Pittsburgh Steel Corporation (12.5 percent), Sharon Steel Corporation (7 percent), and Cleveland-Cliffs (18 percent). J&L owned 75 percent of the J&L-Cliffs Ore Partnership, with the Cleveland-Cliffs Iron Ore Corporation, a subsidiary of Cleveland-Cliffs, responsible for the remaining 25 percent.[15]

In announcing the formation of the J&L-Cliffs Ore Partnership, William R. Roesch, J&L's chairman and president, explained that it would "improve the quality and quantity of J&L's iron ore resources with limited capital investment." J&L made a commitment to purchase 3.4 million tons a year from Cliffs as a replacement for ore from its Tracy mine on the Marquette Range, closed down in 1970, and the Adams mine in Canada, which had been sold. In addition to a guaranteed supply of pellets from Tilden, J&L and Cleveland-Cliffs were to share the profits from the sale of pellets on the open (spot) market.[16]

The purpose of the J&L-Cliffs Ore Partnership was to facilitate borrowing $57.6 million from a group of banks and insurance companies. With the organization of the Tilden mine partnership and expansion of the Empire mine, Cleveland-Cliffs entered a new era of funding using project debt. Before 1970 a partner generally used equity to finance its pro rata share of the capital and development costs and purchased at cost its share of the output. However, between 1970 and 1979 the huge costs associated with pellet projects made equity financing unrealistic. Project funding through loans became the accepted approach to raise the capital necessary for these ventures. The partners paid interest on these loans out of funds generated after the mine became operational. There were several advantages to using project financing instead of a public stock offering to

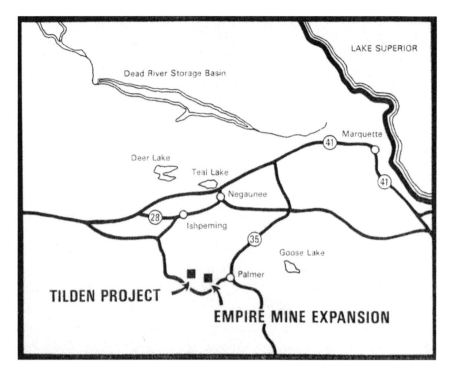

Map showing the location of the Empire mine and the proposed Tilden mine on the Marquette Range, 1973. *(Courtesy of Cliffs Natural Resources, Inc.)*

raise capital. Financing through banks and insurance companies kept the debt off the partners' balance sheets, as long as each owned less than 51 percent. Project financing allowed partners and lenders to renegotiate their agreements as contingencies arose. Proprietary information related to costs could be protected since they were not subject to Securities and Exchange Commission oversight. The group of steel company partners, using their parent companies as collateral, spread the risk. This made financial institutions more willing to invest in this large-scale and expensive technology.[17] Since these were minority partnerships spread among several competing steel companies, no red flags were raised from the antitrust point of view.

For the Tilden project in particular, a favorable rating by both Moody's and Standard and Poor increased competition among potential lenders and allowed the partnership to obtain lower interest rates.[18] Another important advantage from the steel companies' point of view was that project financing allowed them to obtain large tax credits and significant depreciation and depletion deductions as long as the projects were generating income.[19] Between 1970 and 1979 the amount of capital raised by Cleveland-Cliffs and its partners for three pellet plants exceeded $1.2 billion. The company borrowed approximately $260 million to fund these projects.[20] Project financing was an ideal financing vehicle so long as the partners responsible for paying off the debt remained solvent.

First shipment of pellets from the Tilden mine, December 17, 1974. Note the traditional tree planted in the first carload for good luck. *Left to right:* Einer Lindquist, general superintendent, Tilden; Ronald Harma, plant superintendent, Tilden; James Westwater, senior vice president; Maxwell Madsen, chief ore grader (partially visible); Gilbert Dawe, manager–Michigan mines; James Villar, manager–research and development; Walter Nummela, assistant manager–Michigan mines; Robert DeGabriele, manager–engineering and construction; Robert Saari, president of local union 4974, United Steelworkers of America; James Ombrello, operating engineer, Tilden; and Roy Winnen, superintendent, Tilden. *(Courtesy of Cliffs Natural Resources, Inc.)*

In commenting on the expansion of pellet capacity in the Lake Superior region in the 1970s, historian Christopher Hall noted that these large investments in iron ore occurred at a time when the steel industry was postponing improvements in other areas of steelmaking. "Unlike their hesitancy to invest in new technology such as the continuous caster," he said, "the integrated companies plunged into expansion of their iron ore facilities in the 1970s with a vengeance." He pointed out that in 1974 the nationalization of the ore operations of U.S. Steel and Bethlehem in Venezuela seemed to confirm the wisdom of expanding North American mining capacity, but adding many millions of tons of annual capacity at a time when demand was actually declining, in his view, seemed to defy logic.[21] The ability to depreciate these investments and the presumed worldwide scarcity of iron ore may have been the driving factors in these decisions.

Tilden and the Environment

The scale of the pellet projects of the 1970s inevitably had a far greater impact on the environment than underground mining or the company's smaller pellet projects. Moreover, the public's perception of Lake Superior mining companies changed in the 1970s. In the 1950s, when the Reserve and Erie mining companies began mining and processing taconite, they were perceived positively as sources of jobs and economic growth. By the 1960s the negative aspects of open-pit iron mining had become apparent. Duane

A. Smith, in *Mining America: The Industry and the Environment, 1800–1980,* attributed the "great environmental awakening" in part to the impact of Rachel L. Carson's *Silent Spring* and Harry M. Caudill's *Night Comes to the Cumberlands,* both published in the early 1960s. These two books galvanized public opinion and prompted greater government regulation of the mining industry, particularly the strip-mining of coal. Congress passed the Clean Air Act in 1963, the Water Quality Act in 1965, and the National Environmental Policy Act creating the Environmental Protection Agency in 1969.[22]

Mining and processing low-grade iron ore, however, was not like strip-mining of coal. With the exception of Reserve, the processing plants associated with lean-ore beneficiation and pelletizing plants deposited their wastes in self-contained settling ponds, recycling as much water as possible in and out of these ponds. Even the dumping of tailings from Reserve's processing facilities into pristine Lake Superior was tolerated until 1960. When the company announced plans to increase pellet capacity to 10 million tons per year and requested a permit to increase the amount of Lake Superior water used in processing to 502,000 gallons per minute, the United Northern Sportsmen objected. They claimed that fishing in Lake Superior had been affected by Reserve's processing plant. However, the sport fishermen lacked scientific evidence to substantiate the cause of the deterioration of the water quality in Silver Bay, on which the Reserve plant stood.[23]

Over the next nine years, public concern over pollution of the drinking water of communities located along the coast of Lake Superior grew. In 1969 the Sierra Club and other citizen groups brought a class action against the Reserve Mining Company and the State of Minnesota. The suit claimed asbestos-like fibers from the tailings had contaminated the drinking water of communities along Lake Superior as far away as Duluth. The federal judge in *Reserve Mining Co. v. EPA* ruled in March 1975 that the company had violated the Water Pollution Control Act. He ordered the mine's 3,100 workers to stop work. The injunction was reversed on appeal because there was no scientific evidence to establish a direct link between Reserve's tailings and asbestos in the drinking water. Nevertheless, the case remains a landmark in the history of the environmental movement. In 1977 Reserve ended the practice of dumping and three years later completed a system of tailing ponds for handling wastewater.[24]

The high visibility of this case had an impact on all surface iron-mining and pellet projects in the Lake Superior region. In the wake of the Reserve lawsuit, Cliffs encountered public scrutiny of its plans for construction of the Tilden mine. In 1971, in anticipation of objections by environmentalists, Cliffs created an Environmental Affairs Committee, headed by John Wilbur. Wilbur emphasized the necessity of large-scale ventures to allow Michigan pellets to compete against Minnesota pellets and high-quality imported ore. He pointed out that Cliffs' 4,000 employees and their families represented one-third of the total population of the Marquette area. Because of the company's commitment to paying fair wages and generating reasonable profits, economics were "paramount whenever we consider a proposed action and its possible alternatives."[25]

A 1972 article in the *New York Times* described the dilemma of Michigan residents caught between the need to save jobs and the desire to preserve the environment. Emissions from pellet plants, the article said, turned the stripes of skunks pink, covered homes with unsightly red dust, and sacrificed whole forests to the open pit. Nevertheless, with unemployment exceeding 10 percent, residents were reluctant to stand in the way of a $200 million investment in the region. The *Times* article reported that 200 people had turned out for a public hearing, where they learned that the proposed Tilden operation would require thirty-eight tons of water to produce one ton of pellets. The company assured its critics that it planned to invest $10.5 million in pollution-abatement devices, including both smoke precipitators and water-recycling systems.[26] Thereafter, environmental controls became a critical aspect of all Cliffs' pelletizing operations.[27]

Public hearings held in Lansing before the State of Michigan's Natural Resources and Michigan Water Resources commissions considered the company's proposal to dam and divert the middle branch of the Escanaba River to meet the Tilden mine's need for water. Cleveland-Cliffs engineers calculated that processing the Tilden's pellets would use 6,200 gallons per minute, with 93 percent of the water recycled. The reservoir created to provide this water, named the Greenwood reservoir, would have a twenty-six-mile shoreline and thirteen islands. Trees would be cut down prior to the creation of the new lake. To entice public support, the Tilden Mining Company agreed to provide a four-acre site for public access to the reservoir for recreational activities such as swimming, camping, boating, and hunting.[28] The company told the commissioners that six years of careful planning had gone into the project and urged speedy approval. A spokesman for the Ishpeming Chamber of Commerce emphasized that the Greenwood reservoir created by the proposed dam on the Escanaba River would turn "an inaccessible piece of countryside into an exciting piece of lake property," while a representative from the Michigan Building and Construction Trades Council argued that "a few canoeists and a handful of professors who have never been outside a classroom" should not be allowed to turn the Upper Peninsula into Appalachia.[29]

On May 22, 1972, the Michigan Department of Natural Resources approved the company's plans for the water diversion, dam, and reservoir, and accepted its environmental impact statement. The State of Michigan's Public Act No. 293 of 1972, however, required the company to carefully monitor the quality of water discharged into the river from its tailings basins. To comply with the regulations, Cleveland-Cliffs sent water samples each month to state and federal laboratories.[30]

While the diversion of the Escanaba River was accomplished without a great deal of opposition, a proposal to expand the power plant of the Presque Isle Station in Marquette, run by the Upper Peninsula Generating Company, became a cause célèbre when Marquette housewife Julia K. Tibbitts mounted a crusade to protect a beloved park. In 1975 the Upper Peninsula Generating Company asked for permits for two additional coal-burning generating units. This additional generating capacity was needed for the

Presque Isle Power Plant, looking east, after construction of units 9 and 10, late 1970s. Expansion of the power plant and the construction of a coal-unloading dock were opposed by environmental activists at this time. Presque Isle Park is in the background. *(Courtesy of the John L. Horton Collection, Special Collections, Michael Schwartz Library, Cleveland State University.)*

expansion of the Empire and Tilden mines. The company also planned to build two new coal-unloading facilities next to the plant.

The power plant, located at the entrance to the Presque Isle Park, had not encountered public protest when it was constructed in 1955. Nor had subsequent additions of power units. The park, a gift to the city by Peter White about 1900, was a popular hiking, sailing, and picnic spot with breathtaking views of Lake Superior. Tibbitts became a vocal foe of the expansion of the power plant. She also wanted to stop the installation of the new coal-unloading equipment because of the potential for coal dust to pollute the air. Using a family inheritance from investments in Upper Peninsula real estate over several generations, she founded Superior Public Rights, Inc. and hired lawyers to file a class action suit against the power company and the Lake Superior and Ishpeming Railroad, which owned the coal docks. When local newspapers and television stations refused to cover the lawsuit, her grassroots organization paid for twenty-two large advertisements in the local shoppers' newspaper.[31] Tibbitts even purchased ten shares of stock in Cleveland-Cliffs to allow her to attend Cliffs' annual meeting in 1974 to voice her objections.[32]

Although the *New York Times* did not mention Tibbitts, it singled out Cleveland-

Cliffs in 1975 in an article titled "Profit vs. the Land in Upper Michigan." It reported that nineteen Upper Peninsula environmental groups, "expressing a 'new spirit of militancy,'" had formed a coalition to fight what they considered the "exploitation of natural resources" by mining companies. "The U.P. has a love-hate feeling toward Cleveland Cliffs," the paper editorialized. "The company is resented for its dominance, yet needed for its payroll. Usually it has gotten what it wanted, when it wanted."[33] The company went forward with the installation of the new coal-unloading equipment and units 7, 8, and 9 at the Upper Peninsula Generating Company. Its plans included spending $21.6 million to resolve air-quality issues associated with emissions from the generating units at the Presque Isle Station. Tibbitts may have lost her suit, but she succeeded in raising environmental consciousness.

Aggravations of the Tilden Pellet

If the protests of environmentalists in the 1970s were mere distractions for Cleveland-Cliffs, technical bottlenecks produced serious delays in Tilden production. Even after the company celebrated the first shipment of Tilden pellets in December 1974, company engineers encountered serious problems. The project, expected to cost $165 million, exceeded well over $200 million by 1975. Thinking back on the frustrations of the early 1970s, Tom Petersen, who worked as a plant metallurgical engineer at Tilden, recalled: "It was kind of like peeling an onion. You know, we had such difficulties in the beginning, and as we identified problems and solved them, all we did [was see] the next layer of problems. It was getting to be very frustrating, and it took several years for it to start really working right."[34]

Richard Tuthill, in charge of the Tilden pit operations, recalled that Einer Lindquist, general superintendent of the mine, ran what he humorously referred to as "miracle meetings" twice a week to resolve the various problems that stood in the way of reaching production goals. One question they could not answer was why the separation process worked flawlessly at one time and not another. When it was suggested that variations in the mineralogical composition of the crude ore was at the root of the problem, Tuthill recalled saying: "There's no question it's the ore. I have fed this plant magnetite, hematite, limonite, goethite, and probably a few 'ites' that I don't even know about and none of it works. . . . Obviously it's the ore."[35] What was not obvious was how to fix the problem. Finally, a laboratory investigation by gifted researcher Tsu-Ming Han found that poor separation was caused by small amounts of a type of clay that could be seen only microscopically. He developed a shake test to determine when this clay was present. Tsu-Ming Han, born in China in 1924, earned a master's degree in geology at the University of Cincinnati in 1949 and a degree in economic geology at the University of Minnesota in 1952. He also investigated what caused the weakness of the green balls produced at the Robe River plant in Australia and participated in the effort to reduce the phosphorous in the Tilden concentrate—a problem that ultimately proved intractable.[36]

Repair of one of the Allis-Chalmers autogenous grinding mills at the Tilden Pellet Plant, late 1970s. The mine's ore body, consisting predominately of martite, was very hard to grind and equipment wore out quickly. *(Courtesy of the John L. Horton Collection, Special Collections, Michael Schwartz Library, Cleveland State University.)*

Green ball grate feed belt, Tilden, late 1970s. During the pelletizing process, the green (unfired) balls were continuously heated before reaching the grate-kiln where they were fired. *(Courtesy of the John L. Horton Collection, Special Collections, Michael Schwartz Library, Cleveland State University.)*

Edwin B. Johnson (b. 1923) began his career in research at Cleveland-Cliffs in 1948. He was vice president of operations during the construction of the Tilden plant and served as president between 1983 and 1986. "Whether we liked it or not," Johnson said, "we were pushed into the low-grade ores to survive." *(Courtesy of Cleveland Natural Resources, Inc.)*

In the 1970s the company's interest in Tilden was intense, Ned Johnson recalled, because management was convinced that the future of the company depended on it. As senior vice president in charge of operations, Johnson met with mine partners every two or three months to present a detailed review of progress and bolster their confidence. "They had a piece of the action," Johnson said. "You give them the service, and you give them the quality, and you're going to survive."[37] Long-standing traditions of goodwill helped to ride out the difficulties.

Loss of the Republic Float Contract

Preoccupation with Tilden may have diverted executive attention from the problems of the company's marine division in the 1970s. The company had signed a twelve-year contract to transport all of Republic Steel's ore requirements in 1971. The float contract, the largest ever negotiated on the Great Lakes, doubled Cliffs' lake tonnage. The contract included the "float" up Cleveland's Cuyahoga River to Republic's steel plants in the Flats. Because the Cuyahoga was too narrow for full-sized ore boats, Cliffs needed two sizes of vessels to meet the terms of the contract. At this time the company had eight ore boats. The agreement with Republic required Cliffs to purchase two more river-sized boats capable of navigating the Cuyahoga and to contract with Republic for the use of three additional ore boats. Thus, the contract tied up ten of the company's thirteen vessels, leaving only the company's three larger vessels, the *Edward B. Greene*, the *Walter A. Sterling*, and the *Cliffs Victory*, to provide transportation for higher-paying customers. Almost immediately Cliffs began to lose money on the Republic contract. A strike by the United Steelworkers, which included vessel crews, resulted in what company president Harrison called "the most expensive labor settlement in the company's history." With

The Republic Steel Company's Plant Number 1, June 14, 1973. Cliffs' twelve-year vessel contract with Republic Steel included the "float" up the Cuyahoga River to its plants in the Cleveland Flats. Note the ore boat on the left next to the pellet inventory in the ore yard. *(Photo by Cleveland Press photographer Bill Nehez. Courtesy of Cleveland Press Collection, Special Collections, Michael Schwartz Library, Cleveland State University.)*

higher projected costs for both iron ore and vessel operations, Harrison announced the need to raise pellet prices in 1972.[38]

At this time, the marine department operated semiautonomously under senior vice president John Wilbur. Wilbur urged a comprehensive reexamination of the company's marine policy in 1973 and tried to raise awareness among upper management for the need to upgrade Cliffs' fleet. He thought the *Edward B. Greene*, the *Walter A. Sterling*, and the *Cliffs Victory*, built before locks at Sault Ste. Marie were enlarged to accommodate larger vessels, were becoming obsolete.

Rivals like Interlake Steamship (a subsidiary of Pickands Mather) were building 1,000-foot ore boats because their greater efficiency lowered transport costs. Wilbur thought Cliffs needed to take a cold, hard look at vessel earnings and potential return on investment from 1,000-footers. He suggested purchasing Whiskey Island at the mouth of the Cuyahoga. Dock facilities could be constructed on the island for the transfer of ore from large vessels to either barges or conveyor belts for the trip up the Cuyahoga to Republic's mills.[39]

An analysis of the earnings of the marine department by controller Tom Moore confirmed Wilbur's prediction that instead of a profit, Cliffs could expect to lose about $6 million on the Republic contract. With the numbers in hand, Harrison asked Republic officials to renegotiate. "In the Republic float deal I explained that inflation had gotten out of hand so quickly and so severely that the original intent of the agreement had been abrogated," he wrote after his meeting.[40] Apparently Republic Steel was in no hurry to renegotiate this contract. Three years later, Cliffs was still losing money on it.

Wilbur thought the contract was part of a larger problem. He believed the company should either invest in its marine business or shut it down. This was a delicate issue because of the mystique and prestige associated with lake transport. Cliffs' board of directors' meetings took place on the company's flagship in season. Cliffs' employees looked forward to the opportunity to ride up the lakes every five years. But Wilbur argued that Cliffs' partnerships were so successful that it was "difficult to make the case that we now need the float to sell our ore."[41] Rivals Pickands Mather, M. A. Hanna, and Oglebay Norton, in Wilbur's view, had done a better job managing their fleets.

Since Cliffs had assumed that the float contract with Republic Steel would be renegotiated, it came as a surprise when the steel company reopened the bidding in 1976. As historian Mark Thompson wrote, "Cliffs was at a disadvantage in the fierce bidding war that followed, for they didn't yet have a thousand-footer, and the only two self-unloaders in their fleet—the *Edward B. Greene* and *Walter A. Sterling*—were too long to negotiate the tight turns in the winding Cuyahoga."[42] In 1977 Republic awarded the Interlake Steamship Company the contract for ore transport down the lakes. A second contract for the transfer float to the mills on the Cuyahoga went to the American Steamship Company.[43] Cliffs' loss of the Republic contract had repercussions for the port of Cleveland because Republic Steel constructed a transshipment terminal in Lorain, Ohio, where Pickands Mather's 1,000-footers offloaded their pellets into smaller self-unloading vessels for the short trip up the Cuyahoga River to Republic's mills. The loss of the Republic contract, in Pete Hoyt's view, effectively ended Cliffs' ore-transport business.[44] Hoyt headed Pickands Mather. He was the third generation of Hoyt men in the iron-mining and iron-shipping industry. At this time Pickands Mather was flourishing under the corporate umbrella of Moore McCormack Resources, Inc., a large water transportation and natural resources conglomerate. Pickands Mather's Interlake Steamship Company launched its third 1,000-footer in the spring of 1981. To add insult to injury, Interlake named the new boat the *William J. DeLancey*, in honor of the CEO of Republic Steel.[45]

Sam Scovil at the Helm

When Sam Scovil took over as president and CEO in 1976, he reported the highest earnings and revenues in the company's history—the result of capacity production at the Empire and Tilden mines. Rising demand for iron ore by the steel companies and

Samuel K. Scovil (1923–2010), president and CEO, 1974–86. Scovil pushed for the acquisition of rival mining company Pickands Mather in 1986 during the first collapse of the steel industry. This acquisition, which included mines in Minnesota, Canada, and Australia, made the company the largest manager of iron mines in North America. *(Courtesy of Cliffs Natural Resources, Inc.)*

higher transportation rates had also contributed to a strong performance by the marine transportation division, despite questions over the Republic contract. Only the forest products division reported lower revenues—a result of a four-month strike. The outlook looked extremely favorable for the iron ore industry. Ore demand was expected to increase at the rate of about 2.5 percent a year.[46]

The next year, however, Scovil faced a 113-day strike—the first strike since 1959. Although nationwide steel strikes were illegal as a result of the Experimental Negotiating Agreement between the steelworkers' union and the iron and steel industry, it permitted local unions to strike over local grievances. The 16,000 to 17,000 iron mine workers in northern Minnesota and Michigan were dissatisfied with the three-year contract between the steel companies and the United Steelworkers of America reached earlier that year. The miners claimed that they were entitled to the same incentive pay or production bonuses their counterparts in steel production received. The August walkout included employees of Oliver Mining, Oglebay Norton, Pickands Mather, M. A. Hanna, Inland Steel, and the Reserve Mining Company. In anticipation of the strike, the steel companies had stockpiled a four-month supply of ore and pellets. With much of its inventory already delivered, management of the mining companies had the upper hand in negotiations with the striking workers. Nevertheless, the new contract included establishing an incentive wage system for mining.[47]

Scovil made a point of visiting all the company's mining operations after the strike. In a special report to employees he defended what he considered the company's generous wage scale. Average wages in the iron and steel industries, he said, were about 64 percent higher than wages in U.S. manufacturing, and employees' "hidden paycheck"—benefits—was generous by any standard. Cliffs' employees also benefited from its policy of promotion from within, Cliffs' tuition refund plan, which reimbursed employees for up to 75 percent of course fees, and workshops and seminars offered to enhance their skills.[48]

Scovil's priorities were to continue to improve the company's safety record and reduce energy costs, which had escalated during the energy crisis of the early 1970s. To give the company more energy options, the kilns at the Empire, Tilden, and Republic mines were modified to burn coal in addition to natural gas and oil. He pointed out that Cliffs was in the enviable position of owning vast reserves of iron ore—second only to those of U.S. Steel—and they were the key to its future. To tap these reserves Cliffs had invested $137.6 million in its mines and power plants. Its partners invested an additional $238.4 million, bringing the total for 1977 to $376 million.[49]

Due to the strike, the company had lower earnings in 1977. But what concerned Scovil more was the rise in imported steel to a record 19.3 million tons, an increase of 35 percent from 14.3 million tons in 1976. Nevertheless, in the first quarter of 1978, Cliffs' pellet plants were producing at full capacity and new capacity at Tilden and Empire was expected to be brought onstream over the next three years. The company anticipated its annual pellet production worldwide to rise to 41 million tons.[50]

The Empire and Republic mines had record production again in 1978. Problems getting the equipment restarted at Tilden were attributed to the idling of the mine during the 1977 strike. Nevertheless, the company completed Tilden's expansion in 1979, doubling annual capacity from 4 to 8 million tons. Concentrator lines were increased from six to twelve, and a new pelletizing line was set up.[51] This added capacity proved unnecessary when demand for American steel declined precipitously in the early 1980s.

By 1979 it was clear that it was no longer feasible to continue to operate Mather B, the company's last underground mine. The "mighty Mather" at its height had employed several thousand workers. During the ceremony that closed the mine Ned Johnson, senior vice president–operations, paid tribute to the hard work and courage of underground miners. Ernie Ronn, United Steelworkers of America subdistrict director, recalled his career in underground mining and praised the ability of Cliffs' current workforce to respond to the new demands of open-pit mining.[52]

Bethlehem Steel, a major partner in the mine and the Pioneer Pellet Plant, probably drove the decision to close. It had a significant stake at Hibbing Taconite (Hibtac), located between Hibbing and Chisholm on the Mesabi Range. This mine, managed by Pickands Mather, had a capacity of 8.1 million tons when it began pellet production in 1976. In 1977 Bethlehem convinced Republic Steel to become a partner in the Hibtac expansion. Bethlehem preferred Hibbing's pellets because they could be tailored to blast-furnace specifications, while the Mather B pellet needed to be used "pretty much as nature made it."[53] Gilbert Dawe, who spent thirteen years of a long career with Cleveland-Cliffs at the Mather mine as operating engineer, assistant superintendent, and superintendent, recalled the days of mutual respect between partner and operator. The vice president of

A Cliffs' employee monitors a row of balling drums, Tilden Pellet Plant, late 1970s. *(Courtesy of the John L. Horton Collection, Special Collections, Michael Schwartz Library, Cleveland State University.)*

operations for Bethlehem Steel would come up "and we'd sit down and talk right at the mine." Dawe would inform him of plans and programs and expected grade and costs and how much the mine needed for development. "Easy, you know," he said, commenting on the more recent strains in the relationship.[54]

In addition to the higher iron content of taconite pellets (64.4 percent iron compared to the Mather's 60 percent), labor, energy, and the material costs of processing low-grade ore in Minnesota were lower. These factors more than offset Michigan's lower taxes and transportation costs. After joining the Hibbing Taconite partnership, Republic Steel cancelled a joint venture with Cliffs to build a new pellet plant called Cascade East in Palmer, Michigan.[55] Partnership interests in Negaunee Pioneer were transferred to Empire and Tilden.

Another factor that may have influenced the decision was the difficulty of controlling particulate emissions from the Mather's pellet plant. These emissions met standards set by the Michigan Department of Natural Resources, but people living near the plant objected to the red dust produced from the earthy hematite ore of Mather B. Although the company assisted with washing and repainting nearby houses, it was reluctant to install additional emission-control systems to completely eliminate the problem, since the mine was scheduled to close the next year.[56]

Tilden pellets leaving a balling drum, late 1970s. *(Courtesy of the John L. Horton Collection, Special Collections, Michael Schwartz Library, Cleveland State University.)*

Meanwhile, the company was intent on completing its plans for international expansion and diversification. William E. Dohnal, senior vice president–international, managed the expansion of Cliffs' Robe River mine, located in western Australia. It was completed on schedule and under budget in 1979, though production of 13 million tons was far below its annual capacity of 19 million tons of fines and pellets.[57] Labor problems were rife at Robe River's operations, and the mine's Japanese customers failed to take their annual commitment of pellets. In 1980 the escalating cost of fuel forced the company to temporarily shut down its pellet plant, though shipment of iron fines to Japan increased. Reduced shipments in the final quarter of the year were attributed to an ominous decline of steel production in Europe and Japan.[58]

Meanwhile, as part of its diversification strategy, Cleveland-Cliffs in 1978 acquired Tiger Oil International, an oil and gas contract drilling firm, for $59 million. The acquisition seemed to fit well with the acquisitions the company had previously made in oil shale, uranium, and minerals. "This added emphasis on our energy minerals and services sector," Scovil wrote in the Cliffs annual report for 1978, "was a logical move in our program of expansion in fields that are contracyclical to the Company's current businesses."[59] The acquisition required the company to assume additional debt and the burden of long-term equipment leases for offshore and deep-drilling oil rigs. The board

Project engineering department, located in the Ishpeming central office. The group worked on the design and construction of large-scale engineering projects in the 1970s and 1980s, such as the expansions of Empire, Tilden, the Presque Isle Power Plant, and plants in Australia and Canada. *Left to right:* Andy Tarver, Jim Nankervis, Paul Bussone, Roger Day, and Pete Wahlman. *(Courtesy of the John L. Horton Collection, Special Collections, Michael Schwartz Library, Cleveland State University.)*

authorized an additional investment of $70 million in the business in 1980, renamed Cliffs Drilling, bringing total investment to $200 million.[60] After a succession of Cliffs executives, William Calfee took charge of the Houston operations in 1984. Calfee, a graduate of Williams College who was recruited for the Cliffs executive training program after a stint in the U.S. Navy, had previously worked in the ore sales department. He was responsible for both Cliffs Drilling and the small operations that Cliffs retained in oil shale and uranium development in Wyoming. Although the company had thought that diversification into oil drilling would cushion the bottom line from declines in the cyclical steel business, events proved otherwise.

Steel Industry Bankruptcies and Their Impact on Mining

Trouble began with the bankruptcies of McLouth Steel in late 1981 and Wisconsin Steel, a subsidiary of International Harvester, in early 1982. These two companies had taken 40 percent of the Empire mine's output. Though the assumption had never before been tested, Cliffs had thought that if one partner failed, the other partners would save

Aerial view of the Empire mine, 1979, shows the tailing thickener tanks (*four circular tanks on left*) used to clarify 300,000 gallons of water per minute for reuse in the plant. Three additional tanks for the Empire IV expansion were under construction. *(Courtesy of the John L. Horton Collection, Special Collections, Michael Schwartz Library, Cleveland State University.)*

the partnership by buying a larger percentage of the mine. This did not happen because the other steel company partners were also weakened by competition from imported steel. Responsibility fell on Cleveland-Cliffs to keep the mine operating by increasing its ownership stake in the mine and renegotiating its sales contracts. Cliffs' share of Empire's fixed costs leapt from 20 to 60 percent, a $35 million annual increase, and its royalty income fell sharply.[61]

As demand for pellets fell, Cliffs cut back production at the Empire and Tilden mines. Cliffs also temporarily shut down the Adams and Sherman mines in Canada and placed the Republic mine on standby. It permanently closed the Canisteo mine in Minnesota in June 1981, temporarily ending Cliffs' mining presence in Minnesota. Cliffs' competitors were also reeling from steel company bankruptcies. M. A. Hanna's Groveland mine on the Menominee Range was closed in 1980, leaving Cleveland-Cliffs as the last viable mining company in Michigan. Lower demand for iron ore forced the Cliffs marine division to cut back operations, with only four out of the remaining ten vessels operating in 1981.

Cliffs' mining partners in the steel industry, focused on survival, urged the company to cut operating costs. They resisted approving the purchase of new mining equipment and even, in some cases, asked Cliffs to defer basic maintenance. Partnership relations became less congenial as Cliffs tried to keep its mines from closing down permanently. During the summer of 1982 Cliffs had to temporarily shut down all of its mines in the Upper Peninsula. Despite the many signs of the coming crisis in the iron and steel industry, Cleveland-Cliffs had not anticipated the precipitous decline in its revenues in 1982. "We didn't think it was going to be that bad," recalled John Brinzo, who was controller for the company at that time. The unraveling of the integrated steel business was unthinkable. "Of course, it just completely collapsed and the company was in culture shock," he said.[62]

Protected against destabilizing price wars by its oligopolistic structure, the integrated steel industry had operated profitably for most of the twentieth century. However, without price competition the industry lacked incentive for eliminating inefficiencies until it was too late. From the 1950s to the 1970s the high cost of American steel increased the industry's vulnerability as customers—especially the American automobile industry—increasingly turned to lower-cost, higher-quality imported steel.

After World War II American steelmakers were slow to adopt innovations in steelmaking that would have lowered costs. For example, the basic oxygen furnace (BOF), developed in Austria in the late 1940s, converted the pig iron produced in a blast furnace into molten steel. A blast of pure oxygen speeded up melting, drastically shortening the time required to make steel, from at least six hours to about forty minutes. Other advantages included lower installation costs and lower scrap requirements.[63]

Postwar European and Japanese steelmakers had installed the BOF in the 1950s during the rebuilding of their steel industries, often with American foreign aid. The case of Japan was particularly ironic. As the price of scrap escalated in the late 1950s and American supplies dwindled, American policy makers encouraged the Japanese to adopt the new BOF technology because it required less scrap, never dreaming that the efficiency of the BOF and the low cost of labor would allow the Japanese to become major exporters of steel two decades later.[64] The American steel industry belatedly began to install basic oxygen furnaces at an accelerated pace in the late 1960s. By 1970 about 50 percent of the integrated steelmakers' output was being made in basic oxygen furnaces, but costs were still far higher than those for steel produced abroad.[65]

The integrated steel industry also faced domestic competition from minimills, so called because they began with a production capacity of less than 1 million tons. Minimills reprocessed steel by melting low-cost scrap in electric furnaces. At first, minimills produced semifinished billets, suitable for rolling into bars, rods, and small structural shapes—taking the low end of the steel market away from the integrated mills. Nonunion labor and a low-cost philosophy were among the most important factors in keeping costs down. By the 1960s minimills had captured about 5 percent of the American steel market, each with a narrow specialization and supplying a limited geographical area. In the 1970s, the electric-furnace industry increased market share to about 30 percent and moved into the production of higher-priced steel products, such as rods, bars, and medium-sized structural shapes. The growth of this industry threatened not only the integrated steelmakers but also the Lake Superior mining industry because the electric furnace used scrap metal rather than pellets to produce raw steel, bypassing the blast furnace entirely.[66]

The American steel industry was also slow to adopt continuous casting in its rolling mills. Continuous casting, a technology perfected in the 1960s, speeded up steel production, reduced costs, and improved quality. The new technology allowed steelmakers to bypass many of the discrete, labor-intensive steps in steelmaking by casting molten steel directly into slabs, blooms, or billets. However, while steelmakers in the European Community and Japan, as well as the majority of American minimills, adopted continuous casting in the 1960s, this technology did not become widespread in the integrated steel industry until the late 1980s.[67] Even after American integrated mills adopted the continuous caster, they often failed to achieve the efficiencies of the newer mills of their competitors.[68]

The western Europeans and the Japanese had the advantage of building entirely new plants near tidewater ports. The design of their new mills maximized the efficiency of materials handling and ocean transport. Labor costs were also far higher in the United States. To remedy the vicious cycle of strikes accompanied by price increases in steel, labor and management agreed to the Experimental Negotiating Agreement (ENA), designed to prevent strikes in exchange for automatic wage increases. The ENA remained in effect between 1974 and 1983, but it did little to halt the escalation of wages, now tied to the cost of living. During this period United Steelworkers' wages were 100 percent higher than those of the average American industrial worker.[69]

By the early 1970s the integrated steel companies' long-term prospects looked so dubious they began to invest their cash reserves in industries unrelated to steel. Economist William Scheuerman remarked, "In a word, the period between 1969 and 1974 provided a watershed for the industry, for it was during this time that many U.S. integrated steel firms decided that steel was a much too risky business."[70]

In addition to diversifying, the steel companies turned to the government for assistance. They lobbied for tariffs on steel imports, setting off a national debate over protectionism versus free trade. The real problem, however, was not imported steel but the American industry's inefficiency relative to steelmaking abroad. As author Christopher Hall pointed out, "A campaign against imports directed attention away from the companies' own competitive failings, to a scapegoat that (unlike minimills) had no constituency in Washington; built common ground with a union that might otherwise be expected to be hostile to the companies' layoffs; and required the federal government, rather than the companies, to actually do something." The lobbying efforts, spearheaded by the American

Iron and Steel Institute, were successful in obtaining passage of the Voluntary Restraint Agreement (VRA) in 1968, which imposed import quotas on a country-by-country basis, ostensibly to give the domestic steel industry the chance to modernize. The VRA remained in effect until 1974; once it was lifted, American steel prices immediately rose.[71]

In 1974 and 1975 imports actually declined as demand for steel fell during a world-wide recession. Then imports, especially from Japan, began to rise in the second quarter of 1976, and immediately the American Iron and Steel Institute (AISI) protested against "dumping." Dumping was illegal under American trade laws. Foreign firms that were subsidized by their governments were not permitted to sell their goods in the United States below the domestic market price. However, although dumping was often claimed, it was seldom proved.[72] Although Japan voluntarily limited exports to the United States at this time, large volumes of steel from Korea and several other countries continued to enter the United States. This steel sold at $100 per ton below the U.S. list price. Import volume was so great in 1977, economist Father William T. Hogan wrote, that "it shattered the price structure in the United States and was in part responsible for sharp decreases in profits and some bankruptcies."[73] Many steel companies had to report losses in 1977—their first ever.[74]

The American iron ore industry, tied to the steel industry through partnerships, was equally vulnerable. When the Empire mine's expansion from 5.2 million tons per year to 8 million tons was completed in 1980, Cleveland-Cliffs had expected to absorb a significant number of the 850 men laid off from the Mather operation.[75] Instead, a sharp drop in demand for pellets required the layoff of 2,000 employees from the Empire and Republic mines in June 1980. Nevertheless, Cliffs had record profits in 1981, driven by the strong contribution of its new subsidiary, Cliffs Drilling, and continued demand for iron ore by the steel industry.

In 1982 the company lost $10.1 million on iron ore operations—a figure that reflected lower sales and reduced royalty and management fees due to the bankruptcy of Wisconsin and McLouth, and additional write-offs for loans to McLouth. Revenues from the company's oil and gas division also plunged, from $20.6 million in 1981 to $3 million. Even the company's forest products division was not immune to the economy's downturn. The company lost $30.2 million in 1982, its first loss since the Depression.

The company's belt-tightening included layoffs of hourly employees and a 20 percent reduction of salaried staff in Michigan and Cleveland. The company froze salaries and suspended bonuses. With weak demand for iron pellets expected to continue as the ore industry contracted, Cliffs needed new customers to make up for the ones it was losing to bankruptcy. "In this regard," Scovil bravely stated in the company's annual report for 1982, "the geographical location of our domestic mines, our technical innovations, and our modern mining and processing facilities are important reasons for our continued optimism."[76]

In response to the industry crisis, Cliffs reorganized its partnerships, pulling apart the fabric of relationships stitched together by Jack Hollister, senior vice president–sales, Bill

Calfee, and other members of the commercial group over more than two decades. Calfee recalled scratching out the numbers as they negotiated with steel company representatives. "All of these iron ore joint ventures," he said, "had been built up in anticipation of continuing growth of the steel industry. And when the steel industry fell on its hard times and ended up shutting a lot of capacity down, that left the iron ore industry in a position where it was overbuilt to meet the needs of what now was a smaller industry."[77] Competition among the remaining iron ore companies intensified as it became clear that the industry as a whole had excess capacity. Since the steel companies often had partnership interests in several mines managed by different ore companies, rival ore companies endeavored to either force their competitors' mines to shut down or reduce pellet capacity.

Cliffs' first priority was to make sure that the pellets from Empire and Tilden were fully committed. Cliffs transferred the partnership interests in Negaunee Pioneer and the Republic mine to Empire and Tilden, both of which had better cost structures. Jones & Laughlin gave up most of its interest in the Tilden mine and assumed a larger stake in the Empire mine (35 percent), along with Wheeling-Pittsburgh (10 percent) and Inland Steel (40 percent). However, because of the precarious condition of the steel companies, which were already highly leveraged, their creditors insisted that Cliffs guarantee $400 million of its partners' debts before they would approve the ownership changes.[78]

While the outlook for iron ore began to improve, it was not enough to save Cleveland-Cliffs' marine department, which was shut down in 1984. In February 1986 Cliffs sold its underperforming Robe River mine in Australia to Peko-Wallsend, Ltd. for $54 million. While necessary at that time, it was a move the company would later regret. Then Wheeling-Pittsburgh Steel declared bankruptcy in April 1985. Especially serious was the bankruptcy in July 1986 of LTV, the country's second-largest steel producer. LTV's bankruptcy followed a merger in 1984 of Jones & Laughlin, a subsidiary of LTV, with Republic Steel to become the LTV Steel Company, with headquarters in Cleveland. LTV was a major partner in the Tilden and Empire mines.

By November 1986 Cliffs' stock price had plunged from a high of $46 per share in 1981 to $6 per share—a decline of 90 percent. Market capitalization of the company declined from $569 in 1981 to $74 million. With many of Cliffs' partners in bankruptcy, they reneged on their ore contracts. To add insult to injury, some partners began selling their excess pellet inventory, thus becoming Cliffs' competitors. Ore prices, guided for more than fifty years by the "lower lake list price," now rose and fell with demand—sometimes even falling below production costs.[79] As a Harvard case study of Cleveland-Cliffs observed, "1986 was Cleveland-Cliffs' Waterloo." The company declared a loss of $41.7 million.[80]

The precipitous drop of the United States' ore capacity from a peak production of 73 million tons in 1979 to just 38 million tons in 1986 had an impact on all four Cleveland mining companies: Cleveland-Cliffs, Pickands Mather, M. A. Hanna, and Oglebay Norton. It was clear that only the most efficient mines would survive. Hibbing Taconite and Erie (managed by Pickands Mather), Minntac (U.S. Steel), and Empire were considered to have the lowest operating costs. However, Cliffs' higher ownership percentage increased its vulnerability when demand for iron ore fell and partners reneged on their contractual obligations. In addition, it was difficult to sell pellets produced for Cliffs' own accounts on the spot market.[81]

Pickands Mather, the largest pellet producer (after U.S. Steel) in the Lake Superior region, was among the first casualties. Despite the precariousness of Cliffs' own position in 1987, Scovil never wavered in his determination to acquire this former rival. He knew that Pickands Mather's parent company, Moore McCormack, considered Pickands Mather a liability. Referring to his decision to proceed with the acquisition, Scovil said, "I guess I felt that if you had the right setup, why, there was no way the steel business in the United States was going to go completely belly-up." He thought Cleveland-Cliffs and Pickands Mather were better off together, despite the uncertainties of the present. Pickands Mather's assets, from Scovil's point of view, were not only its mining operations in Minnesota, Canada, and Australia, but also its staff of excellent operating people.[82] This acquisition, effective December 30, 1986, involved no cash—only the exchange of Cliffs' oil and gas reserves, valued at $12 million—the equivalent of about 2 million barrels of oil—for all of Pickands Mather's stock. Moore McCormack spun off Pickands Mather's Interlake Steamship subsidiary to James R. Barker and Paul R. Tregurtha. Barker had once headed Pickands Mather's marine department. He stepped down as CEO of Moore McCormack to get back into the marine business.[83]

The acquisition of Pickands Mather by Cleveland-Cliffs joined two venerable Cleveland mining companies founded by the Mather family. Cleveland-Cliffs' new management team worked closely with Bob McInnes, Pickands Mather's former CEO, to integrate the two companies. Henry P. Whaley, head of Pickands Mather's iron and coal operations since 1974 and an authority in the development of Minnesota taconite, became vice president–operations. Cleveland-Cliffs' headquarters' staff moved from the Union Commerce Building on Euclid Avenue into Pickands Mather's offices in the Diamond Shamrock Building at 1010 Superior Avenue.

The Pickands Mather acquisition meant that Cliffs now controlled 40 percent of North America's iron-mining capacity, including mines in Minnesota that were more cost-effective than those it managed in Michigan. Although Michigan enjoyed tax and transportation advantages over Minnesota, production costs were 6 to 7 percent higher because of the higher energy and capital costs associated with processing nonmagnetic ore.[84] The acquisition increased the number of iron mines that Cliffs was operating in North America from four to seven, including the historic Erie mine (renamed LTV Steel Mining Company in 1987), the Hibbing Taconite mine in Minnesota, and the Wabush mine in Newfoundland/Quebec, Canada. The company acquired 36.2 percent of the Savage River mine in Tasmania, Australia, and new Japanese and Australian partners. Cliffs also took over Pickands Mather's coal mines in West Virginia and Kentucky. The

Thomas Moore (1934–2002), president and CEO, 1986–97, took over shortly after the acquisition of Pickands Mather. He turned the company around and fended off two takeover attempts in the 1990s. *(Courtesy of Cliffs Natural Resources, Inc.)*

acquisition changed Cliffs' U.S. industry ranking to number one, with a managed capacity of 41 million tons. The former industry leader U.S. Steel (USX) followed with 30.9 million tons, M. A. Hanna with 23.5 million tons, Oglebay Norton with 3.5 million tons, and Inland and Algoma with 2 million tons.[85] The determination to join two of Cleveland's oldest mining companies at Cliffs' darkest hour demonstrated the vision, fortitude, and optimism of Sam Scovil and laid the foundation for the next consolidation of the mining industry after the turn of the twenty-first century.[86]

As soon as the Pickands Mather deal was struck at the end of 1986, Scovil stepped down, along with Ned Johnson, who had served as company president since 1983. The new president and CEO, M. Thomas Moore, and an executive team that included William Calfee, John Brinzo, and Frank Forsythe, faced the daunting task of turning around the company.

Unlike the two Yale-trained CEOs who preceded him and who were from socially prominent old Cleveland families, Moore had a working-class background. He was born in Brooklyn, New York, and brought up by his mother in Connellsville, the heart of western Pennsylvania coal country. After graduation from the local Indiana (Pennsylvania) State Teachers College in 1956, he was recruited for U.S. Steel's prestigious management-training program at the Duquesne Works. He moved on to the American-Standard Company in Baltimore, where he became plant controller. After completing graduate work in business administration, he went to work at Celanese Corporation in New York as a senior financial and planning analyst. He joined Cleveland-Cliffs as assistant controller in 1966. Moore brought a new level of professional excitement and financial discipline to the company. He was hardworking and ambitious, moving up the executive ranks quickly, to controller in 1968 and senior vice president and chief financial officer in 1983.[87]

Tilden pellets loaded from the
pocket en route to the ore dock in
Marquette, late 1970s. *(Courtesy of
the John L. Horton Collection, Special
Collections, Michael Schwartz Library,
Cleveland State University.)*

Moore's first concern upon taking the helm was saving the Tilden mine. The impact
of the bankruptcy of LTV was especially severe at Tilden, because Cliffs was forced to
carry LTV's 12 percent share. This made Cliffs responsible for 51 percent of Tilden's fixed
costs. Of the remaining partners, only Stelco was considered secure. Algoma was given a
fair chance of survival, while Wheeling-Pittsburgh and Sharon Steel had already filed for
Chapter 11. Indeed, a report issued by Paine Webber and the State of Michigan stated
that Tilden was in a "race for its life."[88]

To generate the cash needed to restructure the Tilden partnership, Cliffs sold its interest in the Presque Isle Power Plant to the Wisconsin Electric Power Company (WEPCO)
for $247.5 million and negotiated a long-term contract with WEPCO to provide power
to its Tilden and Empire mines. This reduced Cliffs' long-term debt. It also spun off
Cliffs Forest Products Company to Timber Products Company for $10 million. The sale
included a sawmill and veneer mill located in Munising, Michigan.[89] In September 1987
the company purchased the Tilden mine's long-term debt at a steep discount, using the
proceeds of a stock offering.

Restructuring the Tilden mine's partnership was accompanied by a $30 million
upgrade to the pelletizing plant to make it capable of producing both magnetite and
hematite pellets. Although Tilden pellets contained only .05 percent phosphorus, these
trace amounts of phosphorus could make steel brittle. Algoma's large ownership stake
in Tilden required the Canadian company to take 3.2 million tons of Tilden's pellets.

Aerial view of the Tilden mine, late 1990s. The pellet plant and pellet storage piles can be seen in the foreground, with the mine's large hematite pit immediately behind the plant. The Tilden plant was modified to process both hematite and magnetite in 1987. The magnetite pit can be seen as a faint V-shaped area on the right beyond the hematite pit. The city of Ishpeming is in the background. *(Courtesy of Cliffs Natural Resources, Inc.)*

Algoma needed magnetite pellets (which contained much lower phosphorus) to mix with the Tilden hematite pellets to reduce the phosphorus load in the blast-furnace charge. Fortunately, the Tilden ore body contained, in addition to its vast reserves of hematite, a thirteen-year reserve of magnetite. The company's geologists suggested mining this magnetite vein and shifting the plant over to run magnetite lines for three months in the winter. This enabled Algoma to reduce its complement of hematite pellets while preserving the terms of the Tilden partnership agreement.[90]

Fluxed pellets gave the company the ability to custom-tailor its pellets to meet the needs of its partners and customers. The addition of a fluxstone mixture of dolomite and limestone to the pellet formula improved the melting characteristics of the blast-furnace charge, lowering fuel costs and improving the efficiency of the process. The Cliffs research department had experimented with fluxed pellets in the 1960s. Cliffs had begun producing them commercially in the mid-1980s for Dofasco at the Sherman and Adams mines in Canada. They came onstream at Tilden and Empire in 1987 and at the Wabush mine in 1989.[91]

Fluxing the Tilden pellet changed the chemistry and improved its softening and

Pellet quality specialists meet with a blast-furnace operator at Bethlehem Steel, 1992. *Left to right:* Bethlehem Steel blast-furnace superintendent Ron Chango; Hibbing Taconite director–total quality process Jack Croswell; Cliffs' manager–technical services Walter Nummela in the control room at the Burns Harbor, Indiana, plant. *(Courtesy of Cliffs Natural Resources, Inc.)*

melting characteristics, making the Tilden pellet attractive to more partners and customers. Steelmakers in the 1980s often added BOF slag, a phosphorus-rich by-product of the basic oxygen furnace process, as a flux to the blast-furnace charge. By using Tilden's fluxed pellets in their blast furnaces, they were able to reduce their BOF slag requirements. More efficient blast-furnace management lowered their costs. Don Gallagher, manager of ore sales at that time, teamed up with Walt Nummela, general manager–technical, to market Tilden's fluxed pellets to several partners and customers, including Acme and McLouth. These steelmakers found the new fluxed pellet so attractive they were willing to trade their complement of magnetite pellets from different mines for Tilden hematite pellets.

While Cliffs' researchers designed Tilden's fluxed pellet to be acceptable and marketable to multiple steel companies, they could also customize the fluxed pellet to satisfy the needs of a particular customer. They worked with Inland Steel's engineers to develop a fluxed pellet to the specifications needed for optimum operation of Inland Steel's Number 7 blast furnace at Indiana Harbor.[92] Number 7, built in 1980, was among the largest and most efficient blast furnaces in the United States.

Cliffs' versatility in responding to the demands of customers and partners for cus-

A representative of Cliffs' sales department inspects the pellet stockpile of the LTV steel mill in Canton, Ohio, 1997. *(Courtesy of Cliffs Natural Resources, Inc.)*

tomized pellets was made possible by the increase in the number of mines managed by Cleveland-Cliffs—a direct consequence of the acquisition of Pickands Mather. Each of Cliffs' mines produced pellets with slightly different chemical characteristics. Working hand in glove with the research department, Gallagher and other members of the sales department could mix and match types of pellets more exactly to blast-furnace specifications. Their growing expertise in this area helped customers achieve optimal performance of their blast furnaces and cut costs. The ability to "mix and match" not only contributed to greater efficiency of the blast furnace, but also reduced pellet transportation, handling, and storage costs. The sales department's philosophy was to know each steel company's needs from the bottom up, personally calling on the blast-furnace superintendent. They examined the grades of pellets and space for stockpiling in the ore yard. Since the sales force also handled dispatch, its members could swap pellets from several different mines for an equal amount from one or two mines. In addition to pellets matched more exactly to desired chemical characteristics, the swap reduced transportation costs and simplified the management of a steel company's ore yard.[93]

The company's improved prospects went hand in hand with the reduction of North American pellet capacity from 90 million tons in 1982 to 55 million tons in 1987, the development of higher-quality pellets, and lower production costs. For these reasons, a study by the Great Lakes Commission found it unlikely that steel plants located in the Midwest would switch to imported ores.[94]

As the American steel industry began to recover in the late 1980s and Cliffs' ore business picked up, the company's stock price rose from $6 in the fourth quarter of 1986 to $10.50 in the first quarter of 1987. Although the company's low valuation at that time might have prompted management to buy back the company's shares, it was reeling from the default of the steel companies on the $400 million debt that Cliffs had guaranteed.[95]

Fending Off the First Takeover Attempt

Buoyed by the improved outlook, the company began once again to formulate a strategic plan that included diversifying into other mineral resources. Then in June 1987 came an unexpected takeover attempt by David Bolger, a real estate investor from Ridgewood, New Jersey. Bolger, who owned only 5.1 percent of Cleveland-Cliffs' stock, claimed that the company was mismanaged and called for the replacement of the board of directors with a new management team to recapitalize the company and create value for the shareholders. He won a petition to hold a special shareholders' meeting at the end of 1987.

Tom Moore was determined to defeat Bolger's bid for shareholder votes. He realized that to win the proxy battle, he needed to give up the company's plans to diversify and present a convincing proposal for refocusing the company on its core business—the mining of iron ore. Returning to its core competency was the time-honored response of the company to adversity. It had worked in the nineteenth century several times and during the Great Depression, but would it save the company in the late 1980s? The company had succeeded in blocking Cyrus Eaton from acquiring control, but few of the "old Cleveland" shareholders with loyalty to management were left. They had been replaced by more speculative shareholders with short-term horizons. It was possible that the company could lose the proxy battle. In addition to abandoning its plan to diversify, Moore announced that the company would sell nonstrategic assets and either recapitalize or put the company up for sale. "This bold plan, which required leveraging the company again, was a calculated risk to retain control of the company's future," he recalled afterward in a speech to the Newcomen Society.[96]

Moore and his management team took their case directly to the shareholders. Management's proposal won the votes of 58 percent of the shareholders, thanks to the eleventh-hour support of a major mutual-fund manager whose company owned a significant block of Cliffs' stock. Moore credited employees and key outside financial, legal, and other advisors for saving Cleveland-Cliffs. In reporting the positive results for management of the proxy battle, the *Wall Street Journal* said the company was betting that "iron ore will turn to gold."[97] Looking back on the difficulties of the previous year in his letter to shareholders, Moore wrote: "The year 1987 was one of the most unusual, controversial, and pivotal years in your company's 137-year history. At the start, there was widespread skepticism about Cliffs' survival prospects in the face of stagnant markets, underutilized production capacity, and a huge debt burden."[98] In 1988 Cliffs purchased 31 percent of its common stock and all of its preferred stock. Shareholders received a distribution of stock in Cliffs Drilling Company and MLX Corporation (originally acquired as a result of the McLouth bankruptcy) and the proceeds of the sale of its timber mills, a total of $230 million in cash and securities.[99]

The steel crisis had caused ten mines in the Lake Superior region to shut down and others to downsize. Nevertheless, Cleveland-Cliffs had managed to win a larger share of a smaller market, and its mines returned to full operation. The industry as a whole, in Moore's view, still had about 10 million tons of excess iron ore capacity, but at least the steel industry appeared to be recovering, and Cliffs' pellet inventories were the lowest they had been in six years.[100]

Moore attributed the rebound of the steel industry to the reimposition of voluntary import quotas by the Reagan administration.[101] Others disagreed. Caterpillar, Inc. refused to support the American Iron and Steel Institute's calls for protection. The steelworkers' union pointed out that 30,000 jobs had been lost during this period. Imports still had about 25 percent market share in 1986. Commenting on the futility of trade protection, historian Christopher Hall wrote, "The integrated companies ran an outstandingly successful political and legal campaign, but they fought against a symptom, not against the problem."[102] The immediate problem was still overcapacity. However, prospects brightened as the value of the dollar declined, President George Bush promised to extend the voluntary import quotas, and demand for steel picked up.

By 1989 Cleveland-Cliffs was inching toward settlement with Wheeling-Pittsburgh and LTV. Stan West negotiated the settlement with Wheeling-Pittsburgh, while John Brinzo served as chair of the unsecured creditors' committee in the LTV proceedings. The settlements netted more than $300 million for Cliffs' shareholders. As the two steel companies emerged from Chapter 11, they agreed to long-term supply contracts at competitive prices, in addition to keeping their equity interests in Cliffs' mines.[103]

Recognition of the value of these sales contracts to the future of the company may have prompted John Brinzo to propose purchasing the outstanding partnership interests in the mines Cliffs managed in North America. He thought that the weakness of the steel companies and the consolidation that was occurring in the iron-mining industry had created an opportunity for Cliffs. In a confidential memo to Moore, he proposed an $800 million joint venture funded by Cliffs and a group of venture capitalists that featured purchase of the steel companies' mining interests. The cost of the "The Great American Iron Ore Partnership" in theory could be partially offset by the equity Cliffs already had in the mines of its bankrupt partners and the low valuation of the mines at that time.[104]

In Brinzo's view, the advantage of owning the mines was the possibility of negotiating long-term sales contracts with the reorganized steel companies. The new partnership would be entitled to receive full depletion tax benefits. (The steel companies previously had received these benefits.) In addition, Cliffs would continue to receive management and royalty fees from its new investment partners, as well as income from its international ventures. The new entity could provide the company with greater flexibility in negotiating contracts directly with the United Steelworkers.[105] The idea (later called the "grand slam") failed to capture the imagination of the board and it was quietly shelved in 1990, only to come to life again after the second collapse of the North American steel industry.

Buoyed by the partial recovery of the steel industry and the company's victory over Bolger, the board of directors turned its attention to growing the company through strategic partnerships—the company's time-honored modus operandi. M. A. Hanna was in the process of divesting its mining assets. It shut down the Butler mine in Minnesota and sold its 15 percent share in the National Pellet mine to National Steel, which owned the other 85 percent. Hanna also owned a controlling interest in the Iron Ore Company of Canada (IOC), the largest iron ore producer in Canada. The Cliffs board considered a proposal to form a partnership with Mitsubishi International Corporation to acquire majority control. After Cliffs withdrew from the deal, Mitsubishi went forward with the acquisition of Hanna's stake in the company.[106]

Second Takeover Attempt

By 1991 the company had largely erased its debt and had accumulated $100 million in cash reserves. At this point the company faced another takeover battle—this time spearheaded by Julian Robertson, a skilled hedge-fund manager from North Carolina who had spent most of his finance career working for Kidder, Peabody. Robertson headed Tiger Management, a $1.65 billion hedge fund that had accumulated 10.46 percent of Cliffs' stock, making Robertson the company's largest single stockholder. Unlike Bolger, who was relatively unknown and lacked credibility in the investor community, Robertson was a well-respected Wall Street insider. At a meeting on March 14, 1991, Robertson and two Tiger managing directors urged Cliffs' management to use the $100 million cash reserve to buy back its stock to increase shareholder value. Moore flatly refused. He recalled that "the company felt it was important to preserve maximum financial flexibility for strategic growth moves as well as protection against potential renewed steel industry problems."[107] Moore wanted to use the company's cash reserves to diversify into the reduced iron business. Minimills now controlled 30 percent of the domestic steel market, and their growth was limited only by their dependence on scrap. He thought Cliffs needed to develop a hot briquetted iron technology to supply this growing segment of the steel industry.

To consolidate his position on enhancing the value of the stock, Robertson insisted that Tiger Management appoint five board members. With all the grief of a second proxy contest to consider, Moore agreed to this demand. Upon reflection and consultation with the board members, some of whom would be required to step down, he reneged on this agreement—an action that infuriated Robertson. In explaining his action to a reporter, Moore said, "They [Tiger Management] had shown no interest in the continuing development of the company's business. When your financial capability is improving, that's not the time to dissipate it. Our goal is not to gradually reduce the share size and liquidate the company, but to build our business. That takes some patience."[108]

Moore's refusal to follow Robertson's recommendations precipitated a second proxy contest focused on the makeup of the board. Robertson considered the board too insular—made up of Clevelanders who served on each other's boards. Indeed, two representatives of old Cleveland families were still on the board: James D. Ireland III, who had served since 1986, and Jeptha H. Wade of Boston, present since 1957. An Ohio securities law permitting cumulative voting enabled Robertson to win the proxy fight. Five men came off the twelve-member board and Robertson installed his five nominees. Remarkably, Moore continued to command the support of the entire board. Even Robertson's appointees—determined to act in Cliffs' best interests—did not press for increasing short-term shareholder value.[109] Although the company fended off both takeover attempts and returned to its core business of iron mining, abandoning the company's plans to diversify had not been by choice. Moore was convinced that the American integrated steel business was not sustainable over the long term. Although returning to its core business in times of economic stress had been the company's traditional response to adversity, times had changed. Competition from imported steel and domestic minimills was only going to increase. However, initially his hands were tied by the new makeup of the board determined by Robertson.

Recovery and Acquistion of Northshore Mining

During the positive economic environment of the Clinton years, the company flourished, despite the underlying weakness of the American steel industry. Moore noted in 1992 that "Cliffs has a 'franchise position' in the ore sales market and many opportunities for new business."[110] Building on the lessons of the first wave of bankruptcies in the steel industry, members of the sales department learned to craft contracts that protected the company from becoming an unsecured creditor in the event that a partner or customer failed. During the Clinton years they had little trouble selling the ore produced for Cliffs' own accounts.

In April 1994 Cliffs again reorganized the Tilden partnership, with Algoma Steel owning 45 percent, Stelco 15 percent, and Cliffs the remaining 40 percent.[111] Algoma Steel closed its mine in Wawa, Ontario, and committed to an increase in tonnage from the Tilden mine, making it possible to bring an additional grinding line onstream.[112] At this time Cleveland-Cliffs operated six mines: Empire and Tilden on the Marquette Range; Hibbing Taconite and LTV Steel Mining (the former Erie mine) on the Mesabi; the Wabush mines in Labrador, Canada; and the Savage River mines in Tasmania, Australia.

An important boost to morale and profits was the acquisition of the Reserve Mining Company, located in Babbitt and Silver Bay, Minnesota. As pointed out in chapter 5, Cliffs had sold its 10 percent ownership stake in Reserve Mining in 1950 after a decade of participation in its development. The mine's bankrupt partners, Armco and LTV, had shut down the mine in 1986 during the first wave of steel bankruptcies. The mine was attractive to Cleveland-Cliffs because it had large iron ore reserves and a mining infrastructure that consisted of a forty-seven-mile railroad to Silver Bay, where large crushing, concentrating, and pelletizing facilities were located. Reserve's port on Lake Superior could accommodate some of the largest ore vessels on the Great Lakes.

The company had tried to acquire the mine during bankruptcy proceedings in New York in 1989, but the bid had gone to Cyprus Amex Minerals of Colorado, a large copper-mining company. Senior vice president–sales A. Stanley West served on the team from Cliffs that represented the company at the bankruptcy court auction. He recalled that the layout of the court resembled a church, with the judge seated on a podium behind the Communion rail. "When he started the proceeding, the Cyprus people were all inside the Communion rail, and we were on the outside, so we knew things weren't looking real good that day," he said.[113] Although Cliffs made the highest bid for Reserve, the judge apparently was swayed by the objections of the State of Minnesota and the mine's owners, who suspected that Cleveland-Cliffs would restrict production in favor of Michigan pellets.[114]

Reserve, renamed Northshore by Cyprus Minerals, restarted pellet production. Although the mine had a capacity of 10 million tons, Cyprus brought onstream only the mine's newer equipment and limited production to 2 to 3 million tons per year. The efficiency of Northshore's operations and its nonunion workforce made the mine extremely cost-effective. Cliffs immediately began to experience the effects of this new competition. Bill Calfee pointed out in a report to the board of directors, for example, that "Northshore's iron ore pricing and other excess ore supplies were imposing downward pressure on iron ore prices."[115] The mine's former owner, Armco Steel, continued to be the mine's major customer. However, Cyprus needed more business to make its operations profitable and had its eye on Weirton Steel. The Weirton plant had been spun off by National Steel in the 1980s. Cliffs also coveted the Weirton account and was perhaps more savvy in the assessing Weirton's needs. Cliffs offered to invest $25 million in Weirton's preferred stock in return for a multiyear contract for pellets. Weirton then asked for the same deal from Cyprus. When Cyprus refused, it lost the Weirton account. Unable to make a profit on Northshore pellets, Cyprus put Northshore up for sale.[116]

In 1994 Cleveland-Cliffs purchased Northshore for $66 million, plus $28 million for working capital. This key acquisition increased Cliffs' pellet-sales capacity by 67 percent. Northshore was the first pellet operation to be owned outright by Cleveland-Cliffs. The acquisition was a harbinger of the "grand slam" strategy that would later eliminate partnerships with the steel companies in favor of mine ownership and multiyear contracts.[117]

The company had a strong finish in the fourth quarter of 1994. Cliffs' board minutes noted, "Pellet inventories at near all time lows; high scrap prices are causing higher pellet consumption; the integrated steel plants are running flat out, and steel and slab imports are at record levels."[118]

Cleveland-Cliffs purchased the Northshore Mining Company (at Babbitt and Sliver Bay, Minnesota) from Cyprus Minerals in 1994. This extremely efficient operation turned out to be a harbinger of the strategy to acquire the mines of bankrupt steel partners after 2001. *(Courtesy of Cliffs Natural Resources. Inc.)*

Although Cliffs had record earnings in 1995, a cloud appeared on the horizon that the positive business climate of the Clinton years could not dispel. McLouth Steel petitioned for protection under Chapter 11 of the U.S. Bankruptcy Code in October. McLouth owned 25 percent of Empire. Cliffs supplied all of McLouth's ore requirements, about 1.5 million tons of pellets annually. This amounted to 13 percent of Cliffs' revenues. Stan West, in an article in *Cliffs News,* warned employees that while McLouth was still operating, it was unclear whether it would survive a second bankruptcy.[119] With a great deal of excess capacity in the steel industry, other marginally capitalized partners, like Sharon, Algoma, and Stelco, were also at risk.

A New Technology Venture

As the financial condition of the integrated steel companies deteriorated, Moore began to implement plans for the development of a new iron product the company could market to the electric-furnace industry. Minimills used low-cost energy to melt scrap, bypassing the blast furnace in the production of steel. However, the growing scarcity of high-quality scrap was beginning to limit their growth. To produce high-quality flat-rolled steel, the electric-furnace industry needed a nearly 95 percent pure iron product as a raw material.

In 1996 Cleveland-Cliffs announced a joint venture with LTV and Lurgi Metallurgie GmbH of Germany to produce hot briquetted iron (HBI) for use in electric furnaces. Cliffs persuaded its major customer, LTV, to become a partner in the HBI venture. LTV committed to a 46.5 percent stake, Lurgi, 7 percent, with Cliffs holding the remaining 46.5 percent. Cliffs planned to act as the sales agent for the U.S. market for briquettes produced by a patented process called Circored. The project was expected to cost $160 million, with Cliffs investing $74 million.

After considering different alternatives, the partners chose the Republic of Trinidad and Tobago in the southern Caribbean as the site for manufacturing hot briquetted iron using the new process. This location seemed ideal, since Trinidad and Tobago had abundant natural gas reserves and government support for the project. The proposed subsidiary would import high-grade ore fines from Brazil, convert them into a high-quality metalized product, and sell it to Nucor, Steel Dynamics, and other large electric-furnace steel producers. Ground for the new plant was broken in 1997 with the expectation that the Trinidad plant would become the centerpiece of a growing business segment. At this time the company also considered the acquisition of companies that supplied ferrous scrap to the electric-furnace industry.

In the midst of these new ventures and uncertainty over the future of the integrated steel industry, Tom Moore announced that he had a serious illness that required him to step down in November 1997. Cliffs' board of directors named John Brinzo president and CEO. Earnings that year were disappointing as the integrated steel companies were squeezed not only by flat-rolled steel produced by the electric-furnace industry but also by cheap imported steel from Japan, Russia, Korea, and Brazil. The next year imports reached a high of 30 percent of domestic steel consumption—a record 41.5 million tons, or 33 percent higher than in 1997.[120] Imported steel was so inexpensive relative to steel produced in the United States that some integrated steel producers installed electric furnaces to produce some grades of steel. Others chose to shut down their blast furnaces and purchase imported slabs to produce flat-rolled steel. "Cliffs faces a highly competitive environment, marked by technological change, growth in electric-furnace steelmaking, and emerging iron making processes," Brinzo wrote in the 1998 annual report. "In this environment, success will go to the innovative and visionary companies."[121] This proved an understatement. The next year, as inexpensive steel from abroad continued to flood the market, the company's steel partners reduced their pellet commitments. As inventory grew, Cliffs stopped producing pellets for its own accounts. Brinzo ruefully admitted that the company had underestimated how bad 1999 would get. Net income fell from $57.4 million in 1998 to just $4.8 million in 1999. After numerous construction delays, the hot briquetted iron plant in Trinidad had started up, but mechanical problems continued to interfere with the production of commercial-grade briquettes. Although the company teetered on the edge of bankruptcy for the next three years, a window of opportunity was opening. Between 2001 and 2005, John Brinzo and his team would pull the company back from the brink, reinventing Cleveland-Cliffs as a merchant company as he executed the grand slam strategy he had proposed a decade before.

7

REINVENTING CLEVELAND-CLIFFS, 2000–2006

In October 2001 the *Wall Street Journal* published an article by Clare Ansberry with the headline "Seizing the Moment: Steelmakers' Troubles Create an Opening for an Iron-Ore Giant—In Sinking Economy, Some Exploit Opportunities Where Others Retrench." Ansberry described how American steel companies had begun to go bankrupt even before the sharp decline of the market after the terrorist attack on the World Trade Center on September 11. The mining interests of these bankrupt steel companies were now attractive bargains. "Standing at this decisive crossroads, Cliffs' Chairman John Brinzo has decided to act," Ansberry wrote, "trying to snap up some of those properties while he can—and while the price is right. 'Time is of the essence,' he says. 'We're trying to walk the line between husbanding resources and taking advantage of an opportunity.'" Ownership of its mines would liberate Cleveland-Cliffs from dependency on partnerships, allowing the company to regain its independence as an ore merchant. "Every good idea has its time," Brinzo told Ansberry. "Now is the time."[1]

Brinzo had proposed this idea in 1990 as the steel companies emerged from their first bankruptcies. Then the acquisitions would have required significant capital. What made Brinzo's strategy to acquire the mining interests of bankrupt partners feasible in 2001 was the even greater distress of the steel industry and the recognition that some companies would have to liquidate.

The bankruptcies of steel partners in the 1980s had forced the company to renegotiate ownership percentages in an effort to keep Cliffs' mines operating at capacity. Then owning large shares in its mines was considered a liability. However, as the amount of ore produced for its own accounts increased, the company's commercial group gained

John Brinzo (b. 1942) took the helm as president in 1997, just before the second wave of steel bankruptcies took place. He saw a "window of opportunity" to buy the ownership interests of the company's steel partners as the steel industry consolidated and positioned Cliffs as a global supplier of iron pellets. After returning the company to strong profitability he retired as chairman of the board in 2006. *(Courtesy of Cliffs Natural Resources, Inc.)*

experience selling this ore through multiyear contracts, becoming skilled negotiators in the process. When the steel industry recovered in the 1990s, the company made a profit on these pellets. Unlike rivals Oglebay Norton and M. A. Hanna, Cleveland-Cliffs developed a "merchant flavor," with members of its commercial group gaining reputations as honest deal makers who drove a hard bargain.[2] Company executives also became thoroughly versed in issues related to the bankruptcy of Cliffs' partners through service on unsecured creditors' committees, often taking leadership roles to protect Cliffs' interests. As in the case of Weirton Steel, Cliffs was sometimes willing to take an equity position or loan capital to ailing steel customers in exchange for multiyear sales contracts.

Building on this expertise in sales and issues related to bankruptcies, Brinzo spearheaded the transformation of Cleveland-Cliffs. He had been a careful steward of the company's finances for three decades. Bill Calfee commented that Brinzo had a keen sense of finance but also "thought like a commercial person."[3] A graduate of Kent State University, Brinzo had started his career at National City Bank. After he completed his MBA at Case Western Reserve University in 1968, he decided that commercial banking did not provide him with enough opportunity to use the ideas and strategies he had learned in business school. He joined Cliffs' five-person financial analysis department in the spring of 1969 where he delved into budgets, analyzed capital expenditures, and did feasibility studies for about two years.

Tom Moore, then company controller, asked him to transfer to Ishpeming to work in the company's central office in Michigan. Previously most financial activities had been carried out in Cleveland, but the company's increasingly complex pellet operations favored decentralization and demanded on-site financial expertise. This organizational restructuring was pushed by Tom Moore. Brinzo, then twenty-nine, moved with his wife and two small sons to Ishpeming. Within a year Brinzo was named controller of the

Empire mine, the company's flagship operation at that time.[4] Brinzo liked the atmosphere of the Upper Peninsula, where generations of families had worked for Cleveland-Cliffs. The area's isolation and its bracing winters and short summers brought labor and management together in a variety of indoor and outdoor activities. Up there the company was the community and the community, the company. Although he returned to Cleveland headquarters after several years, "in terms of enjoying the jobs that I had over the years," he said, "being plant controller of that mine was the best job . . . because there was so much camaraderie, so much teamwork in terms of mine operation."[5] Brinzo had been through the first set of steel company bankruptcies in the 1980s. Like Sam Scovil, who pushed through the acquisition of Pickands Mather at a time of crisis, he was an optimist.

When Brinzo took over as CEO of Cleveland-Cliffs in late 1997, he did not anticipate the second wave of bankruptcies that was about to engulf the integrated steel companies. In his first letter to shareholders, Brinzo pledged to build shareholder value by increasing the company's ore sales capacity. He also announced that the company was close to starting up its reduced iron facility in Trinidad and Tobago. As discussed in the previous chapter, the company expected the new plant to produce hot briquetted iron for the minimills, which by the late 1990s had captured 50 percent of the steel market in the United States. Although minimills represented an important market, the company's core customers remained the integrated steel companies. Brinzo announced at the annual meeting of shareholders in 1998 that 1997 earnings per share had declined 21 percent from the company's record performance the previous year. The closure of the Savage River mine in Australia and the impact of lower sales volume and margin on the company's North American pellet business accounted for this loss of revenue. Although steel imports were a growing concern, Brinzo expected the company to return to nearly full capacity and record production of over 40 million tons of pellets in 1998.[6]

Instead the company continued to struggle with declining orders for pellets. A five-year labor contract, ratified in August 1999, covered about 2,800 workers at the Empire, Tilden and Hibbing mines. The contract stipulated that the union and management make joint decisions on job cuts as the company braced for temporary shutdowns at all of its mines in the wake of a drastic curtailing of pellet orders.[7] In his review of the past year at the May 2000 shareholders' meeting, Brinzo noted that the company had underestimated the full impact of depressed steel demand in 1999. Cliffs' total earnings for the year 1999 had declined to $4.8 million. However, Brinzo expected the mines to return to full capacity in 2000.[8]

The Steel Crisis and a Window of Opportunity

As the company worked to upgrade its pellet operations, the repercussions of the collapse of the Asian economy, followed by worldwide economic stagnation, began to be felt in the Upper Peninsula. In December 1998 Brinzo noted in *Cliffs News* that the company, along with the rest of the iron and steel industry, was experiencing the "dark side of global competition." Using data from the American Iron and Steel Institute, he said that steel imports in the United States had increased 57 percent from the same period in 1997.[9] No longer able to sell their excess steel capacity to European and Asian customers, steel producers in Russia, Korea, and Brazil were selling their steel to customers in the United States at bargain prices. Indeed, prices were often below the cost of producing it. Unable to compete against the flood of imports, U.S. steel companies were hard-pressed to remain profitable. In 1998, in the wake of the East Asia economic crisis, a dozen steel companies and two labor unions brought antidumping trade cases against Japan, Brazil, and Russia. This triggered an investigation by the International Trade Commission.[10]

Cleveland-Cliffs joined a campaign by the steel producers and the United Steelworkers of America to fight illegal dumping. A public letter signed by John Brinzo and the presidents of seventeen steel companies, as well as the president of the United Steelworkers of America, urged President Clinton to "Stand up for Steel." The letter, published in the *Washington Post* and the *New York Times* September 10, 1998, stated that that the industry had restructured and reduced steel production almost 30 percent while investing over $50 billion in capital improvements. At the same time labor had made extraordinary concessions. "Out of that crucible," the letter stated, "we created the modern American steel industry." These efforts to increase productivity and reduce costs now appeared futile, as the North American steel industry struggled to survive.

In addition to asking the Clinton administration to bring antidumping cases against offenders, the iron and steel industry asked the administration to impose tariffs on certain classes of steel.[11] The question of tariffs was controversial. Steel consumers like General Motors and Caterpillar, Inc., for example, purchased low-cost, imported steel because it lowered their manufacturing costs. The *Wall Street Journal* reflected the general view that protection was bad policy and advised the steel industry "to tackle the global market more aggressively and eliminate the moral hazard that has spawned this industry of beggars on the steps of the Capitol."[12] Industry analysts contended that tariffs would contribute to the devaluation of currencies and cause the economies of the developing world to stagnate.[13]

Michigan and Minnesota mining communities held rallies to "Stand Up for Steel." The president of the United Steelworkers of America Local and the general manager of Hibbing Taconite collected about 16,000 letters from residents of the Mesabi Range addressed to President Bill Clinton and U.S. senators Paul Wellstone (D-Minnesota) and Rod Grams (R-Minnesota). They asked for tariff protection against unfairly traded imports and advocated a reduction in Minnesota's taxes on processed ore. In Marquette, the *Mining Journal* published a special edition featuring articles and paid advertisements encouraging everyone to support the effort to save American jobs.[14] Congressman Bart Stupak (D-Michigan) cosponsored H.R. 506, a bill to cap the volume of imports. A rally in Washington, DC, urged the U.S. House and Senate to join the fight to preserve mining and steel industry jobs. Senator Carl Levin (D-Michigan) cosponsored a Senate resolution calling on the president to strengthen enforcement of existing laws against dumping and to present a comprehensive plan for dealing with the problem. "We cannot absorb the world's over capacity at the expense of our domestic steel industry, related industries and their employees," Levin wrote in an article for the Marquette paper. "There can be no free trade without fair trade. Much of this foreign steel was produced with the help of illegal government subsidies and is being sold in our market below the cost of production. Such violations of our trade laws should not be tolerated."[15]

Imported semifinished steel, referred to as "slabs," represented a peril for the iron ore industry as well. When slabs could be purchased more cheaply than the integrated mills could produce them, the temptation was great for steelmakers to eliminate the first step in the steelmaking process: smelting pellets in a blast furnace to produce pig iron. At first the danger that the integrated steel producers would shut down their blast furnaces seemed remote because the mills were operating at capacity. To satisfy the increased demand for steel by the automotive, steel, and construction industries, however, steelmakers purchased imported steel slabs to supplement domestic production.[16]

Nevertheless, even in early 2000, the Cliffs board of directors began to see signs of distress in the steel industry. In January pellet customer Acme Metals, Inc., asked Cleveland-Cliffs for a $5 million loan to help the company keep manufacturing steel.[17] By the beginning of the fourth quarter, Cleveland-Cliffs was directly experiencing the repercussions from dumping of steel slabs. Cliffs' board minutes noted that sales volume from the Tilden mine was at risk because Weirton Steel, an important consumer of Tilden pellets, intended to shut down its blast furnaces in favor of purchasing slabs. The decline in demand for Tilden pellets would make it necessary to shut down the mine for ten weeks.[18]

As imported steel flooded the market, it caused the price of domestic steel to collapse at a time when energy costs were escalating. This produced a domino effect, making it extremely difficult for North American integrated producers, including Bethlehem, Rouge, AK Steel, and Canadian partners Algoma and Stelco, to remain profitable. Wheeling-Pittsburgh, the nation's ninth-largest steel company and longtime partner in the Empire mine, declared bankruptcy in November 2000.

On December 29, 2000, Cleveland-Cliffs learned that LTV had filed for Chapter 11. This was particularly distressing because LTV, the country's third-largest steel producer, owned 25 percent of the Empire mine and was Cliffs' largest customer. LTV immediately closed down its wholly owned LTV Mining Company (formerly Erie Mining Company) in Hoyt Lakes, Minnesota. The closing of the mine, managed by Cleveland-Cliffs, put 1,100 people out of work.[19]

In his presentation to shareholders at the company's annual meeting in May 2001, Brinzo reported that all of the sixteen surviving integrated steel companies in North America were financially distressed, with one-third in Chapter 11. Cliffs had not been producing at capacity since 1998, the company's last financially successful year. Nevertheless, at a time when Cliffs was barely holding its own in a distressed industry, Brinzo saw a window of opportunity opening. "The natural reaction in a period of crisis is to hunker down and wait out the storm," he said. "We can't do that. We must be bold if we are to exploit the opportunities that are created during this period of adversity, and make the changes that are necessary for success in our business." Brinzo had experienced the travails of the steel industry firsthand in the late 1980s when he served as chair of LTV's unsecured creditors' committee. LTV's first reorganization had taken seven years. He believed that excess capacity in the iron and steel industry would drive more industry consolidation. The time had come to implement his plan to expand Cliffs' production capacity through the acquisition of bankrupt mines. "In our view, virtually all the ownership interests in North American mines held by steel companies are for sale," Brinzo told shareholders. "This provides a unique opportunity for Cliffs. Our strategy is to be a bigger, more powerful force in a consolidating industry, and we are actively engaged in evaluating a number of opportunities."[20]

Since Brinzo anticipated an expanding role for Cleveland-Cliffs in the global economy, he hired David H. Gunning to provide the company with greater merger and acquisition expertise. Gunning, a prominent Clevelander, had headed the worldwide corporate practice of the law firm Jones, Day, Reavis and Pogue prior to taking over as CEO of the Capital American Financial Corporation until the company was sold in 1997. He joined Cliffs as vice chairman in charge of business development and within a year was made a member of the board of directors.[21]

Whether the company could capitalize on the opportunity to purchase the bankrupt mines remained an open question. Cliffs at this point had no cash reserves. Although borrowing was difficult, particularly for companies associated with the steel industry, the company succeeded in obtaining a revolving line of credit of $100 million from its banks. In early 2001 it purchased the assets and assumed the liabilities of the LTV Steel Mining Company in a deal engineered by James Trethewey, senior vice president–business development.[22] In addition to the mine's shuttered pelletizing operations, the acquisition included a seventy-four-mile rail line and loading dock on Lake Superior and a 225,000-kilowatt power plant. Cliffs received a vital cash infusion of $50 million from the sale of the mine's power plant.[23] This provided desperately needed capital.

Meanwhile, Brinzo made public his plan to acquire the mines of bankrupt partners. In a July 2001 article in the *Cleveland Plain Dealer,* headlined "Crisis in Steel Industry Creates Mine Ownership Opportunities," Peter Krouse cited the positive effect this strategy would have on Cliffs' long-term prospects.[24] In October the *Wall Street Journal* reported that Brinzo hoped to be able to purchase all North American iron ore mines, with the exception of those owned by USX (U.S. Steel), which were not for sale. "It's an amazing shift," commented Peter Kakela for the *Journal* article.[25] Kakela, a professor of resource development at Michigan State University in East Lansing, was the grandson of a Finn who had found work in the mines in the late 1800s. He had watched the mining industry flourish in the 1970s and contract in the 1980s, providing incisive analysis at each turning point.

Brinzo did not conceal the risk the company faced. In the 2001 annual report he reminded shareholders that the only use for pellets was to smelt them in a blast furnace to produce raw steel. As the number of blast furnaces declined in North America, the company's exposure had grown. In 1970 there were 169 blast furnaces in operation; in 1988, 50. In 2001 the number had shrunk to just 36. Brinzo was betting the remaining larger, more efficient blast furnaces would sustain a smaller North American steel industry.[26] However, the Tilden and Empire mines could be saved only by cutting the number of mining jobs. "I deeply regret what this means to the employees and families who are affected," he told a somber gathering of business and community leaders in Marquette. "Our choice is to make the type of changes that will allow our mines to be competitive, or risk losing them completely." Nevertheless, he remained upbeat. He urged acceptance of what was beyond anyone's control and "to prepare and shape ourselves for these changes."[27] Employees responded by finding ways to conserve energy and improve efficiency in pellet processing. Time lost through injuries was reduced by 15 percent. The sacrifice of mining jobs was accompanied by a 23 percent cut in headquarters staff.

With its back to the wall, Cleveland-Cliffs mounted a new effort to save the domestic iron ore industry. The "Stand Up for Iron Ore" campaign, orchestrated by Dana Byrne, director–public affairs, focused on the threat to the national security of the United States posed by imported iron ore and steel. Without a domestic iron and steel industry, advocates for the industry argued, the United States would be forced to rely on foreign steel for armaments in a time of war. Congressmen James Oberstar (D-Minnesota) and Stupak demanded that the Bush administration take up the cause. They called for an investigation by the U.S. Department of Commerce of the illegal dumping of raw steel into U.S. markets under Section 232 of the Trade Expansion Act of 1962.

In January 2001 Oberstar and Stupak pointed out in a letter to Norman Mineta, President George W. Bush's secretary of commerce, that every ton of imported semifinished steel slabs displaced 1.3 tons of taconite pellets produced in the mines of Minnesota and Michigan. During the decade of the 1990s imports of semifinished steel in the United States had quadrupled from 2 million tons to 8.6 million tons. The United States had also imported 10 million metric tons of iron ore in 2000, with Brazil responsible for 2.4 million. The letter urged Mineta to consider the policy implications of dependence on foreign iron and steel.[28] Monsignor Louis Cappo, a respected Catholic priest who served as chair of the Lake Superior Community Partnership, stood up for the livelihoods of the people he served in northern Michigan and Minnesota. In a full-page letter to President Bush, published in the congressional newspaper *Roll Call,* he questioned whether America, already dependent on a foreign cartel for its oil, could afford to lose its ability to produce another essential raw material. "Can America depend on other nations to supply our needs for steel during a time of future world tension or turmoil?" he asked.[29]

Rouge Steel took an active part in the Stand Up for Iron Ore campaign. Rouge Steel's CEO pointed out in his letter to the Department of Commerce that bankrupt competitors McLouth and Sharon Steel had already shut down their blast furnaces to take advantage of the cheap steel slabs flooding the market. Rouge Steel produced specialized steel for the automotive industry. If illegal dumping could not be stopped, Rouge would be forced to shut down the hot end of its plant and lay off one-third of its workforce, or approximately 1,000 workers. Since Rouge Steel purchased approximately 3.5 million tons of iron ore pellets from the Tilden mine, the closing of its blast furnaces would bring hardship to the people of the Upper Peninsula.[30]

Even representatives of the electric-furnace industry (possibly reacting to the high prices for scrap) joined the campaign to stand up for iron ore. They argued that once the integrated steel industry became completely dependent on foreign slabs, steel producers abroad would raise prices and increase the export of their finished flat-rolled steel products. This unfair competition would cause the demise of the entire North American industry, including the minimills.[31]

Political pressure brought by members of the Lake Carriers' Association, organized labor, several integrated steel producers (WCI Steel, Rouge, and Geneva), and members of the electric-furnace industry (Nucor, Ipsco, Steel Dynamics) at last succeeded in persuading the Department of Commerce to investigate whether the imports of iron ore and semifinished steel had put the national security of the United States at risk. Hearings under Section 232 of the U.S. Trade Expansion Act of 1962 took place in Virginia, Minnesota, and Marquette, Michigan, in July 2001.

Before the hearings in Marquette, about 4,000 people from Ishpeming, Palmer, Negaunee, and Marquette marched from the domed stadium of Northern Michigan University to the University Center, where the hearings were in progress. Those who testified at the hearings included labor and industry representatives, religious leaders, and university presidents as well as local and national politicians. John Brinzo and general managers Paul Korpi and Mike Mlinar of the Empire and Tilden mines emphasized that the remaining thirty-six blast furnaces in the United States were at risk. If they were to shut down, the country would lose the capability to produce steel for its national defense or industrial needs. Representatives of the transportation industry, including the Lake Carriers' Association and the railroads, which formed vital links between the mines and the Lake Superior and lower lake ports and the steel mills in Ohio, Pennsylvania, Indiana, and Illinois, stressed that significant unemployment in their industries would follow the closing of the mines. In his testimony, Senator Levin, chairman of the Senate Armed Services Committee, urged restricting imports through tariffs or quotas to keep the iron ore industry viable in "peace-time and if necessary, in war-time."[32]

Expert witness Peter Kakela argued that government protection of a domestic iron ore

"Stand Up for Iron Ore," Marquette, 2001. To protest the "dumping" of imported steel slabs, the mining communities of the Upper Peninsula marched to Marquette, where hearings were being held by the U.S. Department of Commerce. *Left to right:* Laurie Stupak; then state representative Mike Prusi; Bruce Talus, United Steelworkers; state representative Rich Brown (behind umbrella); U.S. senator Debbie Stabenow (Michigan); U.S. congressman David Bonior (Michigan); U.S. congressman Bart Stupak (Michigan); Mike Carlslon, United Steelworkers; and behind banner in sport coat, state representative Paul Tesanovich. *(Dale Hemmila photo. Courtesy of Cliffs Natural Resources, Inc.)*

supply was not new. The Defense Production Act of 1950 had spurred the development of pelletization by allowing mining companies to rapidly depreciate capital investment in the new processes. He emphasized that in 2001 steel companies no longer depended on American iron ore, but the American iron-mining industry still depended on U.S. steelmakers. And even though 47 percent of the steel manufactured in the United States was being produced in electric furnaces, 53 percent was still being produced in blast furnaces. He predicted that if steelmakers continued to purchase steel slabs, it would force the remaining mines in Minnesota and Michigan (where 98 percent of American pellets were produced) to close. He suggested that at some point in the future, if the price of imports or a war cut off the supply of semifinished steel, it could affect the ability of American heavy industry to respond to military demands.[33] Although the campaign to stand up for iron ore failed to get a favorable ruling from the Department of Commerce on Section 232, the effort raised awareness of the connection between a viable pellet industry and national security.

Less than two months later, the terrorist attack on the World Trade Center on September 11, 2001, heightened these national security concerns. As the economy stalled, the steel industry collapsed. Algoma Steel, which owned 45 percent of the Tilden mine, declared bankruptcy. Bethlehem Steel, which owned 70 percent of Hibbing Taconite, did the same shortly afterward. Cleveland-Cliffs reported its first loss since 1986 at the end of 2001. In a special section of the company's 2001 annual report, titled "Tough Times, Answers to Tough Questions," Brinzo explained that it was clear the North American iron and steel industry had reached a critical turning point: "Cliffs has joined its partners and customers in a war for the survival of North American blast furnaces."[34] Then LTV announced that it planned to liquidate. LTV shut down its rolling mills and blast furnaces in northern Ohio. In Cleveland, LTV's entire 7,000-person workforce joined the rolls of the unemployed. The exodus from the steel company included Brinzo's son Brian, who worked in LTV's accounting office. This was an event that Cleveland-Cliffs' management never expected would happen.

Salvaging the Company's LTV Contracts

LTV's liquidation was particularly serious because the Cleveland-based steel company still owed Cliffs several million dollars for past deliveries of iron ore. To get its investment out of the pellets heaped on LTV's docks, the only solution for Cliffs appeared to be an extreme one: to buy LTV's blast furnaces and begin producing raw steel from Cliffs' pellets. Brinzo asked his management team to explore the idea of purchasing LTV's East Side Works. Since its last bankruptcy in 1984, LTV had invested nearly $1 billion in the modernization of its Cleveland plant, making it an attractive bargain.[35] Studies determined that it might be feasible for Cliffs to produce semifinished steel that could be sold to other steel manufacturers without competing directly with its pellet customers.

While Brinzo was in the process of pitching this proposition to the board, he discovered that a group of local venture capitalists had formed an entity called Cleveland Steel, Inc. and intended to purchase the Cleveland Works. At this point, Cliffs' only cash reserve was the $50 million realized from the sale of the LTV Mining Company's power plant in 2001. Cleveland-Cliffs offered to invest $20 million in the venture, with the proviso that Cleveland Steel would purchase all of its iron pellets from Cleveland-Cliffs.[36]

When Cleveland-Cliffs invested in Cleveland Steel, Brinzo did not know that Wilbur Ross, a bankruptcy specialist, also had designs on LTV's assets. A graduate of Yale and Harvard Business School, Ross had run Rothschild's bankruptcy advisory practice for more than two decades. In November 1997 he set up a private equity fund for Rothschild to purchase and consolidate companies in distressed industries. Commenting on his approach to acquiring ailing companies, Ross told *Fortune* magazine: "It's a Darwinian thing. The weaker parts get eliminated, and the stronger ones come out stronger. Our trick is to figure out which is which, try to climb on to the ones that can be made into the stronger ones, and then try to facilitate the demise of the weaker ones." In 2000 Ross bought the equity fund from Rothschild and founded W. L. Ross and Company. *Fortune* described Ross as a contrarian. "Where others see a scrapheap, he sees a fabulous return on investment. . . . Today, with global markets in recession and the bull market a distant memory, companies are piling up on the rubbish heap. And no one picks through the trash better than Wilbur Ross."[37]

Ross had designs on all of LTV's steelmaking facilities, which included steel mills in Warren, Ohio; east Chicago, Indiana; and Hennepin, Illinois, in addition to the company's Cleveland works. He was convinced that the George W. Bush administration would yield to the mounting political pressure and impose tariffs on steel products.[38] Standing in the way of Ross's plans was Cleveland Steel. When he discovered that Cleveland-Cliffs had invested in Cleveland Steel, he contacted John Brinzo. Ross and Brinzo had known each other since the first LTV bankruptcy when Ross had served as financial advisor to LTV's secured creditors' committee. He inquired whether Cleveland-Cliffs would be willing join a new investment group that he was forming in February 2002. Brinzo said the company was interested, but he admitted that Cleveland-Cliffs had little cash to invest.[39]

Nevertheless, sensing a long-term opportunity, Brinzo obtained the agreement of the board of directors to withdraw from Cleveland Steel—a move that ended the local venture. Cliffs then invested $13 million in the company Ross was forming to buy distressed steel companies. Cliffs received a 7 percent interest in the new entity and Ross's commitment to purchase pellets from Cleveland-Cliffs.[40] Ross paid $80 million cash for LTV's fixed assets, $45 million in working capital for inventory, and assumed about $200 million in environmental and other liabilities. The assets of LTV were reputedly worth about $3 billion.

In an interview published in *Skillings' Mining Review* as front-page news in March 2002, Brinzo listed "positive signs" for an iron-mining industry turnaround: the economy was improving; there was increasing demand for pellets; the imposition of tariffs by the Bush administration was pushing up the price for steel; and LTV was set to begin steel production. He was counting on Ross to purchase domestic pellets rather than imported steel slabs.[41] The next month Ross formed the International Steel Group, or ISG.[42] At that time Cliffs signed a fifteen-year contract to become the sole supplier of pellets to ISG's steel operations, and Brinzo joined ISG's board of directors.[43]

Because LTV had liquidated, ISG did not have to assume the company's legacy obligations to retirees, such as health care, insurance, and pension benefits. As soon as the sale was finalized, Ross named Rodney Mott president and CEO. Mott was a well-respected steel executive with experience in both the integrated steel and minimill industries. He immediately restarted the mills by rehiring union workers. Mott worked out new, more collaborative, work rules with the United Steelworkers of America that included incentives for increased efficiency. A labor agreement reached in December 2002 with the union paved the way for the acquisition of Bethlehem Steel.[44] ISG also benefited from a rise in steel prices after President Bush imposed a 30 percent tariff on many types of imported steel in March 2003, shortly after 30,000 miners and steel plant workers rallied in front of the White House.[45] The controversial tariff, opposed by the World Trade Organization and repealed eighteen months later, nevertheless assisted in steel's recovery.[46]

Cost Reduction and the Restructuring of Cliffs' Partnerships

Cost reduction had been a priority for Brinzo since becoming CEO. This was complicated by the company's aging mines and workforce. All the mines that Cliffs owned or managed had been built between the 1950s and 1970s. Over time the pits had become deeper and wider, increasing haul and lift distances. "The large concentrators and pelletizers in iron ore are relatively inflexible and modernizing 40-year old technology is no simple matter. But we must try," Thomas J. O'Neil, executive vice president–operations, had told workers in 1998.[47] To reduce energy costs at the Empire mine, the company installed a $5.5-million roll press crusher, which was used to break down hard-to-grind ore, and a new filter bay to improve iron recovery. At Tilden a more efficient $3.5-million hyperbaric filter system streamlined production. In a move to reduce maintenance and operating costs at Northshore, the company purchased five 200-ton trucks to replace fourteen 100-ton trucks. Wabush was given the go-ahead to upgrade mining equipment. Cliffs also upgraded software systems throughout the company.[48] As longtime employees retired in Michigan, Cliffs replaced them with younger, more highly trained workers.[49] In early 2000 the company launched a company-wide campaign called ForCE21, the acronym for "For Competitive Excellence in the 21st Century," to improve efficiency and reduce costs. The program sought to change the culture of the company by involving employees more directly in operations from team-based problem solving to environmental stewardship.[50]

To increase the efficiency of the Empire mine, the company asked the State of Michigan to grant a permit to use additional company-owned land for the disposal of waste rock from the mine. This included seventy-nine acres of wetlands. The Upper Peninsula Environmental Coalition and the National Wildlife Federation opposed this expansion because it would destroy wetlands and adversely affect the headwaters of the Escanaba River.[51] After litigation, the company received permits for the waste rock disposal area. As mitigation for the destruction of wetlands, the company was required to meet wetlands banking requirements of the Michigan Department of Environmental Quality (MDEQ) and the U.S. Environmental Protection Agency (EPA) by creating wetlands elsewhere. The wetlands design plan restored or created 670 acres of wetlands within a 2,300-acre preserve near the company's now-abandoned Republic mine. The program took three years to implement, with monitoring continuing until 2004. Republic's gaping open pit gradually filled with water and the former pit and tailing ponds teemed with new life. The nearby wetlands became a habitat for endangered sandhill cranes and peregrine falcons, as well as loons, wood ducks, and great blue herons.

Republic Wetlands Preserve, 2007. Since the late 1970s, environmental mitigation has been an important aspect of mine planning. To obtain a permit from the State of Michigan to use an area that included wetlands for waste rock from the Empire pit, the company agreed to set aside 615 acres for the creation of wetlands at the site of the former Republic mine in 2001. The preserve is regarded as a model of successful environmental mitigation and a valuable asset to the community. *(Courtesy of Cliffs Natural Resources, Inc.)*

The Republic Mine Reclamation and Wetlands Preserve became an example of successful environmental mitigation and a valuable asset to Marquette County. In 2009 the National Mining Congress listed the Republic Wetlands Preserve among its models of reclamation.[52]

As long as its mines were controlled by partnerships, it was difficult for Cleveland-Cliffs to make improvements in mining operations. However, as the distress of the steel industry continued, the company implemented Brinzo's plan to increase ownership. In January 2002 Cliffs acquired bankrupt Algoma Steel's 45 percent interest in the Tilden mine in exchange for assuming the company's liabilities in the mine. Algoma agreed to a fifteen-year sales agreement to make Cliffs its sole supplier of iron ore pellets.[53] The company was unable to swing the same deal with Stelco for its 15 percent interest in each of the Tilden and Hibbing Taconite mines.

In May 2000, to keep Rouge Steel from going bankrupt, Cliffs loaned the ailing steel company $10 million in return for a multiyear contract to purchase all of its iron ore from the Tilden mine. Up to that time, the Eveleth Taconite mine (EVTAC) had supplied Rouge Steel (Rouge owned a 45 percent interest in EVTAC). After the Rouge Steel Company filed for bankruptcy in October 2003, many feared that U.S. Steel intended to purchase Rouge with the intention of closing the plant down. A plan for the purchase of the steel company by the Russian steel giant Severstal won the backing of the United

Auto Workers Local 600. Governor Jennifer Granholm offered an incentive package, crafted by Dana Byrne and shepherded through the Michigan legislature by state senator Don Koivisto (D-Ironwood), that included tax relief. The act provided a credit against Michigan's single business tax equal to $1 for every ton of hematite pellets that Severstal used in its Dearborn blast furnaces. Since the Tilden mine was the only mine in the state producing hematite pellets, the act guaranteed a customer for its pellets. Severstal proved loyal to Cliffs. It not only paid back the loan Cliffs had made to Rouge with interest, but also continued the exclusive sales contract.[54]

In July 2002 the company secured an additional 8 percent interest in Hibbing Taconite from Bethlehem Steel. Then, in August, it increased its ownership in the Wabush mine in Canada by 4 percent. By the end of the year the company had increased its ownership of the Empire mine to 79 percent and had executed a twelve-year pellet sales agreement with Ispat-Inland.[55]

As the steel industry consolidated, Cleveland-Cliffs went from sales contracts with many companies to contracts with the fewer, larger companies emerging from bankruptcy. Instead of royalties and management fees from partners, income now came from profits on long-term sales contracts. "Cliffs is in the midst of remaking itself in a dramatic way to meet the needs of the integrated steel companies operating in the United States and Canada," Brinzo wrote in his letter to shareholders. The company was no longer a caretaker for its partners. It had become a customer-oriented, sales-driven company. As a result of the increase in Cliffs' ownership stake in its mines, production capacity had risen significantly, from 11.8 million tons in 1998 to 19.5 million tons.[56]

The ownership of majority shares of the Empire and Tilden mines allowed the company to combine them into one operation called the Cliffs Michigan Mining Company in the spring of 2003. The consolidation was part of a company-wide effort to reduce costs. It involved a 32 percent reduction in Cleveland headquarters staff, a reduction of the number of corporate officers from twelve to nine, and a 22 percent reduction of salaried workers in Michigan. Merging the two mines kept Empire open, despite the diminishing quality of its ore body. Although located in two different communities, Palmer and Ishpeming, the mines were adjacent to one another. Michael Mlinar, general manager of the Tilden mine, was named general manager of the combined operation. Previously the partnership structures of these mines had prevented the company from having employees share a common dry, mining vehicles, and repair facilities. These changes increased efficiency and reduced operating costs.[57] Several months later the company announced that the 107-year-old Lake Superior and Ishpeming Railroad, a now wholly owned subsidiary, would be integrated into the transportation department of the company.[58]

The rationalization of the company's Michigan operations led to changes in the way the company operated its mines, allowing different mines to specialize in pellet types. Edward C. Dowling, Cliffs' executive vice president–operations, appointed Don Prahl general manager of the Northshore Mining Company. He proceeded to streamline pellet production at Silver Bay by eliminating the line for fluxed pellets. By focusing on pro-

Cliffs Cottage, December 2005. Governor Jennifer Granholm and Dana Byrne, vice president–public affairs, celebrate keeping Rouge Steel's blast furnaces operating. The Michigan legislature offered an incentive package that included a tax credit equal to $1 for every ton of hematite pellets used by Michigan steel mills. *(Courtesy of Cliffs Natural Resources, Inc.)*

duction of high-quality standard pellets, the lines could be speeded up and storage and transport simplified.[59]

China and the Emergence of a Global Company

Even the optimistic and prescient John Brinzo did not anticipate soaring demand for iron ore by China. Industrialization and urbanization in China and India in the 1990s created increased demand for steel. By the late 1990s China had become the world leader in steel production, passing Japan in 1996. China needed iron ore. Though China had more than 8,000 mines, they were located inland in the country's northern and western provinces, while most Chinese steel mills were in the southeast coastal regions and along the middle and lower Yangtze River. High transportation costs and the low quality of domestic ore (about 40 percent iron) contributed to the expense of Chinese steelmaking. To lower the costs of steelmaking, China imported ore from Australia and to a lesser extent Brazil (the world's two largest iron ore producers).[60]

Chinese steel companies especially liked iron pellets produced in North America.

The high iron content and compact structure of pellets meant that they melted faster in the blast furnace than their own softer, less highly beneficiated ore. The use of high-grade pellets also saved energy. Nevertheless, it was hard to imagine a market for Lake Superior iron ore in China because of the prohibitive costs of shipment. Thus, it came as a shock to the American mining community in 2003 to learn that a Chinese steel producer had contracted with U.S. Steel's Minntac mine for delivery of 1 million tons of pellets. These pellets were shipped 2,000-plus miles across Canada by rail for delivery to oceangoing vessels for the long trip across the Pacific.[61]

Chinese steelmakers apparently were so confident that pellets would have an impact on the costs and quality of Chinese steel that they became interested in outright purchase of mines in North America. Getting wind of this, Congressman Oberstar contacted his longtime friend the Chinese ambassador, who put him in touch with representatives of the Laiwu Iron and Steel Group, Ltd. Laiwu Steel Group (successor to the Laiwu Iron and Steel Company, founded in 1970) was China's fourth-largest steel producer. It was headquartered in Laiwu City in the Gangchong district of Shangdong Province. Oberstar knew that the owners of the recently closed Eveleth Taconite mine, located near Chisholm, Minnesota, were desperate to find a buyer. He suggested that Laiwu officials consider purchasing EVTAC. But Laiwu found that the cost to transport pellets from Minnesota to ocean ports made the proposal economically unfeasible. At this point Cleveland-Cliffs learned of negotiations between Laiwu and the Eveleth mining interests. Brinzo, Calfee, and Gunning met with Laiwu officials and presented them with a proposal to purchase EVTAC together. Because Cleveland-Cliffs managed the Wabush mine on the western tip of Labrador (in which it now also held a 38 percent equity interest), the company proposed a solution to the transportation obstacle by swapping Laiwu's complement of 4 million tons of pellets from the EVTAC mine for the same amount of Wabush pellets. The Wabush pellets could be loaded on oceangoing vessels at Pointe Noire at Sept Îles, eliminating the cost of transporting pellets from Minnesota, especially the trip through the Welland Canal (between Lake Erie and Lake Ontario), with its tonnage limit.[62]

Cleveland-Cliffs and Laiwu formed a partnership and purchased EVTAC at an auction in November 2003. The cost was $3 million in cash and the assumption of liabilities, with Cliffs owning 70 percent and Laiwu 30 percent of the mine, renamed United Taconite.[63] Cliffs hired many of the former EVTAC Thunderbird mine's former employees and lost no time in restarting production at its Forbes, Minnesota, Fairlaine plant. Pellets produced at EVTAC went directly to Stelco or Dofasco steel mills on the eastern end of the Great Lakes. Demand for pellets in 2005 allowed Cliffs to reopen a second pellet-processing line at EVTAC that had not operated since 1999. Reopening this mine bolstered the economy of a proud mining community, increased Congressman Oberstar's popularity with his constituents, and benefited the State of Minnesota, which recouped lost tax revenues. At the reopening of the mine Congressman Oberstar expressed his

CEO John Brinzo (*center*) cuts the ribbon at a ceremony celebrating the first carload of pellets shipped from the United Taconite Mine in Eveleth, Minnesota, in early 2004. Brinzo is flanked by Joe Striekar, president, local union 6860, on the left, and Congressman James Oberstar (D-MN) and Simon Shi, vice general manager, Eldon Development, Ltd.–Laiwu Steel Group Ltd., to his right. (Courtesy of Cliffs Natural Resources, Inc.)

wonder at the turn of events. "Few would have imagined that the globalization that contributed to EVTAC's demise would also work to insure its future," he said. "Cheap Brazilian ore was a factor in EVTAC's demise and now, thanks to China's growing steel industry, and the partnership between Laiwu and Cliffs, the future of the mine is insured for at least another ten years." The joy at the ribbon-cutting ceremony was palpable. It was the first time in thirty-four years of working for Cliffs that John Brinzo had seen the traditional evergreen marking the opening of a new mine planted atop the first carload of pellets. This was a turn of events he had never anticipated when the company implemented his strategy for repositioning Cliffs for the twenty-first century. With this acquisition Cliffs' managed-ore capacity rose to 37 million, or about 45 percent of total North American capacity, and Cleveland-Cliffs now became the leading producer of iron ore in America.[64]

As Cleveland-Cliffs emerged as a global company, poised for further growth, the 2003 annual report, titled "Turning Point," presented a detailed and positive picture of the company's prospects. Direct sales of its pellets had increased dramatically because of the relationship that Brinzo had fostered with Wilbur Ross. As ISG acquired new companies, including Acme and Bethlehem Steel in April 2003, and Weirton Steel in February 2004, Ross continued to award Cliffs' multiyear contracts to supply his growing steel empire with pellets. Cliffs made another $13 million investment in ISG to assist in raising capital for these acquisitions. This investment would later pay off handsomely.

Despite the growth in ore sales in 2003 and the company's improved outlook, Cliffs sustained a net loss of $32.7 million, or $3.19 per share.[65] This loss was due in part to the company's write-off of the hot briquetted iron venture in Trinidad and Tobago. The operation was closed down and put up for sale, victim of a flawed technical process and weakness in the international market for the product.[66]

In addition to losses from the HBI venture, in the second quarter of 2003 production at the Tilden mine had to be revised downward from 8 million tons to about 7.2 million tons, the result of the discovery of an "anomaly" in the hematite ore body that compromised iron-recovery rates. It was necessary to "mine through" this bad spot to reach higher-grade ore. Then both Tilden and Empire had to be shut down because the Presque Isle Power Plant flooded when an upstream dam collapsed during a spring storm. This meant layoffs for about 1,200 employees for up to six weeks and the loss of 250,000 tons of pellet production for each week the mines were shut down. No sooner was Tilden up and running when one of its pelletizing kilns had to be shut down because of a cracked riding ring. If this were not enough, the Weirton Steel Corporation, a major ore customer, filed for Chapter 11 in May.[67]

As the company struggled to cope with its operational problems in mid-2003, Brinzo tapped Yale-educated Don Gallagher, vice president–sales, to take over as chief financial officer. By the end of the year the company's cash position was desperately low. To raise capital Gallagher consulted with the company's investment advisors, Hill Street and Associates. They recommended using a convertible preferred stock offering to secure desperately needed funds. The offering, handled by Morgan Stanley in January 2004, raised $175 million. This increased Cliffs' options at a time when the company was stretched to the limit. "We got the cash," Gallagher said. "We could take a breath."[68]

The issue of the high cost of labor came to a head shortly before the company's contract with the United Steelworkers of America was due to expire July 31, 2004. The United Steel Workers union represented about 4,000 employees: the workforce of Tilden and Empire in Michigan and Hibbing Taconite and United Taconite mines in Minnesota. Wabush workers in Canada, covered under a different contract, also threatened to strike.

Although strikes in 1993 and 1999 had been motivated by pension issues, at issue in 2004 was the company's insistence that American workers and retirees pay a greater share of medical insurance costs. Workers also objected to the company's announced intention to contract out more full-time maintenance and support service jobs and to reduce vacation leave. In an open letter to employees published in the local newspapers, Brinzo noted that, while the company was returning to profitability after three years of losses, it was still at risk because of rising employment and legacy costs. Taking into account wages, health care, pension costs, vacation, and other fringe benefits, the company claimed it paid an average of $53.73 per hour to its mine workers. It could not afford a work stoppage that would drive customers to seek other sources of pellets or abandon them altogether in favor of imported steel slabs. "The threat of losing our position in the marketplace," Brinzo wrote, "potentially for good, is a risk that we cannot afford to take." The letter reminded employees that stagnant prices, cash shortages, and the rising cost of labor had driven LTV, Bethlehem, and National Steel into bankruptcy liquidation, wiping out retiree benefits and leaving fewer jobs in the steel industry.[69]

If contract negotiations broke down and the union went on strike, the company threatened to bring in temporary replacement workers. To make sure that workers were aware that this was not an empty threat, the company began to train these workers at its nonunion Northshore plant in Silver Bay, Minnesota. Trailers were moved onto the company's other mining properties in anticipation of housing temporary workers.[70] The company's tactics divided families and neighbors, producing a firestorm of dissention and ill will. State senator Mike Prusi (D-Ishpeming), a former Cliffs' union man who had worked on the Stand Up for Iron Ore campaign, publicly challenged the company's controversial decision to hire replacement workers.[71]

The two sides reached agreement five minutes before the deadline. The company agreed to a 9 percent wage increase and a $220-million contribution to pension plans and retiree health care over the four years the contract was to run. The press called the settlement a "rare victory for labor."[72] In view of the negative publicity the company received for threatening to hire replacement workers, one might question the wisdom of Cliffs' decision. Brinzo thought he had no choice. "I wasn't going to risk an interruption of the flow of pellets to those aging blast furnaces because that may well have resulted in being shut down forever," he explained later to a journalist from *Forbes* magazine.[73] Shortly after the settlement, Cliffs announced plans to expand production at United Taconite, boosting annual capacity from 4.3 million to about 6 million gross tons.[74] It also reported record profits for the second quarter of $32.8 million, a welcome turnaround from the second quarter of 2003, when the company lost $21.2 million.

By the end of 2004 the reinvention of Cleveland-Cliffs was a fait accompli. The title of the 2004 annual report, "Revitalized and Gaining Momentum," though apt, was an understatement. The transformation of Cleveland-Cliffs from minority owner and manager of mines to majority owner and independent merchant of iron pellets had taken just four years. "Essentially every contract that we have today was finalized, improved, or negotiated anew in the past four years," Brinzo wrote.[75] The acquisition of bankrupt iron-mining assets at bargain prices was made possible by the consolidation of the North American steel industry and a reduction of American pellet requirements. Cleveland-Cliffs had emerged as the only independent iron-mining company left in the United States.

In 2004, when Cliffs sold its investment in ISG, the value had risen from $26 million to $238 million. Later that year Ross sold ISG to billionaire steel tycoon Lakshmi N. Mittal for $4.5 billion. Cleveland-Cliffs' multiyear contracts for pellets were transferred to Mittal Steel, headquartered in the Netherlands. "The amazing story of ISG, which arose from the ashes of the LTV Steel bankruptcy, is now legendary and is central to Cliffs' recovery," Brinzo wrote in the 2004 annual report.[76] Mittal Steel, which owned steel mills in Indonesia, Trinidad, the Netherlands, and Kazakhstan, as well as Ispat-Inland in the United States, would combine with the European firm Arcelor Steel to become the world's largest steel company.[77]

While Brinzo had counted on the strategic repositioning of the company to secure a return to financial stability, the extraordinary demand for pellets created by an expanding Chinese steel industry was an unexpected dividend. As world prices for iron ore rose, the company reported record revenues of $1.2 billion and realized a $166.3-million improvement in operating income in 2004. The year ended with earnings per share of $11.80, compared with a loss of $1.60 per share recorded in 2003. Brinzo announced that the company had ended 2004 free of debt and with a reserve of $400 million in cash and marketable securities. The company's cash dividend had been reinstated and the company announced a two-for-one stock split.[78]

In early 2005 the company began another bold initiative—the acquisition of Portman, Ltd., Australia's third-largest iron-mining company, headquartered in Perth. Cliffs' sale of its Robe River mine in the northwestern Pilbara region of western Australia in 1986 and its divestiture of its share of the Savage River mines in Tasmania had left the company without direct exposure to the Asian market. Portman could remedy this. With offices in both Beijing and Tokyo, Portman had sales contracts with fourteen Chinese and two Japanese steel producers. Portman had been exporting iron ore to China since 1994 and to Japan since 1996. The company's two mines, Koolyanobbing and Cockatoo Island, produced direct-shipping fines and lump ore (62–65 percent iron), which required only minimal processing before shipment. Portman was unusual in the degree of remoteness of its operations from populated areas of Australia. Most of the employees and contractors resided in Perth and flew to the mines for two weeks, then returned to Perth and had the following week off. In 2004 Portman had produced 6 million tons of high-grade iron ore and reported record profits of $32.8 million.[79]

Portman had rebuffed mergers with minerals resources companies in the past. The press speculated that iron-mining giants BHP Billiton and Rio Tinto, both with operations in western Australia, intended to make bids for Portman. At a time of rising world ore prices, Cliffs offered $350 million, or Australian $3.40 per share, in January. The company found itself bidding against Credit Suisse First Boston, presumably on behalf of a hedge fund.[80] Cliffs extended the deadline for the tender offer and increased its offer to A$3.85, or about $450 million. An article in *American Metal Market* quoted industry expert Peter Kakela as "amazed" that the company was ready to put that much cash down

Aerial view of the plant at Koolyanobbing, one of two mines acquired in 2004 when Cliffs purchased Portman, Ltd., Australia's third-largest iron-mining company, headquartered in Perth. *(Courtesy of Cliffs Natural Resources, Inc.)*

on Portman, comparing this acquisition to that of EVTAC for just $3 million.[81] By April 2005 the effort led by Brinzo, strongly supported by senior executives David Gunning, Don Gallagher, and Bill Calfee, had succeeded. Cleveland-Cliffs acquired an 81.1 percent stake in Portman for $433 million. It was the largest acquisition in the company's history. Portman's managing director and several board members resigned and Brinzo, Gunning, Gallagher, and Calfee joined the Portman board.[82]

Cliffs' acquisition of Portman contributed to making 2005 an exceptional year, but the company's most important news was its continued profitability. Operating income increased to $375 million, with total revenues of $1.7 billion. The company had a net income of $278 million—more than twice that of the previous year. Cliffs' equity market value reached $2.4 billion, about ten times higher than in 2003. "How did this happen?" Brinzo asked rhetorically in the 2005 annual report. The plan to acquire distressed mining operations had added 10.8 million tons to the company's sales capacity with what was considered a modest investment. The company's six mines now controlled 46 percent of the North American market and Cliffs owned Australia's third-largest iron ore company. Brinzo admitted that the company had been so engrossed in capitalizing on the window of opportunity in North America that it had failed to notice the impact of Chinese demand on the price of iron ore. The rise in global ore prices had given an enormous boost to Cliffs' revenues from ore sales, tempered by a rise in energy prices.[83] A healthy cash reserve allowed the company to invest capital in the Mesabi Nugget Project, an innovative technology to convert iron concentrates into a raw material appropriate for electric furnaces. This was a joint venture with Kobe Steel, Steel Dynamics, Inc., and the State of Minnesota.[84]

Having transformed Cleveland-Cliffs from a mining operator to an iron ore merchant, Brinzo was ready to step down four months shy of his sixty-fifth birthday. In May 2005, in anticipation of Brinzo's retirement, the board chose Joe Carrabba to take over as president and chief operating officer of Cleveland-Cliffs. Carrabba hailed from Ohio. He had attended high school in Frankfort, a small town south of Columbus, and Capital University in Columbus, where he majored in geology. He received his MBA from Frostburg State University in Frostburg, Maryland. Carrabba brought a background of twenty-two years working for Rio Tinto managing limestone, bauxite, and diamond mines in Canada and Australia. His charge was to continue Cleveland-Cliffs' globalization and expansion to include the extraction and marketing of other mineral resources, including coal.

Convinced that the North American iron-pellet industry had reached a point of maturity, Carrabba focused on sustaining that business in Michigan, Minnesota, and Canada while looking for new opportunities in the United States and abroad. Under Carrabba, who became chairman of the board, president, and chief executive officer in December 2006, Cliffs became the mineral resources company that predecessors had envisioned but failed to realize. The company expanded into the metallurgical coal business (essential in smelting iron ore), with mines in Alabama and West Virginia. Carrabba brought the company's resources and talent to bear on sustaining the environment through better management of water resources and air quality issues.

Cliffs' five iron mines in the Lake Superior region (and an additional mine in Canada) continue to be the core of Cliffs' business. Cliffs' annual pellet capacity of 37 million tons is sold to the integrated steel industry of North America largely through long-term sales contracts. In Australia the company produces direct-shipping fines and lump ore, as well as coal, with sales offices in Beijing, Tokyo, and Perth. It owns a 45 percent stake in the Australian Sonoma Project, a coking and thermal coal mine. In Latin America the company in 2007 acquired a minority ownership stake in the Amapá Project, a vast iron reserve in Amapá, Brazil. Through an alliance with Kobe Steel, the company plans to begin manufacturing iron nuggets when the Empire mine is closed in 2010. The iron nugget will open an important new market for Lake Superior iron ore—the electric-furnace industry, the fastest-growing segment of the U.S. steel industry.

To reflect these new initiatives as an international mineral resources company, in 2008 Cleveland-Cliffs changed its name to Cliffs Natural Resources, Inc. It remains headquartered in Cleveland, Ohio. There the old ore boat, the *William G. Mather*, lies

Joseph A. Carrabba took over as chairman, president, and chief executive officer in December 2006. Under Carrabba, Cliffs continued to expand internationally and domestically. To reflect the company's growth as a minerals resources company with iron and coal mines in North America, the Asia-Pacific area, and Latin America, the company changed its name to Cliffs Natural Resources, Inc. in 2008. *(Courtesy of Cliffs Natural Resources, Inc.)*

At the closing bell at the New York Stock Exchange on January 26, 2007, Joseph Carrabba, president and CEO, and other members of the Cliffs family celebrate the company's 160th anniversary and the transition to a new management team. *Left to right:* Ashley Carrabba, Marlene Brinzo, Nelson Chai (executive vice president and chief financial officer, New York Stock Exchange), John Brinzo (former president and CEO), Juanita Carrabba, Joseph Carrabba, Donald Gallagher (president–North American business unit), Maureen Gallagher, Dana Byrne (vice president–public affairs), and Rita Byrne. *(Courtesy of Cliffs Natural Resources, Inc.)*

permanently at anchor on Lake Erie, a symbol of the city's important role in the growth of the iron industry of the Lake Superior region and a tribute to one of Cleveland's oldest companies.

Conclusion: A Capacity for Survival

If you know where to look and are willing to walk a little bit off established paths on the east side of Ishpeming, you can still find the small, water-filled pits where the Cleveland Iron Mining Company initiated mining operations over 150 years ago. Only a few miles to the south are the gigantic pits of Cleveland-Cliffs' Tilden and Empire mines, hundreds of feet deep and thousands of feet wide. The mines are a testament to the longevity of the company and its ability to respond to the changing demands of the integrated steel industry for iron ore.

At the turn of the twentieth century the company survived the consolidation of American steel companies and their integration backward into iron mining. The transition from the Bessemer process to open-hearth steel production occurred just as the company's Bessemer-grade ores were approaching depletion. And the open-hearth process had an additional bonus: a new market for Cliffs' hard, lump ore, a product that had been rapidly going out of favor. Had the Oliver Iron Mining Company in 1897 or 1901 been willing to pay an extra $1.5 million, Cleveland-Cliffs could easily have been absorbed into the U.S. Steel behemoth. Instead it emerged in 1910 as the largest independent producer of iron ore in America.

Executive leadership has played an important role in the company's survival. During the company's early years Samuel Livingston Mather forged the connection between Cleveland and Michigan's Upper Peninsula and proved to skeptical furnace operators in Pittsburgh, Youngstown, and Cleveland that Lake Superior mines could provide them with a high quality and seemingly inexhaustible supply of ore. In the words of an early biographer, among steelmakers the "prosperity and high reputation of the Cleveland Iron Mining Company was due to his personal influence and popularity." Known for his conservatism in business, nevertheless he had "the courage to venture where favoring opportunity led the way."[85]

The reputation of the company for efficient operation persuaded Jeptha Wade to work with Mather on the acquisition of Iron Cliffs in 1890, a company with much larger landholdings and potential ore reserves. Majority ownership of the company's stock enabled members of the Wade and Mather families to control the board of Cleveland-Cliffs for several generations, providing continuity and preventing the takeover of the company by outsiders.

Samuel Mather served as an officer for almost forty years. His son William G. Mather presided over the company for forty-two years. Much like Brinzo a century later, William chose not to hunker down and simply survive during a crisis. When ore prices radically dropped in the 1890s and steel firms integrated backward into iron mining between 1895 and 1905, he expanded the company's operations to other Lake Superior ore ranges through purchase, lease, or partnerships. He sought niche markets for the company's non-Bessemer-grade ore, lowered operating costs, and diversified into charcoal iron, wood chemicals, coal, electric power generation, railroads, and wood products. He expanded the benefits offered to his workforce to ensure labor peace.

Building on the strong base provided by Samuel and William Mather, E. B. Greene and Alexander Brown steered the company through the Depression and World War II. Greene's financial expertise was critical in rescuing the company from the debt it had incurred through involvement in Cyrus Eaton's aborted Midwestern Steel Company. When the company faced declining reserves of direct-shipping ores after World War II, it named Walter Sterling, with a background in mining operations, as president. He assembled the talented cadre of mining engineers that kept underground mining viable and drove the transition to pellets. Later, Stuart Harrison's background in finance and Sam Scovil's gifts as a salesman contributed to the company's ability to convince partners to share the burden of large-scale project financing of new lean-ore mines and associated pellet-processing facilities. Just as Greene, with expertise in finance, had helped carry the company through the Great Depression, Tom Moore and John Brinzo, both skilled in managing the financial end of the business, helped the company survive the collapse and consolidation of its steel company partners—not once, but twice. And Brinzo's strategic vision provided the key to the company's survival and globalization in the early twenty-first century. After serving as operating agent for its steel company partners for almost three decades, at the beginning of the twenty-first century the company returned to its role as an independent iron-mining company and iron ore merchant, bringing the history of the company full circle.

Several of the traditions that emerged in the company's first century turned out to be good survival strategies. Among the Mathers' most important legacies was their insistence that the company retain ownership of its mines and reserves of iron ore wherever possible. In times of economic stress, the company generally chose to sell off assets not directly related to iron ore mining and to return to its core business. The ore properties that the company acquired during its first century were still in its possession in the 1950s and provided the foundation for the company's effort to turn its vast reserves of low-grade iron ore into pellets. Cliffs would use the same approach—sell off peripheral holdings, refocus on iron mining—when faced with the heavy capital demands required for the transition to pellets in the 1950s and early 1960s. The same focus on its core business drove the acquisition of rival Pickands Mather in the late 1980s and the Northshore Mining Company in the 1990s.

A penchant for using partnerships, or joint ventures, to leverage limited capital resources was another long-stranding corporate practice that paid dividends in the long run. The company entered into the first of these at the very birth of its mining enterprise on Lake Superior: an agreement with entrepreneurs from the Jackson Iron Company to share the expense of establishing claims and building a road from their iron mountains to what would become the port of Marquette. William Mather formed numerous partnerships in order to penetrate new iron ranges, open new mines, and expand production to reach the economies of scale necessary to remain competitive with the much larger and better-capitalized integrated steel companies that had invaded iron ore mining. These same partnerships established both personal and business ties between Cleveland-Cliffs and key independent steel companies like Bethlehem Steel and Lackawanna Steel.

After World War II, it was natural for Cleveland-Cliffs, with its long-standing tradition of working with partners, to turn once again to joint ventures to leverage the limited resources it possessed and to develop possible customers for the new product. Because it retained a stake in ore production, even during the era of minority partnerships with steel companies, Cleveland-Cliffs never completely lost its "merchant flavor," continuing to market pellets produced for its own accounts through the 1990s. Cliffs' partnership with Laiwu Steel to purchase EVTAC can be seen as an extension to the global marketplace of the company's tradition of joint ventures.

A third corporate tradition that abetted the company's survival was investment in customers' companies to secure a market for its ores. This tradition can be traced back to the Civil War, when Cleveland Iron agreed to take a portion of Middlesex Furnace's pig iron in exchange for furnishing the furnace with its ores. In the 1920s Cleveland-Cliffs purchased a 50 percent share in the Trumbull-Cliffs blast furnace. Later, this practice helped keep Rouge Steel in business. It fostered the strategic alliance with Wilbur Ross after LTV's collapse and facilitated the partnership with the Chinese company Laiwu.

Finally, Cleveland-Cliffs had a reputation for readily adopting new mining technology once others had proved its value. For example, after the Lake Superior Iron Company erected the first pocket dock for loading iron ore on vessels in 1857, Cleveland Iron introduced an improved version the following year. While not the first company to begin mining underground on the Marquette Range, Cliffs later became the leading underground mining company in the region, introducing the widely used caving system to the industry. It was not the first to use high explosives or the diamond drill on the Marquette Range, but it enthusiastically adopted both relatively early. The same can be said for the use of concrete in mine shafts and steel headframes. The company was the first iron-mining firm to comprehensively electrify its properties using waterpower, but only after William Mather had seen the successful use of hydroelectricity in Swedish mines. Likewise, Cleveland-Cliffs did not pioneer in the beneficiation of low-grade ores or their conversion to pellets. That was first done on a commercial scale on Mesabi taconites, but the company closely watched these developments and then systematically attacked the problem of pelletizing Michigan's lean jasper ores. The company's development of a unique flotation process for treating Tilden's hematite ore was one of the few occasions in the company's history when it forged ahead with the development of a new and untried technology.

Lake Superior's iron ores will not last forever. Cliffs' resilience as a mining company may once again be tested as it seeks to diversify into other natural resources and establish a presence on other continents under its new corporate name. In future times of stress, will Cliffs Natural Resources pull back and return, as did its predecessors, to its core iron ore–mining business? Or will coal by then have become its major focus? What seems unquestionable is the company's history of surviving adversity and its commitment to mining Lake Superior iron ore for as long as it possibly can.

ABBREVIATIONS

AISI: American Iron and Steel Institute Vertical File, Hagley Museum and Library, Wilmington, DE

AM: Archives of Michigan, Lansing, MI

BENT: Bentley Historical Library, Ann Arbor, MI

CCI: Cleveland-Cliffs Iron Company, Cleveland, OH

CCI-Empire: Cleveland-Cliffs Iron Company, Empire mine, Palmer, MI

CCI-HF: Historical Files, Cleveland-Cliffs Iron Company, Cleveland, OH

CCI-LDP: Cleveland-Cliffs Iron Company, Land Department Papers, MS 76–90, Archives of Michigan, Northern Michigan University Repository, Marquette, MI

CEP: Cyrus Eaton Papers, MS 3913, Western Reserve Historical Society, Cleveland, OH

CIMC-CCICP: Cleveland Iron Mining Company and Cleveland-Cliffs Iron Company Papers, MS 86–100, Archives of Michigan, Northern Michigan University Repository, Marquette, MI

CIMC-WR: Cleveland Iron Mining Company Correspondence, MS 3136, Western Reserve Historical Society, Cleveland, OH

DSC-CSU: John L. Horton Collection, Department of Special Collections, Michael Schwartz Library, Cleveland State University, Cleveland, OH

ICP-I: Iron Cliffs Company Papers, RG 65-37, Archives of Michigan, Northern Michigan University Repository, Marquette, MI

ICP-II: Iron Cliffs Company Papers, RG 66-36, Archives of Michigan, North-
    ern Michigan University Repository, Marquette, MI
ICP-III: Iron Cliffs Company Papers, RG 68-102, Archives of Michigan,
    Northern Michigan University Repository, Marquette, MI
LRL: Longyear Research Library, Marquette County Historical Society, Mar-
    quette, MI
MTU: Michigan Technological University Archives and Copper Country His-
    torical Collections, Houghton, MI
PLSMI: *Proceedings of the Lake Superior Mining Institute*
PWP: Peter White Papers, Bentley Historical Library, Ann Arbor, MI
WGMFP: William Gwinn Mather Family Papers, MS 4578, Western Reserve
    Historical Society, Cleveland, OH

NOTES

INTRODUCTION

1. Letter from Ernie Ronn to John Brinzo, Jan. 1, 2001, CCI-HF.
2. Interview with John Brinzo by Virginia Dawson, Oct. 19, 2006.
3. CCI, *Annual Report,* 2006, 2, CCI-HF.
4. U.S. Steel mines ore in the region, but it is an integrated steel company, rather than a mining company.
5. See Harlan Hatcher, *A Century of Iron and Men* (Indianapolis: Bobbs-Merrill, 1950); and Walter Havighurst, *Vein of Iron: The Pickands Mather Story* (Cleveland: World, 1958).
6. David A. Walker, *Iron Frontier: The Discovery and Early Development of Minnesota's Three Ranges* (St. Paul: Minnesota Historical Society Press, 1979); Marvin G. Lamppa, *Minnesota's Iron Country: Rich Ore, Rich Lives* (Duluth, MN: Lake Superior Port Cities, 2004); E. W. Davis, *Pioneering with Taconite* (St. Paul: Minnesota Historical Society, 1964).
7. Kenneth Warren, *Big Steel: The First Century of the United States Steel Corporation, 1901–2001* (Pittsburgh: University of Pittsburgh Press, 2001) and *The American Steel Industry, 1850–1970: A Geographical Interpretation* (Oxford: Clarendon, 1973). Warren focuses on the process end of the industry rather than the raw materials end, in spite of the fact that half of U.S. Steel's valuation at the time of its creation was its iron ore holdings.
8. Paul A. Tiffany, *The Decline of American Steel: How Management, Labor, and Government Went Wrong* (Oxford: Oxford University Press, 1988); Robert P. Rogers, *An Economic History of the American Steel Industry* (London: Routledge, 2009); and Christopher G. L. Hall, *Steel Phoenix: The Fall and Rise of the U.S. Steel Industry* (New York: St. Martin's, 1997).
9. Alfred D. Chandler Jr., *Scale and Scope: The Dynamics of Industrial Capitalism* (Cambridge, MA: Harvard University Press, Belknap Press, 1990).
10. Richard B. Mancke, "Iron Ore and Steel: A Case Study of the Economic Causes and Consequences of Vertical Integration," *Journal of Industrial Economics* 20 (1972): 223 and table, 221.
11. Samuel L. Mather to Jay C. Morse, Apr. 19, 1878, item 2783, CIMC-CCICP.
12. "Cliffs' J. S. Wilbur: Ore Specialist," *Steel,* Oct. 17, 1955, 83.

13. Cleveland Iron Mining Co., *Annual Report of the Directors to the Stockholders . . . for the Year Ending May 17, 1876* (Cleveland: Sanford & Hayward, 1876), esp. 7, 11.

CHAPTER 1

1. For the copper rush to the region, see Robert James Hybels, "The Lake Superior Copper Fever, 1841–47," *Michigan History* 34 (1950); David J. Krause, *The Making of a Mining District: Keweenaw Native Copper, 1500–1870* (Detroit: Wayne State University Press, 1992), 136–43; and Larry Lankton, *Cradle to Grave: Life, Work, and Death at the Lake Superior Copper Mines* (Oxford: Oxford University Press, 1991), 7–10. U.S. Senate, *Report of General Walter Cunningham,* 28th Cong., 2nd sess., Jan. 8, 1845, S. 98, vol. 7, 4, reported almost 900 mining permits granted in the area between November 1844 and July 1845. For "barren and worthless," see U.S. House of Representatives, *Mineral Lands of Lake Superior,* 29th Cong., 1st sess., 1846, H.R. 211, 11; for Clay's remark, see John N. Dickinson, *To Build a Canal: Sault Ste. Marie, 1853–54 and After* (Columbus: Published for Miami University by the Ohio State University Press, 1981), 8, citing *Congressional Globe,* 26th Cong., 1st sess., Apr. 21, 1840, 349–51, 828–29.

2. James E. Jopling, "Dr. Morgan L. Hewitt," 1935 manuscript, LRL; "Dr. Morgan L. Hewitt Is Subject of Paper," *Marquette Mining Journal,* Jan. 18, 1935.

3. Samuel P. Orth, *A History of Cleveland, Ohio* (Chicago: S. J. Clarke, 1910), 1:191–95.

4. This was the name according to T. B. Brooks, "Part I: Iron-Bearing Rocks," in Thomas B. Brooks et al., *Geological Survey of Michigan: Upper Peninsula, 1869–1873,* vol. 1, (New York: Julius Bien, 1873), 28; and A. P. Swineford, *History and Review of the Copper, Iron, Silver, Slate and Other Mineral Interests of the South Shore of Lake Superior* (Marquette: Mining Journal, 1876), 116. Brooks and Swineford were familiar with the pioneers of the region, so their testimony carries some weight, but I have seen no original documents bearing the name Dead River Silver and Copper Company. A handwritten copy of the "Articles of Association" of the Cleveland Iron Company, dated Nov. 9, 1847, in Cleveland-Cliffs corporate headquarters does refer to the location "known as the Dead River Company location" but has that phrase crossed out. Moreover, some of the people involved with Cassels in mineral exploration and the iron ore claim, such as W. A. Adair and George Freeman, were also involved in a Dead River and Ohio Mining Company that in the summer of 1846 attempted to mine copper and silver in the neighborhood of the Dead River near Marquette. Thus, it is possible that the Dead River Silver and Copper Company mentioned by Brooks and Swineford was really the Dead River and Ohio Mining Company. See *Cleveland Herald,* June 25, 1846, 3, and July 1, 1846, advertisement, 2, for the existence of the Dead River and Ohio Mining Company.

5. For instance, when John Burt appeared in Pittsburgh in 1851 to promote the iron deposits of Michigan's Upper Peninsula a prominent furnace owner there told him: "We have an abundance of good ores in Pennsylvania and have no need of your Michigan ores." See John Burt, "Autobiography," c. 1883, 5, LRL.

6. Philo M. Everett, "Recollections of the Early Explorations and Discovery of Iron Ore on Lake Superior," *Michigan Pioneer Historical Collections* 11 (1887): 162–65; Philo Everett to brother, Nov. 10, 1845, LRL; Brooks, *Geological Survey,* 12, quoting from a letter dated Jackson, Mich., Nov. 10, 1845, by Philo M. Everett to Captain G. D. Johnson. The letter is also printed in A. P. Swineford, *Swineford's History of the Lake Superior Iron District,* 2nd ed. (Marquette: Mining Journal, 1871), 12.

7. Everett, "Recollections," 165–66; Swineford, *History and Review,* 95–96; Brooks, *Geological Survey,* 14–20. Until May 1846 claims were filed under a lease-permit system. See Krause, *Making of a Mining District,* 138–40, 179–82, for a description of the system. The system failed to promote mining and was eliminated in favor of purchase in 1847.

8. Jacob Houghton's statement in T. B. Brooks et al., *Geological Survey of Michigan's Upper Peninsula, 1869–1873* (New York, 1873), appendix D, 2:236–37, describes the scene. Jacob Houghton was on Burt's crew. For other accounts of Burt's "discovery," see Swineford, *Swineford's History,* 8–13; Brooks, *Geological Survey,* 9–14; and "Iron Mining on Lake Superior," *Lake Superior Journal,* July 12, 1855. Everett, testifying under oath in an unrelated trial in 1881, asserted he was unaware that Burt's survey

crew had passed through the area when the Jackson party visited its claim in 1845 (*Jeremy Compo v. Jackson Iron Company,* State of Michigan Supreme Court, 43, 54, copy in LRL). Abram V. Berry, president of the Jackson Iron Company, testified the same (*Jeremy Compo,* 83). Everett says the same thing in his "Recollections," 167. Frank B. Stone, *Philo Marshall Everett: Father of Michigan's Iron Industry and Founder of the City of Marquette* (Baltimore: Gateway, 1997), 19, also maintains that Burt's "discovery" did *not* set off a scramble for iron. In the usual struggle over priorities, some, desiring to make Burt's discovery more significant, have suggested that Burt's two Indian guides *may* have told others, and that the person who informed Everett of the mountains of shining ore *may* have picked up the knowledge in this way. See, for instance, John S. Burt, *They Left Their Mark: William Austin Burt and His Sons, Surveyors of the Public Domain* (Rancho Cordova, CA: Landmark Enterprises, 1985), 63. Harlan Hatcher, *A Century of Iron and Men* (Indianapolis: Bobbs-Merrill, 1950), 24–25, suggests, without documentation, that Burt's "discovery" set off a scramble for iron and led to the first mineral claims in the iron region. No evidence supports this assertion. In fact, Burt and his associates initially made little of the "discovery," and it is unlikely the guides would have had any appreciation of the significance of the discovery since the region's Indians made no use of iron ore.

9. Everett, "Recollections," 167–68.

10. Ibid.; and James Harrison Kennedy, "The Opening of the Lake Superior Iron Region," *Magazine of Western History* 2 (Aug. 1885): 348, quoting Berry.

11. Everett, "Recollections," 167–68; Kennedy, "Opening," 348; depositions of Charles Johnson and Edgar Kidney, c. 1850, CCI-HF; "Agreement of Parties Interested in the Iron Location, Apr. 29, 1847," CCI-HF; Burton H. Boyum, "Cliffs Illustrated History," manuscript, c. 1986, 3-8, CCI-Empire, Boyum Collection (in the numbering system used in the Boyum MS, 3-8 means chap. 3, p. 8).

12. "Agreement of Parties Interested in the Iron Location, Apr. 29, 1847"; and Cleveland Iron Co., "Articles of Association," Nov. 9, 1847, CCI-HF. The slightly earlier emergence of the Dead River and Ohio Mining Company, which included some of the people also interested in the iron deposits but who attempted in the summer of 1846 to mine copper, suggests that not all of the group that backed Cassels were willing to give up on the dream of finding copper wealth. See *Cleveland Herald,* June 25, 1846, 3; and July 1, 1846, advertisement, 2, for the Dead River and Ohio Mining Company activities.

13. *Cleveland Herald,* Nov. 10, 1851. See also *American Railroad Journal,* Sept. 24, 1853.

14. For examples of the belief that these new ore deposits were so vast as to be "inexhaustible," see *The Iron Resources of Michigan and General Statistics of Iron. . . .* (Detroit: H. Barns, 1856), 3, 18; Marquette Iron Co., *Prospectus for the Organization of the Marquette Iron Co. of Lake Superior, Michigan* (New York: L. H. Biglow, 1864), 4; Sharon Iron Co., *The Sharon Iron Company, Sharon, Mercer Co., PA, Connected with the Jackson Iron Mountain, Lake Superior* (New York: Baker, Godwin, 1852), 4, 7–8; Cascade Iron Co., *Prospectus of the Cascade Iron Company of Lake Superior* (Philadelphia: W. P. Kildare, 1865), 3; American Iron Mining & Manufacturing Co., *Prospectus Preparatory to the Organization of the American Iron Mining & Manufacturing Co. . . .* (New York: Francis Hart, 1864), 10; *Detroit Free Press,* June 12, 1848; *Mining Magazine,* Aug. 1853, 198; Alexander Campbell, "The Upper Peninsula: An Address on the Climate, Soil, Resources, Development, Commerce and Future of the Upper Peninsula of Michigan [delivered Feb. 6, 1861]," *Michigan Pioneer and Historical Collections* 3 (1881): 251; and J. G. Chamberlain to Cousin Henry, Aug. 6, 1854, J. G. Chamberlain Letters, LRL.

15. Brooks, *Geological Survey,* 189, on 40 percent ores being common in Pennsylvania.

16. J. S. Newberry, *The Iron Resources of the United States* (New York: A. S. Barnes, 1874), 758.

17. Edgar Kidney deposition, c. 1850, and Charles Johnson deposition, c. 1850, CCI-HF; Boyum, "Cliffs Illustrated History," 3-8.

18. Nahum Keys deposition, c. 1850, James E. Peters deposition, c. 1850, and Charles Johnson deposition, c. 1850, CCI-HF; Boyum, "Cliffs Illustrated History," 3-9; Everett, "Recollections," 168–69; Brooks, *Geological Survey,* 28; and Peter White, "The First Fifty Years: The Cleveland Cliffs Iron Company Celebrates Its Semi-centennial at Ishpeming," *Michigan Miner,* Aug. 1, 1900, 13.

19. Cleveland Iron Mining Company of Michigan, *Charter and By-laws of the Cleveland Iron Mining Company of Michigan* (Cleveland: Sanford & Hayward, 1851).

20. Problems with the permit system then used by the U.S. government may also have caused uncertainty

in the minds of the company's early investors, for it required claimants to post a bond of several thousand dollars to lease land for mining purposes for a maximum of nine years, after which everything would revert to the government. This system was replaced in 1847. See Krause, *Making of a Mining District*, 138–40, 179–80.

21. Everett, "Recollections," 170–71; Brooks, *Geological Survey*, 28–29; White, "First Fifty Years," 13. White in 1849 was an employee of the Marquette Iron Company.

22. Brooks, *Geological Survey*, 20–21; A. P. Swineford, *Annual Review of the Iron Mining and Other Industries of the Upper Peninsula for the Year Ending Dec. 1880* (Marquette: Mining Journal, 1881), 5; Kennedy, "Opening," 351–52. White, "First Fifty Years," 12–13, describes his experiences during the Marquette Iron Company's landing.

23. Robert B. Gordon, *American Iron, 1607–1900* (Baltimore: Johns Hopkins University Press, 1996), 14, on the advantages of a bloomery forge over a blast furnace, and 90–100, on the operation of such forges.

24. "Semi-centennial: The Cleveland Iron Mining Company," *Iron Trade Review*, July 26, 1900, 16.

25. Olive Harlow, "Diary," entry for July 10, 1850, LRL, noted production of the first bloom of iron. See also Kenneth D. LaFayette, *Flaming Brands: Fifty Years of Iron Making in the Upper Peninsula of Michigan, 1848–1898* (Marquette: Northern Michigan University Press, 1990), 3–4. White, "First Fifty Years," 13, says the first ore was taken from the Cleveland location in July 1850; "Semi-centennial" says the first ore taken from outcroppings was in 1849.

26. *Lake Superior Journal*, Nov. 12, 1853.

27. Everett, "Recollections," 171; Ralph D. Williams, *The Honorable Peter White* (Cleveland: Penton, 1907), 48–49; Boyum, "Cliffs Illustrated History," 3-11, 3-12; Nancy A. Schneider, "The Struggle to Establish Civilization on the Michigan Iron Range," *Inland Seas* 54 (1998): 24. The depositions, which have been cited in earlier reference notes, are still in the historical files of the Cleveland-Cliffs Iron Company, Cleveland, Ohio.

28. Kennedy, "Opening," 366, 367; Brooks, *Geological Survey*, 29. Outhwaite traveled to the region in 1848, 1850, and 1852. The *Cleveland True Democrat*, July 30, 1850, 3, for example, notes Hewitt as being among a group of Clevelanders returning from the "upper lakes." Even while hesitating to invest in the area, CIMC's investors still refused an offer to sell to the Sharon Iron Company in 1850, as reported in Kennedy, "Opening," 348–49.

29. Michigan Corporation Securities Division, Annual Mining Reports, 1853, "Cleveland Iron Mining Company," MTU. See the document dated Dec. 28, 1852.

30. Boyum, "Cliffs Illustrated History," 3-17. Boyum, a former Cleveland-Cliffs manager in the Upper Peninsula, unfortunately does not indicate the location of the documents to which he refers.

31. Ibid., 3-18.

32. White, "First Fifty Years," 16, 17.

33. Amos R. Harlow to "Friend Fisher" [Waterman Fisher], July 8, 1849, Harlow Family Papers, 1836–50, BENT.

34. On the cost of producing iron and its selling price, see Brooks, *Geological Survey*, 21; White, "First Fifty Years," 17; Williams, *Honorable Peter White*, 53–54; George P. Cummings, "Reminiscences of the Early Days on the Marquette Range," *PLSMI* 14 (1909): 214–15; and George A. Newett, "The Early History of the Marquette Iron Ore Range," *PLSMI* 19 (1914): 302.

35. For the charcoal problem, see White, "First Fifty Years," 16; for examples of internal disputes, see C. A. Linbaugh[?] to A. R. Harlow, June 6 and June 20, 1851, and E. B. Gray to A. R. Harlow, July 15, 1851, file C, item 2768, CIMC-CCICP. See also E. B. Gray to A. R. Harlow, July 15 and Aug. 23, 1851, file C, item 2768, CIMC-CCICP.

36. For trouble with paying employees, see Edmond Maguire and James Joyce to A. R. Harlow, Oct. 8, 1851, and James Higginson to Harlow, Nov. 7, 1851, file C, item 2768, CIMC-CCICP. For Fisher's financial problems, see Everett, "Recollections," 172; and Williams, *Honorable Peter White*, 65–66.

37. The Cleveland Iron Mining Company Blotter Book, entry for Jan. 1857, item 1801, CIMC-CCICP, refers to $100,000 being paid "at time of reorganization" and seems to suggest that figure covered the purchase of the Marquette Iron Company. See also Boyum, "Cliffs Illustrated History," 3-12 and 3-13,

who refers to minutes of a special meeting of CIMC on Apr. 18, 1853. Attempts to locate these minutes were unsuccessful.

38. James Harrison Kennedy, "Samuel L. Mather," in *A History of the City of Cleveland: Biographical Volume* (Cleveland: Imperial, 1897), 118–21; Timothy J. Loya, "William Gwinn Mather, the Man," *Inland Seas* 46 (1990): 120–21.

39. James T. Soutter to Samuel L. Mather, Sept. 22, 1845, and Thomas G. Mather to Samuel L. Mather, Nov. 2, 1845 (on Mather's despondency), folder 568, container 63, WGMFP.

40. While Mather was to devote the vast bulk of his energies to making the Cleveland Iron Mining Company work, the dream of copper wealth died hard. His personal correspondence contains many references to investments in copper-mining companies. See, for example, John Veale to Samuel L. Mather, May 12 and 26, 1853, folder 590, container 67, and W. J. Cadman to Samuel L. Mather, Jan. 12, 1855, folder 596, WGMFP. Mather's personal papers in early 1855 (folder 596) have numerous assessment notices for copper-mining stocks he held.

41. "William J. Gordon," in Maurice Joblin, pub., *Cleveland Past and Present: Its Representative Men* (Cleveland: Fairbanks, Benedict, 1869), 111–13; Nancy A. Schneider, "Breaking the Bottleneck," *Inland Seas* 55 (1999): 127.

42. Cleveland Iron Mining Co., *Act of Incorporation with the Articles of Association and By Laws of the Cleveland Iron Mining Company* (Cleveland: Sanford & Hayward, 1853). The articles of incorporation are dated Feb. 5, 1853 (10). Samuel L. Mather was secretary and treasurer; Morgan L. Hewitt president. For information on the stockholders and their holdings, see Michigan Corporation Securities Division, Annual Mining Reports, for 1854 and 1855, "Cleveland Iron Mining Company," MTU.

43. Michigan Corporation Securities Division, Annual Mining Reports, 1853, "Cleveland Iron Mining Company," MTU. For a list of items acquired from Marquette Iron Company, see "Report of Cleveland Iron Mining Company . . . for July 1853," Cleveland Iron Mining Company, Blotter Book, 1856–61, Jan. 1857, item 1801, CIMC-CCICP.

44. *Lake Superior Journal,* Nov. 12, 1853; Michigan Corporation Securities Division, Annual Mining Reports, 1854, "Cleveland Iron Mining Company," MTU.

45. White, "First Fifty Years," 16–17.

46. *American Railway Times,* Aug. 13, 1857.

47. *Green Bay and Lake Superior Railroad Company Articles of Association* (Cleveland, 1851), 5–12, cited by Saul Benison, "Railroads, Land and Iron: A Phase in the Career of Lewis Henry Morgan," PhD diss., Columbia University, 1953, 68–69, who indicates he possessed the pamphlet; Williams, *Honorable Peter White,* 61.

48. Burt, "Autobiography," 9–10; *American Railroad Journal,* May 29, 1852. The story of Ely's railroad is covered in depth in Benison, "Railroads, Land and Iron"; and John T. Gaertner, *The Duluth, South Shore and Atlantic Railway: A History of the Lake Superior District's Pioneer Iron Ore Hauler* (Bloomington: Indiana University Press, 2009).

49. *New York Weekly Tribune,* July 10, 1847.

50. John N. Dickinson, *To Build a Canal: Sault Ste. Marie, 1853–54 and After* (Published for Miami University by the Ohio State University Press, 1981) is the most complete account of the canal's background and construction. See also Brooks, *Geological Survey,* 23–25.

51. *Lake Superior Journal,* June 3 and July 12, 1855.

52. See LaFayette, *Flaming Brands.*

53. H. K. Beshon to W. F. Carry, Sept. 20, 1855, CCI-HF; Marquette Iron Co., *Prospectus,* 13; Kennedy, "Opening," 362; and Swineford, *Swineford's History,* 11. Jackson Iron Co., *16th Annual Report, June 1869* (New York: C. O. Jones, 1869), 12, comments on early reports that the ore was too hard and "consequently valueless." See also Williams, *Honorable Peter White,* 68–75, 78, 139; *Sharpsville Advertiser,* Jan. 24, Feb. 14, Feb. 21, and Mar. 21, 1877, for a discussion by participants in the process on early problems smelting Lake Superior ores.

54. J. B. Curtis to M. L. Hewitt, Dec. 12, 1853, CCI-HF; John Outhwaite to Samuel L. Mather, Dec. 3, 1856, folder 592, container 67, WGMFP; Benjamin F. French, *History of the Rise and Progress of the Iron Trade of the United States from 1621 to 1857* (New York: Wiley & Halsted, 1858), 156–58; James

M. Swank, *History of the Manufacture of Iron in All Ages . . .* (Philadelphia: James M. Swank, 1884), 250–51; Williams, *Honorable Peter White,* 68; Campbell, "The Upper Peninsula," 253; and Marquette Iron Co., *Prospectus,* 12, 13–18. Some of the tests are reported in *Iron Resources of Michigan,* 14–24.

55. Kenneth Warren, *The American Steel Industry, 1850–1970: A Geographical Interpretation* (Oxford: Clarendon, 1973), 42–43; and John Birkinbine, "The Iron Ore Supply," *Transactions of the American Institute of Mining Engineers* 27 (1897): 519.

56 John Fritz, *Autobiography of John Fritz* (New York: John Wiley, 1912), 74.

57. Boyum, "Cliffs Illustrated History," 3-19, quotes the document, but he does not indicate its location. Since Boyum was employed by Cleveland-Cliffs, one can assume that he probably saw the document someplace.

58. Samuel L. Mather to J. J. St. Clair, Feb. 27 and Mar. 20, 1854, file G, item 2768, CIMC-CCICP.

59. Samuel L. Mather to J. J. St. Clair, June 21, 1854, and Mather to St. Clair, July 31, 1854, file G, item 2768, CIMC-CCICP. See also *Lake Superior Journal,* June 3, 1854, and July 12, 1855.

60. "Iron Region of Lake Superior," *Mining Magazine,* July 1854, 104.

61. *Lake Superior Journal,* June 3, 1854; D. H. Merritt, "History of Marquette Ore Docks," *PLSMI* 19 (1914), 305–6, notes that in the fall of 1854 his father was contracted to construct a dock and that he began working in November 1854 for operation as early as possible in 1855. On his father's death in December 1854, he succeeded to the job.

62. Michigan Corporation Securities Division, Annual Mining Reports, 1853, 1854, 1855, and 1857, "Cleveland Iron Mining Company," MTU.

63. Samuel L. Mather to J. J. St. Clair, Oct. 20, 1854, file G, item 2768, CIMC-CCICP.

64. W. J. Gordon to J. J. St. Clair, Oct. 19, 1855, file HIJ, item 2768, CIMC-CCICP.

65. Brooks, *Geological Survey,* 28–29; Kennedy, "Opening," 353.

66. Joe Russell Whitaker, *Negaunee, Michigan: An Urban Center Dominated by Iron Mining* (Chicago: University of Chicago Libraries, 1931), 17.

67. Samuel L. Mather to J. J. St. Clair, Feb. 27, 1854; see also Mar. 20, 1854, file G, item 2768, CIMC-CCICP.

68. In 1859 Mather referred to J. J. St. Clair, the company's second agent (1853–56), as a "bitter enemy" of the company. Samuel L. Mather to H. B. Tuttle, Feb. 19, 1859, file B, item 2769, CIMC-CCICP.

69. W. J. Gordon to M. L. Hewitt, Aug. 30, 1855, Morgan L. Hewitt Letters, LRL.

70. Hewitt's wife, who suffered severely from hay fever, was another factor. She discovered that her symptoms decreased significantly in the Upper Peninsula.

71. H. B. Tuttle to A. Kent, Jan. 18, Feb. 3, and Feb. 23, 1858, file A, item 2769, CIMC-CCICP.

72. Samuel L. Mather to H. B. Tuttle, Mar. 18, 1859, file B, item 2769, CIMC-CCICP. Tuttle's letter to Mather could not be found, but Mather puts Tuttle's remarks in quotes in responding to them.

73. For example, M. L. Hewitt to H. B. Tuttle, Feb. 10, 1859, file B, item 2769, CIMC-CCICP; and William H. Mulligan Jr., ed., *Historic Resources of the Iron Range in Marquette County, Michigan, 1844–1941* (Marquette: Economic Development Corporation of the County of Marquette, 1991), A-34. The section in the Mulligan volume was authored by Burton Boyum, a Cleveland-Cliffs employee, who presumably had access to company records. He says that the local company agent William Ferguson "repeatedly asked Cleveland for more men." But he does not indicate the location of the documents to which he refers.

74. *Lake Superior Journal,* July 19, 1856; Mulligan, *Historic Resources,* A-34.

75. W. J. Gordon to Wm. Ferguson, July 3, 1856, file K–L, item 2768, CIMC-CCICP.

76. "Instructions for Cleveland Iron Mining Co's Agent at Marquette, April 30, 1857," file M, item 2768, CIMC-CCICP. Less than a month later CIMC informed William Ferguson of his dismissal. The company had previously expressed some discontent with Ferguson's lack of accounting knowledge or skills, and this seems to have been the key issue. The letter informing Ferguson that he would be replaced on June 1 also noted that the company was providing him with $1,000 severance (quite generous for the period) and he was still held in high personal regard. H. B. Tuttle to William Ferguson, May 18, 1857, ibid. Ferguson was to return to work for the company later.

77. For example, Samuel L. Mather to H. B. Tuttle, Mar. 25, 1859, file B, item 2769, CIMC-CCICP.

78. Williams, *Honorable Peter White,* 62, reproduces the Tower Jackson letter.

79. Everett, "Recollections," 172; Benison, "Railroads, Land and Iron," 73–74; *Mining Magazine,* July 1854, 104.

80. They also founded a less important company: the Peninsula Iron Company. For the founding of the Lake Superior Iron Company, see Benison, "Railroads, Land and Iron," 74–75, 81; Burt, "Autobiography," 10; Brooks, *Geological Survey,* 27.

81. Everett, "Recollections," 172–73; *Mining Magazine,* July 1854, 104; John Burt to William Burt, May 30, 1853, William Austin Burt Papers, LRL.

82. *Lake Superior Journal,* Sept. 18, 1855. The decision to use strap rails was apparently made during a visit of the directors of CIMC to the region in spring 1854. See Samuel L. Mather to J. J. St. Clair, Mar. 20, 1854, file G, item 2768, CIMC-CCICP, where Mather notes that the decision of plank or rail for the line from the Jackson mine to the Cleveland mine would be made when the directors went up in the spring. See also Williams, *Honorable Peter White,* 144; and John Hearding, "Early Methods of Transporting Iron Ore in the Lake Superior Region," *PLSMI* 29 (1935): 174–75.

83. *Iron Mountain Railway Company Articles of Incorporation,* Mar. 14, 1855, LRL. The Jackson and Cleveland companies jointly owned the Iron Mountain Railway Company, which ran from the bay in Marquette to the Jackson location. The two or three additional miles of railroad from the Jackson location to the Cleveland Mountain was owned solely by CIMC and called the Mountain Iron Railway. Additional separate CIMC-owned mini-railway companies controlled the side tracks to the Cleveland dock in Marquette and the side tracks at the mines. "Instructions for Cleveland Iron Mining Co.'s Agent at Marquette, April 30, 1857," file M, item 2768, CIMC-CCICP.

84. Kathryn Rudie Harrigan, *Joint Ventures, Alliances, and Corporate Strategy* (Washington, DC: Beard, 2003), 4.

85. Ibid., 4–6; Frederic C. Lane, "Family Partnerships and Joint Ventures in the Venetian Republic," *Journal of Economic History* 4 (Nov. 1944); Henry W. Nichols, "Joint Ventures," *Virginia Law Review* 36 (May 1950). The difference between a partnership and a joint venture has long been cloudy, as Nichols demonstrates.

86. Nichols, "Joint Ventures," 428.

87. *Lake Superior Journal,* May 17, 1856; Williams, *Honorable Peter White,* 144.

88. John Disturnell, *Trip through the Lakes of North America* (New York: J. Disturnell, 1857), 68–69; *Lake Superior Journal,* May 17, June 7, 1856; and *Mining Magazine,* July 1857, 94–95. For the optimistic local outlook in late 1855, see *Lake Superior Journal,* Dec. 1, 1855, which claims CIMC could have sold over 50,000 tons had it been able to get ore to the market, and notes that the company hoped to ship 300 tons a day the following summer.

89. Brooks, *Geological Survey,* 22–23; *Iron Resources of Michigan,* 17; Everett, "Recollections," 173–74.

90. White "First Fifty Years," 17; Williams, *Honorable Peter White,* 144–45.

91. Everett, "Recollections," 172–73; *Mining Magazine,* July 1854, 104; Benison, "Railroads, Land and Iron," 74, 82–83. CIMC and the Ely railroad did not remain implacable enemies. Special arbitrator Charles T. Harvey, constructor of the canal at Sault Ste. Marie, worked to resolve the issues between the Elys and the Iron Mountain Railway (see "Iron Mountain Railroad Report of Charles T. Harvey, Referee, Marquette, Michigan, October 2, 1855," LRL; and Benison, "Railroads, Land and Iron," 98–101). CIMC then worked with the Ely interests through much of the remainder of the 1850s and on into the 1860s on such issues as railroad land grants and an aborted attempt to consolidate the Ely railroad, the Iron Mountain Railway, and other railroads seeking to service the iron-mining area. For this story the best source is Benison, "Railroads, Land and Iron."

92. Rukard Hurd, *Hurd's Iron Ore Manual* (St. Paul, MN: Rukard Hurd, 1911), 54; Whitaker, *Negaunee,* 17. Other sources say rates on Ely's road were half what the Cleveland and Jackson companies had been incurring with the plank road. See *Lake Superior Journal,* Aug. 15, 1857; Brooks, *Geological Survey,* 23; Kennedy, "Opening," 360–61; Everett, "Recollections," 174; and Williams, *Honorable Peter White,* 146.

93. *Lake Superior Journal,* May 17, June 7, 1856.

94. *Mining Magazine,* Feb. 1856, 176.

95. Cleveland Iron Mining Co., Blotter Book, Aug. 15, 1857, item 1801, CIMC-CCICP.

96. Two good accounts of the Panic of 1857 are James L. Huston, *The Panic of 1857 and the Coming of the Civil War* (Baton Rouge: Louisiana State University Press, 1987); and Charles W. Calomiris and Larry Schweikart, "The Panic of 1857: Origins, Transmission, and Containment," *Journal of Economic History 51 (1991).*

97. M. L. Hewitt to H. B. Tuttle, Dec. 11, 1857, CCI-HF.

98. For examples, see Samuel L. Mather to William Ferguson, July 3, 1856, file K–L, item 2768, CIMC-CCICP; Mather to M. L. Hewitt, Apr. 4, 1853, Hewitt Letters, LRL.

99. Peter White, "A Mere Sketch of Iron Money in the Upper Peninsula," *Michigan Pioneer Historical Collections,* 10 (1905).

100. Richard H. Timberlake, "Private Production of Scrip-Money in the Isolated Community," *Journal of Money, Credit and Banking* 19 (Nov. 1987).

101. Williams, *Honorable Peter White,* 151–53. For examples of correspondence dealing with "our currency," see Robert Nelson to H. B. Tuttle, May 1 and 2, 1859, file B, item 2769, CIMC-CCICP. Samuel L. Mather to Robert Nelson, June 25, 1860, folder 1, and Sept. 19, 1860, folder 2, container 1, CIMC-WR.

102. H. B. Tuttle to A. Kent, June 10, 1858, file A, item 2769, CIMC-CCICP.

103. William Donohue Ellis, *The Cuyahoga* (New York: Holt, Rinehart & Winston, 1967), 233; Schneider, "Breaking the Bottleneck," 130.

104. H. B. Tuttle to A. Kent, June 21, 1858, file A, item 2769, CIMC-CCICP.

105. Samuel L. Mather to H. B. Tuttle, Feb. 25, 1859, and May 9, 1859, file B, item 2769, CIMC-CCICP.

106. *Cleveland Leader,* Aug. 7, 1858, quoting the editor of the *Lake Superior Journal.* Jasper is an impure form of quartzite containing significant iron impurities; jaspilite was the term used in the Lake Superior iron districts for the jasper associated with regional iron ores, usually containing specular hematite as the impurity. The terms were often used interchangeably.

107. For example, M. L. Hewitt to H. B. Tuttle, Mar. 18, 1859; Samuel L. Mather to Tuttle, Mar. 25 and Apr. 19, 1859; and Robert Nelson to Tuttle, Mar. 30, 1859, file B, item 2769, CIMC-CCICP.

108. The best account is Merritt, "History." H. B. Tuttle to A. Kent, Mar. 3, 1858, file A, item 2769, CIMC-CCICP, has a description of plans for the new dock. See also Williams, *Honorable Peter White,* 148; and Cummings, "Reminiscences," 215, who wrongly credits CIMC with being the first to adopt the pocket system. Boyum, "Cliffs Illustrated History," 4-12 and 4-13, quotes from documents detailing the initial test of the docks, but again does not provide a location for the documents. Finally, see William F. Armstrong, "Historical Sketch of the Marquette Iron Range," 1932, 23–25, LRL; and Ernest H. Rankin, "Marquette's Iron Ore Docks," *Inland Seas* 23 (1967).

109. "Lake Superior," *New York Times,* Oct. 20, 1859; *New York Times,* Mar. 30, 1860.

110. H. B. Tuttle to Samuel L. Mather, June 15, 1860, item 2724, CIMC-CCICP.

111. Brooks, *Geological Survey,* atlas, plate 12.

112. H. B. Tuttle to Samuel L. Mather, June 25, 1860, item 2724, CIMC-CCICP.

113. M. L. Hewitt to Samuel L. Mather, Apr. 14, 1861, file G–L, item 2724, and Ridgway Burton & Co. to Mather, Apr. 29, 1861, file M–S, CIMC-CCICP.

114. Brooks, *Geological Survey,* atlas, plate 12.

115. Cleveland Iron Mining Co., *Annual Report of the Directors to the Stockholders . . . for the Year Ending May 18, 1864* (Cleveland: Sanford & Hayward, 1864), 3.

116. Peter White to Samuel L. Mather, Dec. 20, 1862, Letterbook, July 29, 1861–Aug. 12, 1863, box 26, PWP. On the rapidly rising price of pig iron, see Peter White, "A Brief Attempt at the History of the Mining Industry of Northern Michigan," *Publications of the Michigan Political Science Association* 3 (Jan. 1899): 148.

117. For the wartime demand for charcoal iron and the belief that linking furnaces to mines was the way to go, see American Iron Mining & Manufacturing Co., *Prospectus,* 5. For organization of the Marquette Iron Company, see Cleveland Iron Mining Co., *Annual Report for 1864,* 5; Marquette Iron Co., *Prospectus,* 3–8, 19–20; Brooks, *Geological Survey,* 36; *Lake Superior Miner,* Apr. 8, 1864; and *Cleveland Daily Herald,* Mar. 28, 1864. The *Marquette Mining Journal* noted, Oct. 8, 1868, that the Marquette

Iron Company was "owned by the stockholders" of CIMC, "though under a separate organization," as does Swineford, *Swineford's History,* 28. In 1879 the Cleveland Iron Mining Company would purchase the Marquette Iron Company outright.

118. Cleveland Iron Mining Co., *Annual Report for 1864,* 5.

119. For example, Jay C. Malone, who became the company's Marquette agent in 1865, wrote to Samuel L. Mather on Jan. 5, 1867 (item 2434, CIMC-CCICP) noting that when he took over it was said that the Cleveland mine was "nearly used up." And the *Marquette Mining Journal,* Oct. 8, 1868, commented that some years ago it had been thought "and confidently asserted" that the Cleveland mine had been exhausted.

120. Everett, "Recollections," 170–71.

121. Robert Kelly, "A Trip to Lake Superior," *PLSMI* 19 (1914): 322, gives an account of an 1853 trip to the region.

122. L. I. Gordon to Friend Pharo, Apr. 2, 1857, 3, LRL.

123. Frank P. Mills to Jay C. Morse, Jan. 30, 1872, item 2772, CIMC-CCICP.

124. A. B. Taylor to Samuel L. Mather, Jan. 28, 1873, item 2774, CIMC-CCICP.

125. Brooks, *Geological Survey,* 251, for employment at mines; ibid., vol. 2, appendix G, for Ishpeming data. See also Judith Boyce DeMark, "Iron Mining and Immigrants, Negaunee and Ishpeming, 1870–1910," in *A Sense of Place: Michigan's Upper Peninsula,* ed. Russell M. Magnaghi and Michael T. Marsden (Marquette: Northern Michigan University, 1997).

126. H. B. Tuttle to Samuel L. Mather, July 14, 18, and 20, 1859, item 2432, CIMC-CCICP.

127. Robert Nelson to Samuel L. Mather, Sept. 15 and Oct. 5, 1861, and Mar. 5, 1862, item 2432, CIMC-CCICP.

128. W. J. Gordon to Samuel L. Mather, June 23, 1862, Letters Recv'd, G–L, item 2724, CIMC-CCICP. For the manpower shortage, see Robert Nelson to Samuel L. Mather, May 1, 1862, Letters from R. Nelson, item 2724, and Nelson to Mather, May 4, 1862, item 2432, CIMC-CCICP.

129. Robert Nelson to Samuel L. Mather, Aug. 11, 1862, item 2432, CIMC-CCICP.

130. William Hart Smith to W. J. Gordon, Mar. 20 and June 29, 1863, and Smith to Samuel L. Mather, Mar. 27, 1863, Letters Recv'd, M–S, item 2724, CIMC-CCICP.

131. William Ferguson to Samuel L. Mather, May 17, 1863, item 2432, CIMC-CCICP; telegram, W. J. Gordon to Samuel L. Mather, May 20, 1863, Letters Recv'd, A–F, 1863, item 2724, CIMC-CCICP; *Lake Superior Journal,* May 15, 1863.

132. Peter White to Samuel L. Mather, June 26, 1863, Letterbooks, 1867–1877, box 26, PWP. The same letter is in White to Mather, June 26, 1863, item 2724, CIMC-CCICP. See also William Ferguson to Samuel L. Mather, May 31, 1863, for the loss of men sent up at company expense, and Ferguson to Mather, June 28, 1863, item 2432, CIMC-CCICP, for another comment on men being frightened off by the draft.

133. William Ferguson to Samuel L. Mather, June 1, 2, and 19, 1864, item 2432, CIMC-CCICP.

134. Frank P. Mills to H. R. Tuttle, Mar. 3, 1860; Robert Nelson to H. R. Tuttle, Mar. 12, 1860, Agent at Marquette, Letters, 1860, item 2724, CIMC-CCICP; Peter White to H. B. Tuttle, Mar. 5 and 12, 1860, Letters T–Z, 1860, CIMC-CCICP; Peter White to Samuel L. Mather, May 4, 1861, Letters Recv'd, 1861, M–S, item 2724, CIMC-CCICP.

135. William Ferguson to Samuel L. Mather, May 17, 1864, item 2432, CIMC-CCICP.

136. William Ferguson to Samuel L. Mather, May 16, 17, 19, 20, and 24; June 5, 1864, item 2432, CIMC-CCICP; "Riot at the Iron Mines," *Chicago Tribune,* May 29, 1864, 3; "Marine Matters at Detroit," *Chicago Tribune,* June 4, 1864, 2; "The Strike of the Iron Miners," *Chicago Tribune,* July 20, 1865, 1 (which has comments on the 1864 strike); and Bradley A. Rodgers, *Guardian of the Great Lakes: The U.S. Paddle Frigate Michigan* (Ann Arbor: University of Michigan Press, 1996), 97, 104. Samuel L. Mather's brother Henry was in Marquette in this period and referred to "riotous demonstrations." For his observations on the strike, see Henry R. Mather to Samuel L. Mather, May 24, and June 3, 9, and 11, 1864, folders 619 and 620, container 71, WGMFP.

137. Frank P. Mills to Jay C. Morse, July 11, 1865, item 2770, CIMC-CCICP.

138. Rodgers, *Guardian of the Great Lakes,* 93–105, is the best source, but see also F. A. Roe, "The United

States Steamer *Michigan* and the Lake Frontier during the War of the Rebellion," *United Service: A Monthly Review of Military and Naval Affairs* 6 (Dec. 1891). Roe was commander of the *Michigan* in 1865. See also Samuel L. Mather to Jay C. Morse, June 6; July 10, 12, 14, and 18, 1865; F. P. Mills to Jay C. Morse, July 7, 11, 12, 15, and 17, 1865; J. Outhwaite to Jay C. Morse, July 15, 1865, item 2770, CIMC-CCICP; *Portage Lake Mining Gazette* (Houghton, MI), July 15, July 29, 1865; and "The Strike of the Iron Miners," *Chicago Tribune,* July 20, 1865.

139. N. H. W., "Among the Iron Mines," *Ishpeming Iron Home,* Sept. 5, 1874.

140. William Ferguson to W. J. Gordon, Apr. 7, 1864; Ferguson to Samuel L. Mather, Apr. 7 and 26, May 17 and 25, 1864, item 2432, CIMC-CCICP.

141. Frank P. Mills to R. Nelson, Sept. 8, 1862, item 2724, CIMC-CCICP.

142. Jay C. Malone to Samuel L. Mather, undated, probably around Mar. 26, 1867, item 2434, CIMC-CCICP. M. L. Hewitt wrote to Mather on Nov. 6, 1873, that Mills was "the best man that we have." File G–L, item 2724, CIMC-CCICP.

143. Cleveland Iron Mining Co., *Annual Report of the Directors to the Stockholders of the Cleveland Iron Mining Company for the Year Ending May 17, 1871* (Cleveland: Sanford & Hayward, 1871), 8; Samuel L. Mather to Jay C. Morse, Aug. 31, Sept. 13, 1871, item 2771, CIMC-CCICP. For attempts to retain another important manager, Joseph Sellwood, see Mather to D. H. Bacon, Nov. 7 and 8, 1881, item 2789, CIMC-CCICP. A short biographical note on Mills is in *PLSMI* 15 (1910): 212–13.

144. Peter White to T. B. Meigs, Aug. 3, 1863, Letterbooks, 1861–1867, box 26, PWP. White, "First Fifty Years," 17; Marquette Iron Co., *Prospectus for the Organization of the Marquette Iron Co. of Lake Superior, Michigan* (New York: L. H. Biglow, 1864), 8; Cascade Iron Co., *Prospectus of the Cascade Iron Company of Lake Superior* (Philadelphia: W. P. Kildare, 1865), 13; Cleveland Iron Mining Co., *Annual Report for 1864,* 3–4.

145. Brooks, *Geological Report,* atlas, plate 12.

146. *Lake Superior Mining Journal,* Dec. 2, 1864.

147. J. Ward & Co. to Cleveland Iron Mining Co., July 18, 1862, file A–F, item 2724, CIMC-CCICP.

148. *Lake Superior News and Journal,* Jan. 16, 1863; see also *Cleveland Daily Herald,* Mar. 28, 1864.

149. *American Journal of Mining,* May 19, 1866. The *Cleveland Leader* had noted as early as Sept. 29, 1858, that the coming of Lake Superior ore would change the whole character of the iron trade in the Northwest.

CHAPTER 2

1. *Lake Superior Mining Journal,* June 17, 1864. See also *Beard's Directory and History of Marquette County, with Sketches of the Early History of Lake Superior, Its Mines, Furnaces, etc.* (Detroit: Hadger & Bryce, 1873), 233. J. S. Newberry, *The Iron Resources of the United States* (New York: A. S. Barnes, 1874), 759, 771, 774, noted that over 200 furnaces along the Ohio-Pennsylvania border used Lake Superior ores and that Pittsburgh had become the most important iron market in the United States. See also *Bulletin of the American Iron and Steel Association,* Oct. 29, 1873, 475.

2. William T. Hogan, *Economic History of the Iron and Steel Industry in the United States* (Lexington, MA: Lexington Books, 1971), 1:20, 196.

3. Calculated from *9th Census of the United States* (Washington, DC: Government Printing Office, 1872), 3:760, table 13.

4. John B. Pearse, *A Concise History of the Iron Manufacture of the American Colonies up to the Revolution and of Pennsylvania until the Present Time* (Philadelphia: Allen, Lane & Scott, 1876), 167–68, 229.

5. Thomas Dunlap, ed., *Wiley's American Iron Trade Manual . . .* (New York: John Wiley, 1874), 503, for Port Henry; for other New York mines, which produced generally 20,000 tons or less annually, and New Jersey mines, see Samuel H. Daddow and Benjamin Bannan, *Coal, Iron, and Oil; or, The Practical American Miner. . . .* (Pottsville, PA: Benjamin Bannan, 1866), 541–43.

6. Robert B. Gordon, *A Landscape Transformed: The Ironmaking District of Salisbury, Connecticut* (Oxford: Oxford University Press, 2001), 85. See also William Ralston Balch, comp., *The Mines, Miners and Mining Interests of the United States in 1882* (Philadelphia: Mining Industrial Publishing Bureau, 1882), 130.

7. Dunlap, *Wiley's American Iron Trade Manual,* 481–92.

8. "Frank P. Mills, Sr.," *PLSMI* 15 (1910); *Counties of White and Pulaski, Indiana, Historical and Biographical* (Chicago: F. A. Battey, 1883), 275.

9. Samuel L. Mather to H. B. Tuttle, Mar. 25, 1859, file B, item 2769, CIMC-CCICP.

10. *Lake Superior News and Journal,* June 5, 1863. The various pits in which the company conducted mining operations were usually given separate names: No. 3 mine, Incline mine, Schoolhouse mine, Sellwood's Pit. Some sources, however, simply referred to the "Cleveland mine" and in that term included all operations on the company's property, even if noncontiguous.

11. A. P. Swineford, *Appendix to Swineford's History of the Lake Superior Iron District: Being a Review of Its Mines and Furnaces for 1873* (Marquette: Mining Journal, 1873), 26.

12. William Ferguson to Samuel L. Mather, Apr. 19, 1864, item 2432, CIMC-CCICP.

13. Samuel L. Mather to Jay C. Morse, Sept. 14 and 20, Oct. 28, 1865, item 2770, CIMC-CCICP; Jay C. Malone to Mather, Aug. 8, 1867, item 2434, CIMC-CCICP.

14. Jay C. Malone to John Outhwaite, Nov. 11, 1867, item 2434, CIMC-CCICP.

15. Frank P. Mills to Jay C. Morse, Jan. 4, 1873, item 2774, CIMC-CCICP.

16. "Tariff and the Iron Mines," *New York Times,* Dec. 25, 1893.

17. Cleveland Iron Mining Co., Statement of Condition, Inventory, etc., May 19, 1873, Annual Statements, item 2749, CIMC-CCICP.

18. Swineford, *Appendix,* 27.

19. Henry S. Drinker, *A Treatise on Explosive Compounds, Machine Rock Drills and Blasting* (New York: John Wiley, 1883), 189–212; and Harold Barger and Sam H. Schurr, *The Mining Industries, 1899–1939: A Study of Output, Employment and Productivity* (1944; repr., New York: Arno, 1972), 119, for accounts of the history of rock drills.

20. Robert Nelson to H. B. Tuttle, Aug. 23, 1859, item 2769, CIMC-CCICP; W. J. Gordon to Tuttle, Aug. 23, 1859, ibid. As early as 1857 William Ferguson, then supervising the company's mine, had indicated he would "like to see a good horse or steam drill at work in the new mine," believing it would keep down strikes. See Ferguson to H. B. Tuttle, Apr. 9, 1857, item 2431, CIMC-CCICP.

21. Larry D. Lankton, *Cradle to Grave: Life, Work, and Death at the Lake Superior Copper Mines* (Oxford: Oxford University Press, 1991), 81–86; Eric Twitty, *Blown to Bits in the Mine: A History of Mining and Explosives in the United States* (Ouay, CO: Western Reflections, 2001), 45–48.

22. For a brief history of explosives in mining, see Drinker, *Treatise,* 65–73; Twitty, *Blown to Bits,* 1–34; and Arthur Pine Van Gelder and Hugo Schlatter, *History of the Explosives Industry in America* (1927; repr., New York: Arno, 1972), 321–26.

23. *Marquette Mining and Manufacturing News,* Dec. 19, 1867.

24. Thomas B. Brooks, et al., *Geological Survey of Michigan: Upper Peninsula, 1869–1873,* vol. 1 (New York: Julius Bien, 1873), 269; C. C. Cowpland, "A Resume of the Early History of the Explosives Industry in Marquette County," *PLSMI* 29 (1935): 22; and Kenneth D. Lafayette, "Nitroglycerine: Destructive Servant," *Marquette Mining Journal,* Apr. 17, 1975.

25. The accident at the Jackson mine is mentioned in Dunlap, *Wiley's American Iron Trade Manual,* 467.

26. Brooks, *Geological Survey,* 265.

27. Swineford, *Swineford's History,* 27. Burton H. Boyum, "Cliffs Illustrated History," manuscript, c. 1986, 5-18, CCI-Empire, claims that CIMC was the first to use dynamite in its mine since its ores were harder than others, but cites no document to support this claim.

28. Lankton, *Cradle to Grave,* 94–96.

29. "Statement of Condition, Inventory, etc., May 18, 1870," Annual Statements, item 2749, CIMC-CCICP. Swineford, *Swineford's History,* 28, called them the best lot of horses in the region.

30. Lankton, *Cradle to Grave,* 41–57.

31. *Lake Superior Mining Journal,* Dec. 8, 1865.

32. Jay C. Malone to Samuel L. Mather, Aug. 8, 1867, item 2434, CIMC-CCICP.

33. *Lake Superior Mining Journal,* Oct. 10, 1867; Brooks, *Geological Survey,* 277.

34. Cleveland Iron Mining Co., *Annual Report of the Directors to the Stockholders . . . for the Year Ending May 19, 1869* (Cleveland: Sanford & Hayward, 1869), 5, 6.

35. "Statement of Condition, Inventory, etc., May 19, 1873," Annual Statements, item 2749, CIMC-CCICP.

36. "Estimates of Ore Costs, 1870," folder 1115, container 17, WGMFP.

37. For example, Samuel L. Mather to Wm. Ferguson, July 14, 1856, file K–L, item 2768, CIMC-CCICP. H. B. Tuttle to A. Kent, May 31, 1858, file A, item 2769, CIMC-CCICP. For another example, see Mather to Jay C. Morse, May 22 and June 22, 1865, item 2770, CIMC-CCICP.

38. George W. Tifft, Buffalo Furnace and Machine Shop, to H. B. Tuttle, May 8, 1860, file T–Z, item 2724, CIMC-CCICP. See also a letter of Mar. 6, 1860, ibid.: "[I]t is greatly for your interest to so arrange our business that there will not be any delay in loading vessels at Marquette."

39. Samuel L. Mather to J. J. St. Clair, June 21, 1854, file G, item 2768, CIMC-CCICP.

40. For instance, H. B. Tuttle to A. Kent, June 21, 1858, file A, item 2769, CIMC-CCICP; Jay C. Malone to J. C. Baldwin, Escanaba, May 18, 1868, item 2434, CIMC-CCICP; and Samuel L. Mather to Jay C. Morse, Sept. 10 and Nov. 11, 1872, item 2772, CIMC-CCICP.

41. H. B. Tuttle to Samuel L. Mather, June 23, 1860, item 2724, CIMC-CCICP.

42. F. P. Mills to Jay C. Morse, Aug. 21, 1866, item 2770, CIMC-CCICP.

43. See, for example, H. B. Tuttle to A. Kent, June 26, 1858, file A, item 2769, CIMC-CCICP.

44. C. L. Burton to Samuel L. Mather, July 20, 1860, file A–F, item 2724, CIMC-CCICP.

45. George W. Tifft to Samuel L. Mather, Aug. 15, 1862, and July 24, 1862, item 2724, CIMC-CCICP; Mather to Rbt. Nelson, June 10 and 24, 1862, item 2769, CIMC-CCICP.

46. Samuel L. Mather to Jay C. Morse, Sept. 5, 1865, item 2770, CIMC-CCICP; Mather to Morse, July 1, 1872, item 2772, CIMC-CCICP.

47. John O. Greenwood, *The Fleets of Cleveland-Cliffs, Detroit and Cleveland Navigation . . .* Fleet Histories Series 6 (Cleveland: Freshwater, 1998), 2, 16; John B. Mansfield, *History of the Great Lakes* (Chicago: J. H. Beers, 1899), 446.

48. "Marine Department," 37, in Cleveland-Cliffs Iron Co., *The Cleveland-Cliffs Iron Company: An Historical Review of This Company's Development and Resources Issued in Commemoration of Its Seventieth Anniversary, 1850–1920* (Cleveland: Cleveland-Cliffs, 1920); Greenwood, *Fleets of Cleveland-Cliffs,* 1–2, 16–17; Mansfield, *History of the Great Lakes,* 446, 448; and Harlan Hatcher, *A Century of Iron and Men* (Indianapolis: Bobbs-Merrill, 1950), 110. One of the letterboxes in the Cleveland-Cliffs Collections contains a number of letters from Mark Hanna at Rhodes & Co. to Jay C. Morse, Cleveland Iron Mining Company's agent in Marquette, dealing with the formation of the Cleveland Transportation Company, in which Morse had invested. See file H, item 2773, CIMC-CCICP. These letters began around July 16, 1872.

49. See Samuel L. Mather to Jay C. Morse, Aug. 28, 1872, item 2772, CIMC-CCICP, where he notes, "We are organizing a company to build 4 barges & 4 consorts; expect to have two of each out and ready for opening of navigation next spring." Cleveland Iron people took out $265,000 in stock and Hanna's group $235,000 out of the $500,000 subscribed. A Sept. 26, 1872, letter notes that the company had been formally organized that day and that Morse had subscribed to $20,000 in stock.

50. Samuel L. Mather to Jay C. Morse, Aug. 19, 1872, item 2772, CIMC-CCICP. See also Mather to Morse, Jan. 5, 1874, item 2776, CIMC-CCICP.

51. A. P. Swineford, *History and Review of the Copper, Iron, Silver, Slate and Other Mineral Interests of the South Shore of Lake Superior* (Marquette: Mining Journal, 1876), 143.

52. Dunlap, *Wiley's Iron Trade Manual,* 468.

53. *Bulletin of the American Iron and Steel Association,* Oct. 29, 1873, 475.

54. Ibid., Jan. 23, 1873, 164.

55. Samuel L. Mather to Jay C. Morse, Apr. 18, May 7 and 9, and June 20, 1873, item 2774, CIMC-CCICP.

56. See M. John Lubetkin, *Jay Cooke's Gamble: The Northern Pacific Railroad, the Sioux, and the Panic of 1873* (Norman: University of Oklahoma Press, 2006) for more details on the origins of the Panic of 1873.

57. Samuel L. Mather to Jay C. Morse, Sept. 22, Oct. 7, 14, 25, and Nov. 13, 1873, item 2774, CIMC-CCICP.

58. Samuel L. Mather to Jay C. Morse, May 4, 1874, Dec. 28, 1874.

59. Samuel L. Mather to Jay C. Morse, Apr. 13 and 14, May 6 and 7, July 6 and 8, 1874, item 2776, CIMC-CCICP.

60. Samuel L. Mather to Jay C. Morse, Oct. 17, Nov. 18, 1874, item 2776, CIMC-CCICP.

61. Jay C. Morse to Samuel L. Mather, Oct. 6, 25, 28, and 29, 1873, item 2774, CIMC-CCICP, for reductions in force and wages. Also see Samuel L. Mather to Jay C. Morse, Oct. 25, 1873, ibid.

62. Peter White to Samuel L. Mather, Dec. 3, 1873, Letters, Nov. 19, 1873–Dec. 27, 1874, box 26, PWP.

63. Jay C. Morse to Samuel L. Mather, Oct. 29, 1873, item 2443, CIMC-CCICP.

64. E. B. Isham to C. J. Canda, Nov. 6, 1873, box 41, ICP-III.

65. *Bulletin of the American Iron and Steel Association,* Oct. 15, 1875, 308.

66. William Kelly, "Address of President William Kelly," *PLSMI* 6 (1900): 17.

67. *Marquette Mining Journal,* Aug. 29, 1874.

68. Samuel L. Mather to Jay C. Morse, July 29, 30, 1874, item 2776, CIMC-CCICP.

69. *Marquette Mining Journal,* July 25, Aug. 1, 8, 1874; *Ishpeming Iron Home,* July 23, 1874; and Russell M. Magnaghi, *The Way It Happened: Settling Michigan's Upper Peninsula* (Iron Mountain, MI: Midpeninsula Library Cooperative, 1982), 59–63.

70. Samuel L. Mather to Jay C. Morse, Mar. 30, 1874, item 2776, CIMC-CCICP.

71. Jay C. Morse to Samuel L. Mather, Apr. 7, 1874, item 2776, CIMC-CCICP; see also Mather to Morse, Apr. 6, 7, and 13, 1874, ibid.

72. Samuel L. Mather to Jay C. Morse, May 5, 7, and 8, 1874, item 2776, CIMC-CCICP.

73. Samuel L. Mather to Jay C. Morse, Apr. 19 and May 9, 1878, item 2783, CIMC-CCICP.

74. Samuel L. Mather to Jay C. Morse, Apr. 19, May 9 and 22, 1878, item 2783, CIMC-CCICP. Apparently Mather's Cleveland Transport charters were for $1.50 a ton, Marquette to Cleveland; freight rates on independent vessels had dropped to $1.10 by May. The Lake Superior Iron Company did not have a captive lake shipping company like Cleveland Iron, but in 1873, in an attempt to bring shipping under control, Lake Superior Iron entered into a number of long-term charter arrangements with vessel owners, only to be caught in the same bind as the Cleveland Iron Mining Company. See, for example, Lake Superior Iron Co., *Annual Report, December 31, 1874* (Boston: Rand, Avery, 1875), 9, 10; see also the company's *Annual Report,* Dec. 31, 1875, 7.

75. Samuel L. Mather to Jay C. Morse, Aug. 14, 1876, item 2779, CIMC-CCICP. See also Mather to Morse, Nov. 6, 1875, item 2777, CIMC-CCICP, which seems to suggest that because Cleveland Iron was caught with expensive vessel contracts it urgently needed to get the cost of mining down to make up the difference. See also C. H. Hall to Morse, Jan. 29, 1875, item 2778, CIMC-CCICP, and Mather to Morse, Aug. 10, 1876, item 2779, CIMC-CCICP.

76. Lake Superior Iron Ore Association, *Lake Superior Iron Ores, 1938* (Cleveland: Lake Superior Iron Ore Association, 1938), 322; Henry Raymond Mussey, *Combination in the Mining Industry: A Study of Concentration in Lake Superior Iron Ore Production* (New York: Columbia University Press, 1905), 74–75.

77. Samuel L. Mather to Jay C. Morse, Jan. 15 and Feb. 8, 1878, item 2783, CIMC-CCICP; R. S. Fay (Secretary, Lake Superior Iron Company) to Morse, Aug. 30, 1878, item 2782, CIMC-CCICP. See also Fay to Morse, Apr. 16, 1878, item 2782, CIMC-CCICP, on Menominee ore driving prices down. In October Mather confessed to Morse that he feared the Menominee iron range and wished Cleveland Iron had an interest in some of its deposits (Mather to Morse, Oct. 30, 1878, item 2782, CIMC-CCICP).

78. T. B. Brooks, "Report on the Iron Interests of Lake Superior in Relation to the Lands of the Lake Superior Ship Canal Iron & RR Co.," Dec. 15, 1877, 11–21, folder 2, box 1, Report on Menominee Range, 1877, RG 81-29, Records of the Department of State, AM, outlines the advantages that the Menominee region mines would have over Marquette region mines.

79. Wm. G. Mather to D. H. Bacon, Nov. 19, 1885, folder 6, container 1, MS 3136, CIMC-WR.

80. "New Iron Fields: Increased Production Forcing Prices Down," *New York Times,* July 5, 1885, referring to the opening of the Vermilion Range.

81. Mussey, *Combination,* 76; James E. Jopling, "The Marquette Range—Its Discovery, Development and Resources," *Transactions of the American Institute of Mining Engineers* 27 (1897): 552; A. P. Swineford,

Annual Review of the Iron Mining and Other Industries of the Upper Peninsula for the Year Ending Dec. 1882 (Marquette: Mining Journal, 1883), 24.

82. Brooks, *Geological Survey,* 244–46.

83. Ibid., 274. Brooks, writing in the early 1870s, was critical of the inefficiency of mining operations on the Marquette Range in a number of areas; see ibid., 244–49.

84. For early dislike of soft hematite, see, for example, *Marquette Daily Mining Journal,* Jan. 31, 1893; Ralph D. Williams, *The Honorable Peter White* (Cleveland: Penton, 1907), 157, 160; Samuel L. Mather to Jay C. Morse, Sept. 12, 1871, item 2771, CIMC-CCICP (comments on the waste occurring when shipping); and Mather to Morse, July 15, 1876, item 2779, CIMC-CCICP ("there is a prejudice against all soft or Hematites, & all the reasoning in the world, wont [*sic*] change them").

85. Samuel L. Mather to Jay C. Morse, Dec. 2, 1871, item 2771, CIMC-CCICP.

86. For example, Samuel L. Mather to Jay C. Morse, Dec. 15 and 17, 1871, item 2771, CIMC-CCICP.

87. For a very good account of the rise of the Bessemer process, see Jeanne McHugh, *Alexander Holley and the Makers of Steel* (Baltimore: Johns Hopkins University Press, 1980), 71–210. For the adoption of the process in America, see either McHugh or Thomas J. Misa, *A Nation of Steel: The Making of Modern America, 1865–1925* (Baltimore: Johns Hopkins University Press, 1995), 5–39.

88. Local furnace operators had been mixing hard and soft ores for some years, producing a high-quality charcoal pig iron. It is not clear whether local furnace operators or furnace operators on the lower lakes first recognized that this practice could be used in the context of producing Bessemer-grade pig.

89. Samuel L. Mather to Jay C. Morse, May 15, 1875, and July 27, 1875, item 2777, CIMC-CCICP.

90. Samuel L. Mather to Jay C. Morse, Jan. 15, 1875, item 2777, CIMC-CCICP.

91. Samuel L. Mather to Jay C. Morse, Apr. 19, 1878, item 2783, CIMC-CCICP.

92. Boyum, "Cliffs Illustrated History," 5-17, citing a letter from mine superintendent Fred Mills.

93. Swineford, *History and Review,* 156.

94. Cleveland Iron Mining Co., *Annual Report of the Directors to the Stockholders . . . for the Year Ending May 17, 1876* (Cleveland: Sanford & Hayward, 1876), esp. 7, 11.

95. For problems with the company's inclined pits, see, for example, *Marquette Mining Journal,* May 17, 1879; Samuel L. Mather to Jay C. Morse, May 15 and 19, 1879, item 2784, CIMC-CCICP. A number of additional letters in the 1879 letterbook refer to rock fall problems in the company's other mines.

96. *Ishpeming Iron Ore,* Oct. 25, 1879.

97. For example, Samuel L. Mather to Don H. Bacon, Dec. 3 and 21, 1885, folder 6, container 1, CIMC-WR. Mather a month earlier had referred to the company's soft hematite mine as "our main standby now." See Mather to Bacon, Nov. 30, 1885, ibid.

98. Charles D. Lawton, "The Cleveland Cliffs Iron Company and Its Mine," *Michigan Miner,* Oct. 1, 1900, 11; see also Kelly, "Address," 21; and H. H. Stoek, "Marquette-Range Caving Methods," *Mines and Minerals* 30 (Nov. 1909): 193.

99. Albert L. Waters, "A Thesis on the Modern Methods of Prospecting for Iron Ores in the Lake Superior Region," thesis, Michigan College of Mines, 1893, 27–36.

100. Cleveland Iron Mining Co., *Annual Report for 1869,* 10.

101. Brooks, *Geological Survey,* 200, 204, 259.

102. Samuel L. Mather to J. C. Morse, Dec. 8, 1877, item 2780, CIMC-CCICP.

103. Samuel L. Mather to Jay C. Morse, July 15, 1878, item 2783, CIMC-CCICP.

104. For an example, see James E. Jopling, "A Brief History of the Cleveland-Cliffs Iron Company," *Michigan History* 5 (1921): 159–60.

105. Swineford, *Swineford's History,* 27.

106. Samuel L. Mather to H. B. Tuttle, Aug. 23, 1859; and W. J. Gordon to Tuttle, Aug. 23, 1859, item 2769, CIMC-CCICP. The company had sent Robert Nelson, its mining superintendent, to Niagara to look at some steam drills.

107. For the Lake Superior Company's early use, see C. H. Hall to Rand Drill Co., Mar. 20, 1883, in Swineford, *Annual Review for 1882,* letters of testimony for "Little Giant" drill at end of the volume. For Iron Cliffs Company experimentation see [indecipherable], Barnum Mine, to T. J. Houston, Aug. 28, 1874, vol. 5, Correspondence Book no. 19, ICP-II. For the continued dominance of hand drilling, see Brooks, *Geological Survey,* 266n.

108. John C. Fowle to Alexander Maitland, Dec. 13, 1883, vol. 17, ICP-II. For a very good description of problems with early rock drills in the nearby copper district, see Larry D. Lankton, "The Machine under the Garden: Rock Drills Arrive at the Lake Superior Copper Mines, 1868–1883," *Technology and Culture* 24 (1983): 10–22.

109. Cleveland Iron Mining Co., *Annual Report for 1876,* 11, for expression of some disappointment with the drills' performance. See G. N. McKibbin, Ntl. Drill & Compressor Co., to Jay C. Morse, Sept. 11, 1877, file A, item 2781, CIMC-CCICP, for a rather vague indication of possible resistance on the part of hand drillers. For resistance from miners to mechanized rock drills in the nearby copper district, see Lankton, "Machine under the Garden," 22–34.

110. [N.?] W. Horton, Rand Drill Co., to J. C. Morse, Sept. 29, 1880, item 2786, CIMC-CCICP.

111. [N.?] W. Horton, to Jay C. Morse, July 25, Aug. 13 and 26, 1879, item 2785, CIMC-CCICP.

112. A. P. Swineford, *Annual Review of the Iron Mining and Other Industries of the Upper Peninsula for the Year Ending Dec. 1880* (Marquette: Mining Journal, 1881), 8.

113. Lankton, *Cradle to Grave,* 85–87. Compressed air drills were not as common in western mining until around 1890 (Twitty, *Blown to Bits,* 48).

114. Don H. Bacon, "The Development of Lake Superior Iron Ore," *Transactions of the American Institute of Mining Engineers* 27 (1897): 343.

115. F. P. Mills to H. W. Horton, Rand Drill Co., Aug. 22, 1889, item 2450, CIMC-CCICP.

116. William Baxter Jr., "The Electric Light in Mining Operations," *Engineering and Mining Journal,* July 4, 1896.

117. M. C. Bullock to Jay C. Morse, Apr. 15, May 12, Oct. 27, 1879, item 2785, CIMC-CCICP.

118. F. P. Mills to M. C. Bullock, Mar. 9, 1880, item 2450, CIMC-CCICP; *Ishpeming Iron Agitator,* May 22 and June 5, 1880; Samuel L. Mather to Jay C. Morse, May 24, 1880, item 2787, CIMC-CCICP; F. C. Stanford, "The Electrification of the Mines of the Cleveland-Cliffs Iron Company," *PLSMI* 19 (1914): 189. We have been unable to determine where the first use of electric lights for mining occurred.

119. D. H. Bacon to Jay C. Morse, Oct. 19, 1880, item 2786, CIMC-CCICP; Mather to Jay C. Morse, Oct. 4, 1879, item 2784, CIMC-CCICP; and Frank P. Mills, "Cost of Mine Lighting," *Transactions of the American Institute of Electrical Engineers* 2 (1885), paper 3.

120. *Marquette Mining Journal,* Apr. 5, 1890.

121. F. P. Mills to Jay C. Morse, Apr. 15, 1881, item 2788, CIMC-CCICP; *Ishpeming Iron Ore,* June 4, 1881; and Michigan, Commissioner of Mineral Statistics, *Annual Report of the Commissioner of Mineral Statistics of the State of Michigan for 1880* (Lansing: W. G. George, 1881), 164. See J. M. Gannon to Jay C. Morse, Dec. 9, 1878, item 2782, CIMC-CCICP, for Iron Cliffs' earlier adoption of the telephone.

122. Swineford, *Annual Review for 1880,* 12–13.

123. Kelly, "Address," 20.

124. F. P. Mills to Pound Manuf. Co., Aug. 17, 1888; F. P. Mills to J. Sellwood, Sept. 6, 1888; F. P. Mills to Vulcan Iron Works, Sept. 8 1888, item 2451, CIMC-CCICP.

125. D. H. Bacon to Jay C. Morse, Mar. 22, 1881, item 2778, CIMC-CCICP; F. P. Mills to Wm. G. Mather, Aug. 25, 30, 1890, item 2455, CIMC-CCICP.

126. See Jay C. Morse to D. H. Bacon, June 3, 1885, item 2791, CIMC-CCICP, for coal comparison. See F. P. Mills to Wm. G. Mather, July 16, 1888; F. P. Mills to A. D. Morse, July 16, 1888; and F. P. Mills to Reid Burner Co., Aug. 6, 1888, item 2451, CIMC-CCICP; and *Ishpeming Iron Ore,* June 16, 1888, for the petroleum tests.

127. F. P. Mills to B. F. Sturtevant, June 18, 1890, item 2455, CIMC-CCICP.

128. Samuel L. Mather to Jay C. Morse, Mar. 16, 1876, item 2779, CIMC-CCICP (drills); Mather to Morse, Aug. 2, 1880, item 2787, CIMC-CCICP (lubricants); Mather to Morse, July 26, 1881, item 2789, CIMC-CCICP (labor).

129. See *Marquette Mining Journal,* July 17 and 31, 1869, on the accident. Correspondence between Samuel L. Mather and Samuel Mather, in which the former compliments the latter on the quality of his information on company mining operations is in folder 1, container 1, Samuel Livingston Mather Family Papers, MS 3735, Western Reserve Historical Society.

130. Walter Havighurst, *Vein of Iron: The Pickands Mather Story* (Cleveland: World, 1958), 30–35.

131. Ibid., 15–47.

132. Ibid., 17–43.

133. Ibid., 35.

134. Timothy J. Loya, "William Gwinn Mather, the Man," *Inland Seas* 46 (1990); William M. Milliken, *A Time Remembered: A Cleveland Memoir* (Cleveland: Western Reserve Historical Society, 1975), 111–18, 140–44, 167–68; John N. Ingham, *Biographical Dictionary of American Business Leaders* (Westport, CT: Greenwood, 1983), 863–65.

135. Rheumatism is mentioned in Samuel L. Mather to D. H. Bacon, Feb. 19, 1884, and in Mather to Jay C. Morse, Aug. 6, 1884, folder 5, container 1, CIMC-WR. Problems with eyesight are mentioned in William G. Mather to Bacon c. June 28, 1886, folder 1, container 2, CIMC-WR.

136. Samuel L. Mather to D. H. Bacon, Aug. 19, 1886, folder 2, container 2, CIMC-WR. The gradual transfer of more and more responsibility by Samuel L. Mather to William G. Mather can be observed in the Cleveland Iron Mining Correspondence between 1884 and 1886.

137. See, for example, F. P. Mills Jr. to Wm. G. Mather, Oct. 20 1888 (service), item 2451, and W. Potter to Jay C. Morse, Sept. 26, 1881, item 2788, CIMC-CCICP.

138. A. Kidder to Wm. G. Mather, Jan. 28, Feb. 4, 11, 28, 1889, CCI-HF.

139. Wm. G. Mather to D. H. Bacon, Nov. 19, 1885, folder 6, and Apr. 27, 1886, folder 7, container 1, CIMC-WR.

140. Samuel L. Mather to D. H. Bacon, Sept. 4 (crazy) and Sept. 8 (wild), 1886, folder 2, container 2, CIMC-WR. See also, on problems getting vessels, Fred A. Morse to D. H. Bacon, July 10, 1886, and Wm. G. Mather to D. H. Bacon, Aug. 14 and 16, 1886, folder 1, container 2, CIMC-WR.

141. Wm. G. Mather to D. H. Bacon, Dec. 21, 1886, folder 3, container 2, CIMC-WR.

142. Wm. G. Mather to D. H. Bacon, Jan. 20, 1886, folder 7, container 1, CIMC-WR.

143. Samuel L. Mather to M. L. Hewitt, Sept. 26, 1888, Hewitt Letters, LRL; "Heavy Shipments and Profitable Business of the Owners," *New York Times,* Oct. 9, 1888.

144. Greenwood, *Fleets of Cleveland-Cliffs,* 2–3.

145. John Birkinbine, "Notes upon American Iron Ore Deposits," in Iron and Steel Institute (London), *The Iron and Steel Institute in America in 1890* (London: E. & F. N. Spon, 1890), 393.

146. *Congressional Globe,* 34th Cong., 1st sess., June 3, 1856, pt. 3, appendix, 11–12.

147. *American Railroad Journal,* Dec. 6, 1856.

148. Saul Benison, "Railroads, Land and Iron: A Phase in the Career of Lewis Henry Morgan," Ph.D. diss., Columbia University, 1953, 156–58.

149. Brooks, *Geological Survey,* 40–41; Benison, "Railroads, Land and Iron," 199–201.

150. *Congressional Globe,* 37th Cong., 2nd sess., 1861–62, 3049; *Lake Superior Miner,* Oct. 1, 1864; Benison, "Railroads, Land and Iron," 202.

151. Benison, "Railroads, Land and Iron," 215.

152. J. W. Foster, *The Geology and Metallurgy of the Iron Ores of Lake Superior, Being a Report Addressed to the Board of Directors of the Iron Cliffs Company* (New York, 1865); J. P. Kimball, *Notes on the Iron Ores of Marquette, Michigan: Extracts from a Report to the Iron Cliffs Mining Co., June 1, 1864* (Marquette: n.p., 1864), BENT; Benison, "Railroads, Land and Iron," 225–26.

153. Brooks, *Geological Survey,* 8–39.

154. Iron Cliffs Co., *Minute Book, 1864–80,* Mar. 1, 1866, CCI.

155. Brooks, *Geological Survey,* 31–32; Kenneth D. LaFayette, *Flaming Brands: Fifty Years of Iron Making in the Upper Peninsula of Michigan, 1848–1898* (Marquette: Northern Michigan University Press, 1990), 13, 14–15. For negotiations on leasing, see Iron Cliffs Co., *Minute Book, 1864–80,* Mar. 8, 1866, CCI.

156. Iron Cliffs Co., *Minute Book, 1864–80,* Sept. 28, 1866, and Jan. 24, 1867, CCI.

157. For talk of erecting a foundry, see Edward Kirk-Talcott to Mess. Platt & Thom, Sept. 21, 1866, and T. B. Brooks to W. N. Moon, Sept. 27, 1866, and Jan. 25, 1867, Iron Cliffs Letterbook, folder 1, box 1, RG 81-29, AM.

158. Iron Cliffs Co., *Minute Book, 1864–80,* Mar. 5, 1867, CCI.

159. "Financial Statement to Dec. 31, 1866 [Iron Cliffs]," folder 1, box 19, ICP-III.

160. E. S. Green to H. Benedict, Milwaukee, folder 3, box 39, ICP-III.

161. T. B. Brooks to Edmund H. Miller, Dec. 6, 1867, folder 11, box 4, ICP-I.

162. T. B. Brooks to Executive Committee, Mar. 19, 1867, folder 11, box 4, ICP-I.

163. For examples of the charcoal supply problem, see E. B. Isham to Wm. Barnum, Apr. 14 and 17, 1873, folder 4, box 40, ICP-III.

164. Lafayette, *Flaming Brands,* 50–51.

165. T. B. Brooks to Edmund H. Miller, Jan. 24, 1868, folder 11, box 4, ICP-I.

166. T. B. Brooks to Executive Committee, Dec. 5, 1866, box 1, folder 1, RG 81-29, AM.

167. T. B. Brooks to Edmund H. Miller, July 4, 1868; Brooks to Charles J. Canda, June 24, 1868; Brooks to Canda, July 4, 1868, folder 11, box 4, ICP-I.

168. T. J. Houston to Charles J. Canda, Mar. 6, 1877, "Historical Letter Box," CCI-Empire.

169. *Negaunee Mining News,* Mar. 26, 1868.

170. Charles J. Canda to E. B. Isham, Sept. 21 and Oct. 12, 1868, folder 5, box 21, ICP-III.

171. Gordon, *Landscape Transformed,* 63, 66, 83; Charles Rufus Harte, "Connecticut's Iron and Copper," *60th Annual Report of the Connecticut Society of Civil Engineers* (1944).

172. Charles J. Canda to E. B. Isham, Dec. 1, 1868, folder 5, box 21, ICP-III.

173. W. H. Barnum to E. B. Isham, Dec. 15, 1869, "Historical Letter Box," CCI-Empire.

174. Iron Cliffs Co., *Minute Book, 1864–80,* July 28, 1870, and Aug. 13, 1873, CCI.

175. E. B. Isham to Charles J. Canda, Nov. 6, 1873, folder 1, box 41, ICP-III; and Isham to Canda, Nov. 6, 1873, folder 6, box 21, ICP-III.

176. Iron Cliffs Co., *Minute Book, 1864–80,* Aug. 31 and Oct. 24, 1876, CCI.

177. Iron Cliffs Co., *Minute Book, 1864–80,* Jan. 25, 1879, CCI.

178. For the mix Houston was using, see T. J. Houston to Charles J. Canda, Dec. 4, 1875, folder 2, box 9, ICP-I. For Barnum's order to use only hard ores, see Houston to Canda, Mar. 6, 1876, folder 6, box 9, ICP-I; and Canda to Houston, Mar. 8, 1876, folder 2, box 29, ICP-III.

179. T. J. Houston to Charles J. Canda, Mar. 25 and 27, and Apr. 19 and 29, 1878, folder 1, box 11, ICP-I.

180. T. J. Houston to Wm. Barnum, Apr. 13, 1876, folder 1, box 10, ICP-I; James A. Root to Wm. Barnum, Mar. 3, 1876, folder 1, box 2, ICP-I. Brooks in 1868 had insinuated much the same regarding dealing with labor. For this, see Brooks to Charles J. Canda, June 24, 1868, folder 11, box 4, ICP-I.

181. *Ishpeming Iron Ore,* Feb. 15, 1890; Michigan, Commissioner of Mineral Statistics, *Mines and Mineral Statistics [1889]* (Lansing: Robert Smith, 1890), 127–28.

182. Michigan, Commissioner of Mineral Statistics, *Annual Report of the Commissioner of Mineral Statistics of the State of Michigan for 1882* (Lansing: W. G. George, 1883), 211.

183. Michigan, Commissioner of Mineral Statistics, *Mines and Mineral Statistics [1888],* by Charles D. Lawton (Lansing: Darius D. Thorp, 1889), 139–40; see also Michigan, Commissioner of Mineral Statistics, *Mines and Mineral Statistics [1887],* by Charles D. Lawton (Lansing: Thorp & Godfrey, 1888), 25, 29.

184. For the decision to close the company store, poor returns from exploration work, the Barnum's exhaustion, and the change in policy on leases, see "Annual Statement of the Iron Cliffs Company for the Year Ending November 30, 1887," with follow-on information in the 1888 and 1889 statements, CCI-HF. For the appointment of the committee to consider the condition of Iron Cliffs and possible sale, see Iron Cliffs Co., *Minute Book, 1880–89,* Apr. 15, 1889, CCI.

185. Samuel L. Mather to D. H. Bacon, Mar. 6, 1882, and Mar. 1, 1882, item 2790, CIMC-CCICP.

186. Samuel L. Mather to D. H. Bacon, June 7 and 14, 1886, folder 7, container 1, CIMC-WR. In the winter of 1886–87 CCI crews drilling through the ice of Lake Angeline, just south of Ishpeming, discovered large rich deposits under the lake bed; the exploitation of this find eliminated concerns about the imminent depletion of the company's ore deposits.

187. Samuel L. Mather to M. L. Hewitt, Jan. 12, 1889, Hewitt Letters, LRL. In the Lake Superior iron district second-class ores were originally those with less than 60 percent iron content, but later the term also referred to ores that did not have the proper physical structure or contained too much phosphorus.

188. Raphael Pumpelly, *Report on the Mining Industries of the United States (Exclusive of the Precious Metals), with Special Investigations into the Iron Resources of the Republic . . .* (Washington, DC: Government Printing Office, 1886), 75, table 23.

189. Wm. G. Mather to D. H. Bacon, Apr. 27, 1886, folder 7, container 1, CIMC-WR.

190. Samuel L. Mather to D. H. Bacon, June 14, 1886, folder 7, container 1, CIMC-WR.

191. Samuel L. Mather to D. H. Bacon, Oct. 13, 1886, folder 2, container 2, CIMC-WR.

192. Samuel L. Mather to D. H. Bacon, Oct. 13, 1886, folder 2, container 2, CIMC-WR; Mather to Bacon, Nov. 13 and 17, and Dec. 1, 1886; Mather to Bacon, Dec. 4, 1886, folder 3, container 2, CIMC-WR.

193. *Ishpeming Iron Ore,* Feb. 15 and 22, 1890; *Marquette Mining Journal,* Feb. 8, 1890. There are some miscellaneous documents relating to the acquisition of Iron Cliffs stock in item 2756 (a large Hollinger box of Miscellaneous Papers), CIMC-CCICP, in a packet labeled "Iron Cliffs Company. Misc. papers concerning the purchase of I.C. Co. stock with statements of legal and miscellaneous expense account purchase. 1890." For surprise at the deal and the low price paid for the stock, see Michigan, Commissioner of Mineral Statistics, *Mines and Mineral Statistics [1889],* 127. The commissioner also noted that there was a general feeling of satisfaction locally with the purchase, Iron Cliffs' policies on leasing being considered "illiberal."

194. Cleveland-Cliffs Iron Mining Co., *Minute Book, 1891–1906,* May 16, July 10 and 11, 1891, CCI.

195. Mussey, *Combination,* 98–99, and Michigan, Commissioner of Mineral Statistics, *Mines and Mineral Statistics [1891]* (Lansing: Robert Smith, 1892), 71, quote a circular to stockholders outlining the expected advantages from the merger.

196. Mussey, *Combination,* 99.

CHAPTER 3

1. Notice from Frank P. Mills, Agent, dated Oct. 10, 1890, in folder 560, container 63, WGMFP.

2. "Cottage Built with Care," *Marquette Mining Journal,* Aug. 11, 1975.

3. "The Ore Trade Dead-Lock," *New York Times,* Mar. 12, 1893. See also "Iron Mines Shutting Down," *New York Times,* July 2, 1893.

4. For a detailed account of the Panic of 1893 and its aftermath, see Douglas Steeples and David Whitten, *Democracy in Desperation: The Depression of 1893* (Westport, CT: Greenwood, 1998).

5. CCI, *Annual Report of the President of the Cleveland-Cliffs Iron Company,* 1893, CCI-HF. For much of the period covered by this chapter, CCI annual reports were not publicly available because the company's stock was not publicly traded and the number of stockholders was only around 200–300. Until the 1920s annual reports were short, unpaginated documents with the word *Private* printed at the top of the first page. Annual reports are usually dated sometime in the first half of the year following the financial year that they cover. My citations refer to the financial year, rather than to the date the report was prepared. Thus, Mather's report dated Apr. 13, 1894, I cite above as *Annual Report of the President of the Cleveland-Cliffs Iron Company,* 1893, since it covers the 1893 fiscal year. Subsequent reports will be cited as CCI, *Annual Report.*

6. William Kelly, "Address of President William Kelly," *PLSMI* 6 (1900): 15.

7. Ibid., 17.

8. Michigan, *Eleventh Annual Report of the Bureau of Labor and Industrial Statistics,* 1894, 426, 430–32. See also "Iron Mines Shutting Down," *New York Times,* July 2, 1893.

9. *Ishpeming Iron Ore,* Aug. 12 and Oct. 28, 1893; Jan. 20, 1894.

10. CCI, *Minute Book, 1891–1906,* CCI.

11. "Ore Mines to Shut Down," *New York Times,* June 29, 1893.

12. *Ishpeming Iron Ore,* Oct. 28, 1893.

13. CCI, *Annual Report,* 1894, CCI-HF.

14. Henry Raymond Mussey, *Combination in the Mining Industry: A Study of Concentration in Lake Superior Iron Ore Production* (New York: Columbia University Press, 1905), 128.

15. CCI, *Agents' Annual Reports and Statistics,* 1895, sec. 5, 10, item 2061, CIMC-CCICP.

16. Florence Peterson, *Strikes in the United States, 1880–1936,* Department of Labor Bulletin 651 (Washington, DC: Government Printing Office, 1938), 21.

17. *Ishpeming Iron Ore,* Nov. 29 and Dec. 20, 1890.

18. Michigan, *Thirteenth Annual Report of the Bureau of Labor and Industrial Statistics,* 1896, 356–57.

19. For the Pinkerton Agency bill and samples of the forms labor recruits signed, see "Strike Expense 1895" and "Strike 1895," in item 2756, CIMC-CCICP. See also F. P. Mills to Wm. G. Mather, Nov.

26, 1895, in "Strike 1895," ibid., for clear indications that the Pinkerton activities were coordinated by Cleveland-Cliffs.

20. For White's letter, see Peter White to John Rich, Aug. 26, 1895, part 34, Peter White Papers, microfilm, Peter White Library, Marquette.

21. Herbert J. Brinks, "Marquette Iron Range Strike, 1895," *Michigan History* 50 (1966), is the most comprehensive account of the strike. See also, however, "Marquette Iron Range Strike: Miners' History of the Strike," and "Marquette Iron Range Strike: Mine Owners' History of the Strike," in Michigan, *Thirteenth Annual Report of the Bureau of Labor*, 1896, 350–61, 361–63. A file containing a large number of letters from Frank Mills Jr., the company's mining superintendent, and Samuel Redfern, the company's land agent, is in the "1895 Strike" file, box 2, Boyum Collection, CCI-Empire. These document the company's difficulties in bringing in outside labor and the tenseness of the situation, especially in Negaunee, in early September. Finally, see Frank B. Lyon to W. S. Green, Adjutant General, Michigan National Guard, Sept. 30, 1895, in Michigan, Adjutant General, *Report of the Adjutant General of the State of Michigan for the Years 1895–1896* (Lansing: Robert Smith, 1896), 41–43.

22. "Marquette Iron Range Strike: Miners' History of the Strike," 360.

23. Michigan, *Thirteenth Annual Report of the Bureau of Labor and Industrial Statistics*, 1896, 258; Michigan, *Twenty-first Annual Report*, 1904, 112.

24. *Marquette Daily Mining Journal*, Aug. 9, 12, 21, 23, 24, 25, 27, and 30, 1897. See also John P. Beck, "Homegrown," *Michigan History* 79 (1995), on attempts by Marquette iron miners to organize the Northern Mineral Mineworkers Union in the late 1890s.

25. For cooperation between regional mining companies, see M. M. Duncan to Wm. G. Mather, Aug. 21, 1897, and copy of notice to employees signed by the mining superintendents of regional companies dated Aug. 21, 1897, both found in "Labor Unions" file, box 2, Boyum Collection, CCI-Empire.

26. Wm. G. Mather to M. M. Duncan, Feb. 6, 1899, "Labor Unions" file, box 2, Boyum Collection, CCI-Empire.

27. *Marquette Daily Mining Journal*, Apr. 3, 4, 5, 6, 10, 1899; Beck, "Homegrown," 24; Burton H. Boyum, "Cliffs Illustrated History," manuscript, c. 1986, 10-20–10-22, CCI-Empire. See also the correspondence between Duncan and Mather in March 1899 in "Labor Unions" file, box 2, Boyum Collection, CCI-Empire.

28. "A Crisis Impending," *Marquette Mining Journal*, Apr. 4, 1899; "Crisis This Morning," *Marquette Mining Journal*, Apr. 5, 1899.

29. "Lake Superior Iron Mines," *Iron Trade Review*, Aug. 17, 1899, 10, for efforts to organize the three ranges.

30. Gertrude G. Schroeder, *The Growth of Major Steel Companies, 1900–1950* (Baltimore: Johns Hopkins University Press, 1953), 36.

31. "The Carnegie-Oliver-Rockefeller Agreement," *Iron Age*, Feb. 18, 1897, 32–33; Richard B. Mancke, "Iron Ore and Steel: A Case Study of the Economic Causes and Consequences of Vertical Integration," *Journal of Industrial Economics* 20 (1972); Joseph Wall, *Andrew Carnegie* (New York: Oxford University Press, 1970), 587–615; and James Howard Bridge, *The Inside History of the Carnegie Steel Company* (New York: Aldine, 1903), 260–68. For the bid to buy Cleveland-Cliffs, see "Mining in Michigan," *New York Times*, Dec. 5, 1897.

32. "An Iron-Ore Combination," *New York Times*, Jan. 21, 1895; "The Iron Ore Combination," *New York Times*, Jan. 11, 1896; "The Iron-Ore Pool," *New York Times*, May 19, 1896; "The Ore Pool Must Go," *New York Times*, Feb. 12, 1897; and "Iron Ore Men Organize," *New York Times*, Mar. 27, 1897.

33. On the problems of the pool see, for example, "Lake Iron Ore Matters," *Iron Age*, Apr. 15, 1897, 15–16; and "Lake Iron Ore Matters," *Iron Age*, Apr. 22, 1897, 14–15; as well as Mancke, "Iron Ore and Steel," 223–24, 227–28.

34. "Lake Iron Ore Matters," *Iron Age*, Mar. 18, 1897, 9.

35. "Sale of Mesaba Mines," *New York Times*, Apr. 14, 1901; "The Course of the Iron Markets in 1896," *Engineering and Mining Journal*, Jan. 2, 1897, 12–13; "Lake Superior Iron Mines: Many Changes in Ownership under the Consolidations," *Iron Trade Review*, Nov. 29, 1900; "Background of Domestic Ore Holdings," in U.S. Federal Trade Commission, *Report of the Federal Trade Commission on the Con-*

trol of Iron Ore for the Antitrust Subcommittee of the Committee on the Judiciary, House of Representatives (Washington, DC: Government Printing Office, 1952).

36. For the broader merger mania in American industry in this period, see Naomi Lamoreaux, *The Great Merger Movement in American Business, 1895–1904* (Cambridge: Cambridge University Press, 1985).

37. Mancke, "Iron Ore and Steel," 220–21.

38. H. O. Young, *Supplementary Report on the Facts, by H. O. Young, a Member of the Special Committee to Investigate the United States Steel Corporation,* 61st Cong., House of Representatives, 1912, 15–20, outlines the background of U.S. Steel's acquisition of Oliver. See also Mussey, *Combination,* 117–25; and Schroeder, *Growth,* 37–38. Young, *Supplementary Report,* 80, gives a figure of around 50 percent of all ore. Edwin C. Eckel, "Iron Ore, Pig Iron, and Steel," in *Mineral Resources of the United States, 1906* (Washington, DC: Government Printing Office, 1907), 85, claimed that U.S. Steel controlled two-thirds to three-fourths of the iron ore of the Lake Superior iron region. Schroeder, *Growth,* says 46 percent.

39. "Steel Trust Buys Mines," *New York Times,* Apr. 5, 1901 ($8 million offer); "Sale of Mesaba Mines," *New York Times,* Apr. 14, 1901; and "Wall Street Topics," *New York Times,* Apr. 20, 1901, for the $9.5 million option.

40. U.S. Federal Trade Commission, *Report of the Federal Trade Commission,* 9.

41. For discussions of the turn-of-the-century consolidation of the steel industry and iron ore supplies, see Mussey, *Combination;* William T. Hogan, *Economic History of the Iron and Steel Industry in the United States* (Lexington, MA: Lexington Books, 1971), 1:235–302; Kenneth Warren, *The American Steel Industry, 1850–1970: A Geographical Interpretation* (Oxford: Clarendon, 1973), 104–7, 123; "Lake Superior Iron Mines"; and Schroeder, *Growth,* 125–30. The consolidation also impacted furnace companies, some of which were forced out of business since ore supplies were tied up by the larger steel companies. For an Upper Peninsula example, see Samuel Redfern to Wm. G. Mather, Dec. 7, 1899, folder 1, box 64, ICP-III. For an example of the buying mania, see "Buying Up Iron Mines," *New York Times,* July 30, 1899.

42. CCI, *Annual Report,* 1897, CCI-HF.

43. Kelly, "Address," 14.

44. Crowell & Murray, Inc., *The Iron Ores of Lake Superior,* 7th ed. (Cleveland: Penton, 1930), 85.

45. CCIC, *Annual Report,* 1898, CCI-HF.

46. William G. Mather, "Some Observations on the Principle of Benefit Funds and Their Place in the Lake Superior Iron Mining Industry," *PLSMI* 5 (1898): 10–12.

47. Frank L. Palmer, *Spies in Steel: An Expose of Industrial War* (Denver: Labor, 1928), accuses the Oliver Iron Mining Company of holding the Mesabi Range "in the deadly grip of a great fear." For a general account of methods used by industrialists to keep labor under control, see Rhodri Jeffreys-Jones, "Profits over Class: A Study in American Industrial Espionage," *Journal of American Studies* 6 (1972).

48. See Wm. G. Mather to M. M. Duncan, Apr. 3 and 4, 1899, "Labor Unions" file, box 2, Boyum Collection, CCI-Empire.

49. For an overview of the welfare capitalism movement nationally, the best sources are Stuart D. Brandes, *American Welfare Capitalism, 1880–1940* (Chicago: University of Chicago Press, 1976); and Andrea Tone, *The Business of Benevolence: Industrial Paternalism in Progressive America* (Ithaca, NY: Cornell University Press, 1997). For an overview of the movement in the Lake Superior region (mainly Minnesota), see Arnold R. Alanen, "Companies as Caretakers: Paternalism, Welfare Capitalism, and Immigrants in the Lake Superior Mining Region," in *A Century of European Migrations, 1830–1930,* ed. Rudolph J. Vecoli and Suzanne M. Sinke (Urbana: University of Illinois Press, 1991).

50. Mather, "Some Observations," 14; *Ishpeming Iron Ore,* Sept. 29, 1928. Mather traveled widely in Europe both in his youth and as an adult, often visiting mining districts. That European mining operations were somewhat in advance of American operations in paternalistic practices is clear from William P. Blake, "Provision for the Health and Comfort of Miners—Miners' Homes," *Transactions of the American Institute of Mining Engineers* 3 (1874–75). Mather could also have picked up some ideas on paternalism from the nearby Lake Superior copper district.

51. Mather outlined his arguments in Mather, "Some Observations," 10–20.

52. CCI, *Directors' Trip, September 1916,* 16, folder 124, container 18, WGMFP. The company employed slightly over 3,000 people in mining operations at this time.

53. W. H. Moulton, "Industrial Relations Work of the Cleveland-Cliffs Iron Company," in Cleveland-Cliffs Iron Co., *The Cleveland-Cliffs Iron Company and Its Extensive Operations in the Lake Superior District* (Cleveland: Cleveland-Cliffs, 1929), 48–49; Cleveland-Cliffs Iron Co., *The Cleveland-Cliffs Iron Company: An Historical Review of This Company's Development and Resources Issued in Commemoration of Its Seventieth Anniversary, 1850–1920* (Cleveland: Cleveland-Cliffs, 1920),61–62.

54. W. H. Moulton, "The Sociological Side of the Mining Industry," *PLSMI* 14 (1909): 97; "Swanzy District" and "The Cleveland-Cliffs Iron Company," in *Iron Mining and Agriculture,* souvenir edition of the *Marquette Mining Journal,* Mar. 1, 1913 (hereafter cited as *Iron Mining and Agriculture*); James E. Jopling, "A Brief History of the Cleveland-Cliffs Iron Company," *Michigan History* 5 (1921): 164–65; Harlan Hatcher, *A Century of Iron and Men* (Indianapolis: Bobbs-Merrill, 1950), 226–27; Warren H. Manning, "Villages and Homes for Working Men," *Western Architect* 16 (Aug. 1910); and Arnold R. Alanen, "The Settlements of Marquette County, Michigan," in *Historic Resources of the Iron Range in Marquette County, Michigan, 1844–1941,* ed. William Mulligan (Marquette: Economic Development Corporation of the County of Marquette, 1991), 21–25.

55. CCI, *Agents' Annual Reports and Statistics,* 1910, vol. 1, sec. 3, 93–94, item 2093, CIMC-CCICP; *Ishpeming Iron Ore,* June 19, 1915, and Nov. 15, 1917.

56. Moulton, "Sociological Side," 96; "The Cleveland-Cliffs Iron Company," in *Iron Mining and Agriculture; Cleveland-Cliffs Iron Company* (1920), 60, for the house and garden prizes; W. H. Moulton, "Sanitation for the Mine Locations," *PLSMI* 18 (1913): 40–41, for garbage pickup.

57. Moulton, "Sociological Side," 88; *Cleveland-Cliffs Iron Company* (1920), 60–61.

58. Wm. G. Mather to Austin Farrell, Mar. 13, 1899, *Agents' Annual Reports and Statistics,* 1899, sec. 5, item 2063, CIMC-CCICP.

59. Moulton, "Sociological Side," 91–92; CCI, Ore Mining Department, *Annual Report of the General Manager,* 1930, 535, item 2001, CIMC-CCICP.

60. CCI, Ore Mining Department, *Annual Report of General Manager,* 1918, 665, 736, item 1989, CIMC-CCICP.

61. Moulton, "Sociological Side," 92–93; "The Cleveland-Cliffs Iron Company," in *Iron Mining and Agriculture; Cleveland-Cliffs Iron Company* (1920), 59–60; Moulton, "Industrial Relations Work," 49–50.

62. CCI, *Agents' Annual Reports and Statistics,* Mine Agent, 1914, 594–95, item 2106, CIMC-CCICP; CCI, *Cleveland-Cliffs Iron Company* (1920), 60.

63. CCI, *Minute Book, 1906–1911,* Dec. 31, 1908, CCI; *Ishpeming Iron Ore,* May 22, 1909; Moulton, "Sociological Side," 94–96.

64. CCI, Ore Mining Department, *Annual Report of the General Manager,* 1938, 612–14, item 2009, CIMC-CCICP. See also CCI, *Minute Book, 1920–1931,* Jan. 4, 1922, CCI.

65. "Landscaping on an Iron Range: The Cleveland-Cliffs Iron Co. Believes in the Utility . . . ," *Explosives Engineer* 2 (Aug. 1924). For an example, see CCI, *Agents' Annual Reports and Statistics,* 1913, sec. 1, 82, 191, item 2103, CIMC-CCICP.

66. CCI, *Agents' Annual Reports and Statistics,* 1900, sec. 1, 57, item 2064, CIMC-CCICP; Moulton, "Sociological Side," 91, and "Sanitation for Mine Locations," 38; and John S. Mennie, "A New Change House at the Cliffs Shafts Mine," *PLSMI* 9 (1903).

67. See, for example, Mark Aldrich, *Safety First: Technology, Labor, and Business in the Building of American Work Safety, 1870–1939* (Baltimore: Johns Hopkins University Press, 1997), 2–5, 76.

68. J. Parke Channing, "Mine Accidents: Address of the Retiring President," *PLSMI* 3 (1895): 38, 39, for Michigan iron mines; Aldrich, *Safety First,* 300–301, for bituminous mining.

69. Aldrich, *Safety First,* 111, 166, and 275, for example.

70. The Solicitors Annual Report, in CCI, *Agents' Annual Reports and Statistics,* 1909, vol. 2, sec. 4, item 2092, CIMC-CCICP, indicates that the level of activity in terms of written opinions and reports from that unit doubled between 1904 and 1908. Although most of this activity involved property issues, some involved injury litigation. For examples of CCI's solicitor's involvement in influencing workman's compensation legislation, see CCI, *Agents' Annual Reports and Statistics,* 1911, sec. 1, 15, item 2098,

CIMC-CCICP; and, especially, "Stenographer's Minutes of Meeting of Heads of Departments, June 12, 1912," item 2584, CIMC-CCICP, which contains on pp. 54–60 the report of a Workman's Compensation Law Committee and a discussion of Michigan's law.

71. Wm. G. Mather to M. M. Duncan, Jan. 29, 1900, in CCI, *Agents' Annual Reports and Statistics,* 1899, sec. 1, item 2063, CIMC-CCICP.

72. For example, see the Master Mechanic's Report, CCI, *Agents' Annual Reports and Statistics,* 1907, vol. 2, sec. 3, 7, 8, 9, item 2087, CIMC-CCICP.

73. CCI, *Agents' Annual Reports and Statistics,* Mine Agent, 1914, 551–55, item 2106, CIMC-CCICP.

74. The company printed safety rule books in three languages: English, Finnish, and Italian. See CCI, *Agents' Annual Reports and Statistics,* 1913, sec. 1, 381, item 2103, CIMC-CCICP. For the ethnic breakdown of CCI's workforce in 1901, see CCI, *Agents' Annual Reports & Statistics,* 1901, vol. 1, sec. 4 1/2, 57, item 2067, CIMC-CCICP, which breaks down the 1810 employees of the company in November 1901 by nationality and mine. The leading ethnic groups were 24.7 percent Scandinavian, 21.9 percent Finnish, 19.2 percent English, and 10.6 percent Polish. Irish, French, and Italians composed between 5 and 8 percent of the workforce each. Americans were only 2.5 percent, and many of those were in the general office or laboratory.

75. Supervisor remuneration for safety performance was through a "safety bonus" system instituted in 1929. See CCI, Ore Mining Department, *Annual Report of the General Manager,* 1929, 2, item 2000, CIMC-CCICP; and William Conibear, "Development of Safety in the Iron Mines of Michigan," *PLSMI* 29 (1936): 180.

76. Aldrich, *Safety First,* 221–23.

77. "The Cleveland-Cliffs Company," in *Iron Mining and Agriculture;* "Will Conibear First Mine Safety Inspector," *Marquette Mining Journal,* June 29, 1950 (special Cleveland-Cliffs' centennial issue); William Conibear, "Safety Work at the Cleveland-Cliffs Iron Company," in *The Cleveland-Cliffs Iron Company and Its Extensive Operations,* 52–55).

78. CCI, *Agents' Annual Reports and Statistics,* 1913, sec. 1, 35, item 2103, CIMC-CCICP.

79. For fatality rate data, see CCI, Ore Mining Department, *Annual Report of the General Manager,* 1945, 491–92, item 2016, CIMC-CCICP. See also Conibear, "Development of Safety," 180–84. For a firsthand account of how CCI's safety program worked, see also Charles J. Stakel, *Memoirs of Charles J. Stakel: Cleveland-Cliffs Mining Engineer, Mine Superintendent, and Mining Manager* (Marquette: Marquette County Historical Society, 1994), 111–17.

80. William Conibear, "System of Safety Inspection of the Cleveland-Cliffs Iron Company," *PLSMI* 17 (1912): 110.

81. Harring is quoted in a letter sent from General Manager S. R. Elliott to Wm. G. Mather, Mar. 9, 1933, in CCI, Ore Mining Department, *Annual Report of the General Manager,* 1932, 3, item 2003, CIMC-CCICP. The year before Elliott had noted that CCI's accident record for the previous two years had been "cited frequently" by the Bureau of Mines as "the most outstanding safety record in the entire mining industry of the country" (CCI, Ore Mining Department, *Annual Report of the General Manager,* 1931, 1, item 2002, CIMC-CCICP).

82. S. R. Elliott to E. B. Greene, Feb. 13, 1940, in CCI, Ore Mining Department, *Annual Report of the General Manager,* 1939, 2, item 2010, CIMC-CCICP. See also E. M. Libby, "The Company Surgeon," *PLSMI* 15 (1910), for additional information on company physicians.

83. CCI, Ore Mining Department, *Annual Report of General Manager,* 1918, 729–30, item 1989, CIMC-CCICP; Hatcher, *Century,* 245; *Ishpeming Iron Ore,* June 30, 1917, and Sept. 28, 1918.

84. CCI, Ore Mining Department, *Annual Report of General Manager,* 1945, 629–30, item 2016, CIMC-CCICP.

85. F. C. Stanford, "Training of Workmen for Positions of Higher Responsibility," *Transactions of the American Institute of Mining Engineers* 59 (1918) (Stanford was CCI's chief electrician); CCI, *Agents' Annual Reports and Statistics,* Mine Agent, 1914, 622–36, item 2106, CIMC-CCICP; *Ishpeming Iron Ore,* July 31, 1915; "Negaunee, Michigan," *Engineering and Mining Journal,* Aug. 3, 1912, 230; "Final Report, First Conference: Upper Peninsula County Mine Inspectors, Upper Peninsula Company Safety Engineers, Lake Superior Mining Division Staff of the United States Bureau of Mines, Houghton, Michigan, April 23–25, 1919," 20–22, typescript, MTU.

86. For example, CCI, Ore Mining Department, *Annual Report of the General Manager,* 1921, 620, item 1992, CIMC-CCICP, which reported eighty-seven classes in English attended by 418 people at the clubhouse in CCI's model town of Gwinn. See also Moulton, "Sociological Side," 90, and *Cleveland-Cliffs Iron Company* (1920), 62.

87. The story of CCI's experiments with agriculture are told in detail in Terry S. Reynolds, "'Quite an Experiment': A Mining Company Attempts to Promote Agriculture on Michigan's Upper Peninsula, 1895–1915," *Agricultural History* 80 (2006).

88. CCI, *Agents' Annual Reports and Statistics,* Mine Agent, 1914 (pension dept. report), 616, item 2106, CIMC-CCICP.

89. "A Bond of Interest," *Harlow's Wooden Man* 13, no. 5 (1978): 30. See also *Marquette Mining Journal,* July 7, 1943. For Stakel's impressive role in a variety of community activities, as well as his use of company employees to secure election as Republic Township supervisor, see Stakel, *Memoirs.* The anonymous author of "A Bond of Interest" declared: "While all mining managers shared a deep-seated concern for welfare and the community, none gave of himself more than did Charles J. Stakel." Stuart R. Elliott, Stakel's predecessor as mining manager, was similarly active: see "S. R. Elliott, Retiring . . . ," *Marquette Mining Journal,* July 7, 1943.

90. For contemporary accounts, see Claude T. Rice, "Labor Conditions at Calumet and Hecla," *Engineering and Mining Journal,* Dec. 23, 1911, 1235–39; Rice, "Copper Mines at Lake Superior—III," *Engineering and Mining Journal,* Aug. 3, 1912, 217–21; and "The Upper Peninsula of Michigan," *Harper's New Monthly Magazine,* May 1882, 892–902. For a historian's recent synthesis, see Larry D. Lankton, *Cradle to Grave: Life, Work, and Death at the Lake Superior Copper Mines* (Oxford: Oxford University Press, 1991).

91. CCI's spectrum of paternalistic programs compares well with the much better known exemplars of American welfare capitalism described in Tone, *Business of Benevolence.*

92. See, for example, Palmer, *Spies in Steel,* for an account of the more heavy-handed Oliver operations.

93. Wm. G. Mather to M. M. Duncan, Dec. 23, 1907, and Duncan to Mather, Dec. 27, 1907, "Wages, Labor" file, box 2, Boyum Collection, CCI-Empire.

94. H. R. Harris, Austin Farrell, C. V. R. Townsend, and M. M. Duncan to Wm. G. Mather, July 12, 1915, "Wages, Labor" file, box 2, Boyum Collection, CCI-Empire.

95. CCI, *Agents' Annual Reports and Statistics,* Mine Agent, 1913, 91–92, item 2103, CIMC-CCICP; CCI, *Agents' Annual Reports and Statistics,* Mine Agent, 1914, 106 (Negaunee mine), item 2106, CIMC-CCICP.

96. CCI, *Annual Report,* 1897, CCI-HF.

97. William Gwinn Mather, *Diary,* 1899, entries for Feb. 7, 18; Apr. 20, 21; and May 18, 1899, folder 9, container 1, WGMFP. Reports the previous year that negotiations were pending for the acquisition of the mine by CCI were denied: see "Lake Superior Iron Mines," *Iron Trade Review,* Apr. 28, 1898, 10.

98. CCI, *Minute Book, 1891–1906,* May 1, 1899, Oct. 5, 1899, and Jan. 2, 1894, CCI. In keeping with the company's tradition of finding partners, the Webster mine was purchased in partnership with Pickands Mather (Jan. 2, 1900). These transactions are also summarized in CCI, *Annual Report,* 1899, CCI-HF. In 1925 CCI exchanged its quarter interest in the Lake Superior Iron Company for a tract of land adjoining its Maas mine (*Annual Report,* 1925).

99. CCI, *Minute Book, 1891–1906,* Jan. 16. Jan. 30, and Feb. 15, 1905, CCI, *Annual Report,* 1905, CCI-HF.

100. See, for example, Thomas Pellow to W. C. Boone, Jan. 12, 1905, folder 4, box 1, RG 81-29, Records of Department of State, AM. Pellow, the secretary of the Jackson Iron Company, informed Boone that there had been no market whatsoever for the company's ore the previous year.

101. *Marquette Mining Journal,* Mar. 9, 1905.

102. CCI, *Minute Book, 1891–1906,* Mar. 31, 1902, CCI; CCI, *Annual Report,* 1903, CCI-HF; "New Lease on Negaunee Mine," *New York Times,* Apr. 3, 1902.

103. CCI, *Annual Report,* 1912, CCI-HF. The ore deposits were "very deep," so the mines, especially the Athens, would not become productive until some years later. By 1919 the Athens was in operation, and CCI had extended its ownership to 52.4 percent (CCI, *Annual Report,* 1919, CCI-HF). See also Jopling, "Brief History," 165.

104. "Annual Report to Secretary of State's Office, Michigan, 1905," in CCI, *Agents' Annual Reports and Statistics,* 1905, vol. 3, sec. 2 (Reports to Outside Parties), item 2082, CIMC-CCICP.

105. CCI, *Annual Report,* 1907 and 1908, CCI-HF. See also *Ishpeming Iron Ore,* Jan. 5, 1907, for a description of the value of the diamond drill for exploration.

106. CCI, *Minute Book, 1891–1906,* Apr. 8, 1901, CCI; CCI, *Annual Report,* 1901, CCI-HF; Jopling, "Brief History," 163; and *Maas v. Longstorf; Lonstorf v. Maas et al.,* nos. 2132 and 2133, Circuit Court of Appeals, Sixth Circuit, Mar. 5, 1912 (Lexis 1193).

107. CCI, *Annual Report,* 1903, CCI-HF.

108. "Lake Superior Iron Ores," *Mines and Minerals* 31 (Dec. 1910).

109. This is mentioned in CCI, *Minute Book, 1920–1931,* Aug. 3, 1928, CCI.

110. For exploration efforts, see CCI, *Agents' Annual Reports and Statistics,* 1910, vol. 1, sec. 3, 14–23, item 2093; and CCI, *Agents' Annual Reports and Statistics,* 1913, sec. 1, 350–51, 355, 362, item 2103, CIMC-CCICP. For taking a lease on the mine, see CCI, *Annual Report,* 1915, CCI-HF. See also "Cleveland-Cliffs' Spies-Virgil Iron Mine Has Important Future," *Skillings' Mining Review,* July 28, 1923.

111. See CCI, *Minute Book, 1891–1906,* May 6, 1901, CCI; and CCI, *Annual Report,* 1901, CCI-HF, for the acquisition. See "Lake Superior Iron Ore Notes," *Mines and Minerals* 31 (July 1911), for the comment on the mine being one of the finest.

112. Jopling, "Brief History," 164.

113. CCI, *Agents' Annual Reports and Statistics,* 1902, vol. 1, sec. 1, 34–47 (quote from 37), item 2069, CIMC-CCICP; CCI, *Annual Report,* 1902, CCI-HF.

114. CCI, *Agents' Annual Reports and Statistics,* 1903, vol. 4, sec. 1, 37, 41, item 2076, CIMC-CCICP.

115. "Michigan Iron Mine Purchase," *New York Times,* July 20, 1902; CCI, *Annual Report,* 1905, CCI-HF.

116. Mine Department, "Principal Property & Equipment Added since 1917," folder 124, container 18, WGMFP.

117. See folder 124, container 18, WGMFP; U.S. Federal Trade Commission, *Report of the Federal Trade Commission,* 24.

118. CCI, *Annual Report,* 1919, CCI-HF.

119. CCI, *Annual Report,* 1926, CCI-HF. In 1927 CCI purchased the mine and changed its name to the Pontiac mine. See also CCI, *Minute Book, 1920–1931,* June 16, 1927, CCI.

120. CCI, *Annual Report,* 1929, CCI-HF. In 1929 also, CCI leased the Dean mine on the Mesabi and took over management of the Alexandria mine of the Donner Steel Company.

121. Kenneth Warren, *Bethlehem Steel: Builder and Arsenal of America* (Pittsburgh: University of Pittsburgh Press, 2008), 61, 62.

122. For later examples of joint ventures in nonferrous mining, see Robert H. Ramsey, *Men and Mines of Newmont: A Fifty-Year History* (New York: Octagon, 1973), 54, 57, 63–64, and elsewhere.

123. *Cleveland-Cliffs Iron Company* (1920), 29.

124. Alfred D. Chandler Jr., *The Visible Hand: The Managerial Revolution in American Business* (Cambridge, MA: Harvard University Press, Belknap Press, 1977), 503.

125. CCI, *Annual Report,* 1910, CCI-HF.

126. CCI, *Agents' Annual Reports and Statistics,* 1893, sec. 7 (Vessel Operating Statistics), item 2059; CCI, *Agents' Annual Reports and Statistics,* 1894, sec. 7 (Vessel Operating Statistics), item 2060, CIMC-CCICP.

127. CCI, *Annual Report,* 1899 and 1900, CCI-HF.

128. CCI, *Minute Book, 1891–1906,* Oct. 18, 1904, CCI; CCIC, *Annual Report,* 1905, CCI-HF.

129. CCI, *Directors' Trip, September 1916,* 25, folder 124, container 18, WGMFP.

130. CCI, *Minute Book, 1920–1931,* Dec. 7, 1923, CCI.

131. William D. Rees (Treasurer, Lake Superior Iron Co.) to Peter White, Dec. 18, 1895, "Papers December 1895," box 10, PWP, indicates that part of the disagreement was over the route planned for the railroad.

132. CCI, *Minute Book, 1891–1906,* Aug. 30 and Oct. 8, 1895, CCI. The quote is from CCI, *Annual Report,* 1895, CCI-HF. See also "Lake Iron Ore Matters," *Iron Age,* Apr. 22, 1897, 14–15.

133. CCI, *Annual Report,* 1896, CCI-HF.

134. CCI, *Annual Report,* 1915, CCI-HF. When the LS&I Rail*way* consolidated with CCI's wholly owned Munising, Marquette & Southeastern Railway Company in 1924 to form the LS&I Rail*road* Company, CCI's ownership stake rose to 75 percent (CCI, *Annual Report,* 1923, CCI-HF).

135. H. R. Harris, "Transporting Ore from Mine to Docks," in *The Cleveland-Cliffs Iron Company* (1929), 39–41.

136. For the discovery of the deposit, see Michigan, Commissioner of Mineral Statistics, *Mines and Mineral Statistics [1887],* 27. For accounts of how the deposit was exploited, see CCI, *Agents' Annual Reports and Statistics,* 1892, sec. 1, letter F. P. Mills to Wm. G. Mather, Jan. 16, 1893, unpaginated, and sec. 2, 3, item 2058, CIMC-CCICP; F. C. Stanford, "The Electrification of the Mines of the Cleveland-Cliffs Iron Company," *PLSMI* 19 (1914): 189; Kelly, "Address," 19; and *Ishpeming Iron Ore,* Sept. 12, 1891. Bela Gold, William S. Pierce, Gerhard Rosegger, and Mark Perlman, *Technological Progress and Industrial Leadership: The Growth of the U.S. Steel Industry, 1900–1970* (Lexington, MA: Lexington Books, 1984), 295–96, discuss early use of electric locomotives underground. See also N. Yaworski, O. E. Kiessling, C. H. Baxter, Lucien Eaton, and E. W. Davis, *Technology, Employment, and Output Per Man in Iron Mining* (Philadelphia: Works Projects Administration, 1940), 144.

137. CCI, *Agents' Annual Reports and Statistics,* 1894, 15–16, 19, item 2060, CIMC-CCICP.

138. CCI, *Agents' Annual Reports and Statistics,* 1900, vol. 2, 10, item 2066, CIMC-CCICP.

139. CCI, *Agents' Annual Reports and Statistics,* 1904, vol. 1, sec. 1, 6, item 2077; and CCI, *Agents' Annual Reports and Statistics,* 1905, vol. 3, sec. 1, 11, item 2082, CIMC-CCICP. CCI estimated the cost per ton for tramming using men and mules at $0.187 versus $0.047 using electric tramming. John F. Berteling, "Notes on Mining the North Palms Orebody," *PLSMI* 25 (1926): 62, noted that, in general, the introduction of technology could liquidate labor costs with less economic disturbance than cutting wages; see also Engineering and Mining Journal, *Handbook of Mining Details* (New York: McGraw-Hill, 1912), 4, 7.

140. "Executives of Cleveland-Cliffs Company," *Marquette Mining Journal,* June 29, 1950 (special CCI commemorative issue), suggests this and claims that in 1909 Mather "issued instructions for the underground electrification of the company properties" and in 1910 "proceeded with a program for the development of hydro-electric power."

141. CCI, *Agents' Annual Reports and Statistics,* 1909, vol. 2, sec. 2, 7, 9, 10, item 2092; CCI, *Agents' Annual Reports and Statistics,* 1910, vol. 3, sec. 1, 2, 3, 4, 8, 9, 12, 13, item 2095, CIMC-CCICP.

142. CCI, *Agents' Annual Reports and Statistics, 1910,* vol. 3, sec. 1, 13, item 2095, CIMC-CCICP.

143. Samuel Redfern to Wm. G. Mather, Jan. 5, 1902, folder 4, box 64, ICP-III. See also Redfern to Mather, Sept. 29, 1901, folder 5, box 50, ibid.

144. See CCI, *Agents' Annual Reports and Statistics,* 1908, vol. 3, sec. 2, 12–13, item 2090, CIMC-CCICP, for trip to look at best practices. See also CCI, *Minute Book, 1891–1906,* Apr. 14, 1911, CCI; and CCI, *Annual Report,* 1910 and 1911, CCI-HF. This was a million-dollar project, but instead of seeking a partner, perhaps because there was no obvious one available and because the company intended on using the bulk of the power for its own mines, CCI undertook this alone.

145. "The Scratch Pad," *Marquette Mining Journal,* Oct. 3, 1938.

146. CCI, *Directors' Trip, September 1916,* 6 (table), folder 124, container 18, WGMFP.

147. CCI, Ore Mining Department, *Annual Report of the General Manager,* 1918, 642, item 1989, CIMC-CCICP.

148. CCI, *Annual Report,* 1919, CCI-HF. Alexander N. Winchell, *Handbook of Mining in the Lake Superior Region* (Minneapolis: Byron & Learned, 1920), 66–69, has a brief description of CCI's early hydro-power system. See also CCI, *The Cleveland-Cliffs Iron Company* (1920), 33–34.

149. CCI, *Minute Book, 1920–1931,* Jan. 4, 1926, CCI.

150. "Cleveland-Cliffs Chemical 'Lab' Largest in District," *Marquette Mining Journal,* June 29, 1950 (CCI commemorative edition).

151. Jopling, "Brief History," 164.

152. CCI, *Agents' Annual Reports and Statistics,* 1907, vol. 2, sec. 3, 5, item 2087, CIMC-CCICP.

153. Winchell, *Handbook of Mining,* 69, 71–72, 75; Stakel, *Memoirs,* 57–62.

154. CCI, *Agents' Annual Reports and Statistics,* 1913, sec. 1, 77–78.

155. CCI, *Annual Report,* 1896, CCI-HF. The numerous advantages of the caving method known as top slicing in terms of reduced timber, increased safety, increased extraction of ore, and improved grade are outlined in Roland D. Parks, "Recent Developments in Methods of Mining in the Marquette Iron Range," *PLSMI* 26 (1928): 146. H. H. Stoek, "Marquette-Range Caving Methods," *Mines and Minerals* 30 (Nov. 1909): 193, noted that the caving system "now prevails generally" on the Marquette Range.

156. For general information see Gold et al., *Technological Progress,* 293–94. For CCI use, see *Ishpeming Iron Ore,* Mar. 6, 1915.

157. CCI, *Agents' Annual Reports and Statistics,* 1900, vol. 2, sec. 3, introductory letter, H. R. Harris to Wm. G. Mather, Jan. 17, 1901, item 2066, CIMC-CCICP.

158. Examples of this abound in the annual reports of the mining department for each mine in the period. A typical one would be that for the Lake mine in 1917: CCI, Ore Mining Department, *Annual Report of the General Manager,* 1917, 12–27, item 1988, CIMC-CCICP.

159. For the proportion of production by the two methods, see American Iron and Steel Association, *Annual Statistical Report,* 1940, 14–15. For accounts of the factors behind the replacement of the Bessemer process by the open-hearth steelmaking process, see Thomas J. Misa, *A Nation of Steel: The Making of Modern America, 1865–1925* (Baltimore: Johns Hopkins University Press, 1995), 74–83; Robert P. Rogers, *An Economic History of the American Steel Industry* (London: Rutledge, 2009), 17–21; Jeanne McHugh, *Alexander Holley and the Makers of Steel* (Baltimore: Johns Hopkins University Press, 1980), 274–315; and Hogan, *Economic History,* 2:402–12.

160. Stakel, *Memoirs,* 119–20.

161. Lucien Eaton, "Mining Hard Ore at Cliffs Shaft Mine," 12, in *The Cleveland-Cliffs Iron Company* (1929).

162. J. F. Wolff, "The Economics of Underground Mining," *Skillings' Mining Review,* Apr. 6, 1946.

163. CCI, *Annual Report,* 1895, CCI-HF. In *Annual Report,* 1896, Mather noted that the company made "slight if any" profit on its non-Bessemer grades of ore. Yet in 1893 77 percent of the company's mine output and 88 percent of its ore deliveries were non-Bessemer grades (*Annual Report,* 1897, "Comparison" and "Deliveries of Ore" tables). See also CCI, *Minute Book, 1891–1906,* Nov. 3, 1894, CCI, on the unprofitability of the company's "large bodies" of non-Bessemer ores in times of economic stress.

164. Wm. G. Mather to Samuel Redfern, Feb. 6 and 19, 1891, folder 7, box 8, Jackson Iron Co. Papers, RG 76-91, AM.

165. Samuel Redfern to W. L. Wetmore, May 16, 1892, folder 2, box 62, ICP-III.

166. Richard H. Schallenberg, "Evolution, Adaptation and Survival: The Very Slow Death of the American Charcoal Iron Industry," *Annals of Science* 32 (1975): 352–56.

167. Mather's growing interest in charcoal iron is reflected by a paper he delivered in 1903: William G. Mather, "Charcoal Iron Industry of the Upper Peninsula of Michigan," *PLSMI* 9 (1903).

168. "Austin Farrell," *PLSMI* 25 (1926).

169. James E. Jopling, "Charcoal Furnaces of Michigan," *PLSMI* 25 (1926): 251.

170. "Great Activity in the Iron Fields," *New York Times,* Dec. 15, 1895. "A New Double Furnace to Start," *New York Times,* Mar. 18, 1896, referred to it simply as "the largest in the country using charcoal."

171. CCI, *Agents' Annual Reports and Statistics,* 1896, 17, item 2062, CIMC-CCICP.

172. CCI, *Annual Report,* 1897, CCI-HF. The "Comparison" table in this report indicates that production of pig iron had risen from a bit over 5,000 tons in 1893 to nearly 37,000 tons in 1897, reflecting the new furnace at Gladstone.

173. For a description, see "Pioneer Furnace No. 2," *PLSMI* 9 (1903); and Cleveland-Cliffs Iron Co., *Description of the Exhibit of the Cleveland-Cliffs Iron Co. and Pioneer Iron Co. at the Louisiana Purchase Exposition . . . St. Louis, 1904* (n.p.: [1904?]). See also CCI, *Minute Book, 1891–1906,* Dec. 31, 1900, and Jan. 31, 1901, CCI; and CCI, *Annual Report,* 1901, CCI-HF.

174. The Carp went out of blast in 1908. Because of its poor condition, it was never put into operation again.

175. CCI, *Directors' Trip, September 1916,* 27, folder 124, container 18, WGMFP; CCI, *The Cleveland-Cliffs Iron Company* (1920), 58; *Cliffs Chemical Company v. Wisconsin Tax Commission,* Supreme Court of Wisconsin, June 20, 1927 (Lexis 299).

176. Samuel Redfern to Wm. G. Mather, May 30, 1891, folder 3, box 62, ICP-III. Later letters indicate continued problems with charcoal supply. See Redfern to Mather, Sept. 19, 1891; and Redfern to Mather, May 2, 1892, folder 2, box 62, ICP-III. Securing adequate supplies continued to be a concern. For just one example, see CCI, *Agents' Annual Reports and Statistics,* 1899, sec. 5, 15, 17, 19, item 2063, CIMC-CCICP.

177. CCI, *Minute Book, 1891–1906,* May 16, 1900, CCI; and CCI, *Annual Report,* 1900, CCI-HF. Charcoal iron furnaces required the company to control large amounts of timber. In 1904 alone the company's wood operations cleared 5,154 acres of land, with the projection being that when all three furnaces ran full-time the timber land required annually would be 8,200 acres. CCI, *Annual Report,* 1904, CCI-HF.

178. CCI, *Minute Book, 1891–1906,* May 29 and June 12, 1902, CCI.

179. U.S. Department of Commerce, Bureau of Corporations, *The Lumber Industry,* vol. 1, pt. 3 ("Landholdings of Large Timber Owners") (Washington, DC: Government Printing Office, 1914): 188–92, 242–43. John D. Black and L. C. Gray, "Land Settlement and Colonization in the Great Lakes States," USDA Bulletin 1295, Mar. 23, 1925, 13, mention an anonymous Michigan company that owned over 1.5 million acres of land, totaling 14.2 percent of the entire Upper Peninsula. That company was Cleveland-Cliffs.

180. CCI, *Agents' Annual Reports and Statistics,* 1910, vol. 2, sec. 3, 3, item 2094, CIMC-CCICP. For attempts to attract wood manufacturers, see, for example, Samuel Redfern to Wm. G. Mather, Jan. 4, 1902 (basket manufacture); Redfern to United States Leather Co., Jan. 6, 1902 (tannery); Redern to Amos Rosenberg & Son, Feb. 10, 19, 27, 1902 (hardwood flooring); Redfern to William Musgrave, Feb. 19, 1902 (tannery), folder 4, box 64, ICP-III; Redfern to Wm. Hanna, Sept. 23, 1902 (handle factory), folder 1, box 64, ICP-III; Redfern to Wm. G. Mather, Mar. 24, 1902 (tannery, pulp mill), folder 3, box 65, ICP-III.

181. CCI, *Minute Book, 1891–1906,* June 4, 1902, CCI; and "Proposed Large Pulp and Paper Mills," *New York Times,* May 8, 1902. In 1916 CCI owned only 12 percent of the common and 20 percent of the preferred stock. See CCI, *Directors' Trip, September 1916,* 27, folder 124, container 18, WGMFP.

182. CCI, *Minute Book, 1906–1911,* Oct. 1, 1907, CCI. By 1916 CCI owned 62 percent of the company's stock. See CCI, *Directors' Trip, September 1916,* 27, folder 124, container 18, WGMFP.

183. For information on plans to erect the sawmill, see Samuel Redfern to Marinette Iron Works Co., Apr. 5, 1894; Redfern to D. C. Perscott, Apr. 11, 1894; and Redfern to Austin Lovell, June 11, 1894, folder 4, box 62, ICP-III. See also CCI, *Agents' Annual Reports and Statistics,* 1908, vol. 1, 6, item 2088, CIMC-CCICP; Charles A. Symon, ed., *Alger County: A Centennial History, 1885–1985* (Munising, MI: Bayshore, 1986), 157; and CCI, *Annual Report,* 1919, CCI-HF. The sawmill burned in 1926, and the company did not rebuild it. By this time the company also operated a mill to manufacture railroad ties.

184. Michigan, *Twenty-first Annual Report of the Bureau of Labor and Industrial Statistics,* 1904, 121. A single coke-fired blast furnace in 1906 might easily produce 100,000 tons a year and large steel firms operated dozens of furnaces.

185. Peter Temin, *Iron and Steel in Nineteenth-Century America: An Economic Inquiry* (Cambridge, MA: MIT Press, 1964), 266–67.

186. Schallenberg, "Evolution, Adaptation and Survival," 355–56.

187. Lowthian Bell, "On the American Iron Trade and Its Progress during Sixteen Years," in Iron and Steel Institute (London), *The Iron and Steel Institute in America in 1890* (London: E. & F. N. Spon, 1890), 132.

188. F. B. Gaylord to J. C. Holt, Apr. 16 and Oct. 21, 1897, folder 2, box 69, ICP-III.

189. CCI, *Agents' Annual Reports and Statistics,* 1901, vol. 2, sec. 9, item 2068, CIMC-CCICP; CCI, *Annual Report,* 1904, CCI-HF.

190. "Stenographer's Minutes of Meeting of Heads of Departments, Cleveland, June 12, 1912," 27, 33, item 2584, CIMC-CCICP, for Farrell's remarks. See also "Summary of Net Profit from Operations" in CCI, *Agents' Annual Reports and Statistics,* 1912, sec. 2, item 2099, CIMC-CCICP.

191. CCI, *Annual Report,* 1914, CCI-HF.

192. CCI, *Minute Book, 1891–1906,* Apr. 2, 1900, and May 6 and July 2, 1901, CCI.

193. CCI, *Minute Book, 1891–1906,* Oct. 11, 1902, CCI; CCI, *Annual Report,* 1903, CCI-HF. The company's later purchase of equity in blast furnaces in the Cleveland area led to the sale of its lands at Toledo in 1925 (CCI, *Annual Report,* 1925, CCI-HF; and CCI, *Minute Book, 1920–1931,* Oct. 8, 1924, and Jan. 6, 1925, CCI).

194. CCI, *Minute Book, 1891–1906,* Oct. 4, 1909 (quote), Oct. 30, 1909, and Nov. 27, 1909, CCI. See also CCI, *Minute Book, 1920–1931,* Dec. 9, 1920. In 1943 CCI increased its ownership share to 50 percent (CCI, *Annual Report,* 1943, CCI-HF). See also *Cleveland-Cliffs Iron Company* (1920), 68.

195. CCI, *Annual Report,* 1917, CCI-HF; "Principal Property & Equipment Added since 1917," Coal Operations, folder 124, container 18, WGMFP; *Cleveland-Cliffs Iron Company* (1920), 65, 68.

196. CCI, *Annual Report,* 1922, CCI-HF. By 1937 the company owned coal docks in Green Bay, Escanaba, Port Huron, and Duluth (*Annual Report,* 1937).

197. A. D. Carlton, "Furnishing Coal for an Iron ore Company," in CCI, *The Cleveland-Cliffs Iron Company* (1929), 56–57; also *McMorran v. Cleveland-Cliffs Iron Co.,* Supreme Court of Michigan, docket no. 78, Jan. 7, 1931 (Lexis 726), and *Cleveland-Cliffs Iron Company v. Department of Revenue,* docket no. 31, Supreme Court of Michigan, Dec. 6, 1950, 7–8 (Lexis 303).

198. CCI, *Minute Book, 1891–1906,* Aug. 31, 1910, Sept. 30, 1910, CCI; CCI, *Annual Report,* 1915, CCI-HF; CCI, *Directors' Trip, September 1916,* 27, folder 124, container 18, WGMFP.

199. "New 600-Ton Blast Furnace Plant," *Blast Furnace and Steel Plant,* 9 (Oct. 1921); and Tom M. Girdler, *Boot Straps: The Autobiography of Tom M. Girdler* (New York: Charles Scribner, 1943), 197.

200. CCI, *Minute Book, 1920–1931,* May 17 and Aug. 22, 1921, CCI. CCI originally subscribed to $4 million of the company's preferred stock and $250,000 of its common stock, while Trumbull subscribed to only $250,000 of common stock, giving both 50 percent of voting control. Trumbull later upped its investment in preferred stock. See also CCI, *Annual Report,* 1920, CCI-HF; CCI, *Minute Book, 1920–1931,* Feb. 18, 1925, CCI; and *Cleveland-Cliffs Iron Company* (1920), 71.

201. CCI, *Minute Book, 1920–1931,* May 11, 1923, CCI; CCI, *Annual Report,* 1923, CCI-HF.

202. CCI, *Minute Book, 1920–1931,* June 20, 1923, CCI.

203. See a series of letters from C. G. Heer to Wm. G. Mather between 1924 and 1928 in folders 55 and 56, container 7, WGMFP. See especially Heer to Mather, Apr. 6, 1928, folder 56: "I can tell you now . . . that the foundry business is a dirty business." For interest in securing a foothold in the Chicago area see also CCI, *Minute Book, 1920–1931,* June 20, 1923, CCI.

204. CCI, *Minute Book, 1920–1931,* Jan. 4, Aug. 19 and 25, 1926, CCI.

205. CCI, *Annual Report,* 1929, CCI-HF.

206. CCI, *Annual Report,* 1932, CCI-HF.

207. CCI, *Annual Report,* 1917, CCI-HF.

208. Stakel, *Memoirs,* 69.

209. CCI, Ore Mining Department, *Annual Report of the General Manager,* 1917, introductory letter, M. M. Duncan to Wm. G. Mather, Jan. 1, 1918, item 1988; CCI, Ore Mining Department, *Annual Report of the General Manager,* 1918, 739, 756, item 1989, CIMC-CCICP.

210. H. R. Harris, Austin Farrell, C. V. R. Townsend, and M. M. Duncan to Wm. G. Mather, July 12, 1915, "Wages, Labor" file, box 2, Boyum Collection, CCI-Empire.

211. For example, CCI, Ore Mining Department, *Annual Report of the General Manager,* 1917, 105, 136, 191, 312, item 1988, CIMC-CCICP.

212. Ibid., 92, 181.

213. Stackel, *Memoirs,* 69–73.

214. R. W. Jenner, "Cliffs Dow Chemical Company: A Narrative of Background, Formation and Operation," 5, folder 5, R. Wesley Jenner Papers, MS 8, Northern Michigan University Archives, Marquette. Jenner was president of Cliffs-Dow from 1960 to 1968 and had worked at the plant since the late 1930s.

215. CCI, *Annual Report,* 1915, 1916, 1919, CCI-HF.

216. CCI, *Annual Report,* 1924, CCI-HF.

217. CCI, *Annual Report,* 1921, 1923, CCI-HF. See also CCI, *Minute Book, 1920–1931,* May 17, 1921,

CCI (for price cuts, mine and furnace shutdowns, and the need for "drastic measures" "both by reduction of output and wages"); and CCI, Ore Mining Department, *Annual Report of the General Manager,* 1921, opening letter, 70, item 1992, CIMC-CCICP.

218. CCI, *Annual Report,* 1921, CCI-HF.

219. Ibid.

220. CCI, *Annual Report,* 1922, CCI-HF.

221. CCI, Annual *Report,* 1923, CCI-HF.

222. CCI, Annual *Report,* 1925, CCI-HF.

223. F. L. Childs pointed out these problems as early as 1911. See CCI, *Agents' Annual Reports and Statistics,* 1911, sec. 3, 2, 4, 11, 28, 34, 36, item 2098, CIMC-CCICP. For additional descriptions of the company's problems with freight rates in its wood chemicals operation, see CCI, *Agents' Annual Reports and Statistics,* 1910, vol. 3, sec. 4, 2, 22, 24, 42–43, 45, item 2095, CIMC-CCICP. Companies like General Chemical and Grasselli Chemical radically cut prices on acetic acid and worked together to try to deny CCI entry to the field when it began to produce that product. See CCI, *Agents' Annual Reports and Statistics,* 1910, vol. 3, sec. 4, 22, 30.

224. R. W. Jenner, *Hardwood Distillation in Michigan: Background for Cliffs-Dow Chemical Company* (Marquette: Cliffs-Dow, n.d.), 1.

225. CCI, *Annual Report,* 1923, CCI-HF.

226. CCI, *Minute Book, 1920–1931,* Sept. 16, 1925, CCI.

227. CCI, Ore Mining Department, *Annual Report of the General Manager,* 1925, 79, 87, 418, 420–21, item 1996, CIMC-CCICP; CCI, *Minute Book, 1920–1931,* Nov. 17, 1926, CCI. The Nov. 17 entry in the *Minute Book* simply notes that reports were made to the board about the accident, but it does not reveal the contents of the reports. The Dec. 15, 1926, entry provides a bit more detail on general manager of mining Elliott's report that the work of reclaiming the mine was too hazardous to the life of employees and should be abandoned. Elliott also recommended, besides surrendering the leases on the Barnes-Hecker and adjacent Morris mines, that the company make payments to the families and relatives of those lost beyond the provisions of the Michigan Compensation Law, a recommendation that the company adopted. See Steckel, *Memoirs,* 99–110, on the accident. See also Robert M. Neil, "The Barnes-Hecker, 1926: Michigan's Worst Mine Disaster," *Mining History Journal* 15 (2008); and Thomas G. Friggens, *No Tears in Heaven: The 1926 Barnes-Hecker Mine Disaster,* 2nd ed. (Lansing: Michigan History Magazine, 2002).

228. CCI, *Annual Report,* 1928, CCI-HF; CCI, *Minute Book, 1920–1931,* Oct. 9, 1928, CCI. This cost the company over $700,000 for acquiring surface lands and moving houses.

229. By 1925, for example, 99 percent of the miners at the Cliffs Shaft mine were American Citizens (CCI, Ore Mining Department, *Annual Report of the General Manager,* 1925, 30, item 1997, CIMC-CCICP). By the early 1940s, 75 to 85 percent of the mining workforce was American-born, even if of foreign ethnicity. See, for example, CCI, Ore Mining Department, *Annual Report of the General Manager,* 1942, 127, 229, 303, item 2013, CIMC-CCICP.

230. S. R. Elliott, "Athens System of Mining," *Transactions of the American Institute of Mining Engineers* 66 (1922).

231. S. R. Elliott, J. E. Jopling, R. J. Cenneour, and E. L. Derby, "Mining Methods of Marquette District, Michigan," *Transactions of the American Institute of Mining Engineers* 72 (1925): 133; CCI, Ore Mining Department, *Annual Report of the General Manager,* 1925, 106, 108, item 1996, CIMC-CCICP.

232. CCI, Ore Mining Department, *Annual Report of the General Manager,* 1936, 127, item 2007, CIMC-CCICP.

233. Ford E. Boyd, "Slushing Practice in the Mines of the Oliver Iron Mining Company," *PLSMI* 23 (1923): 68.

234. For the experiment with the underground steam shovel, see CCI, Ore Mining Department, *Annual Report of the General Manager,* 1917, 32, item 1988, CIMC-CCICP.

235. Ibid., for early CCI interest in the device. An early version of the air-powered scraper tested at the Negaunee mine had insufficient power. CCI technical personnel suggested several possible improvements in the course of experiments. See also Gold et al., *Technological Change,* 294–95; and Boyd,

"Slushing Practice," 69, for early attempts to use the technology and a brief comment on the resistance of labor to the technology.

236. Harold Barger and Sam H. Schurr, *The Mining Industries, 1899–1939: A Study of Output, Employment and Productivity* (1944; repr., New York: Arno, 1972), 124, 126–27; Yaworski et al., *Technology, Employment, and Output,* 152–53.

237. Robert S. Lewis, *Elements of Mining* (New York: John Wiley, 1933), 279, 317.

238. CCI, Ore Mining Department, *Annual Report of the General Manager,* 1925, opening letter (unpaginated) and 107, 141–42 (Morris-Lloyd mine), 176, 179–80 (Barnes-Hecker), 203, item 1996, CIMC-CCICP. For later reports of the increases in productivity due to scrapers, see Lucien Eaton, "Present Scraping Practices at the Cliffs Shaft Mine," *PLSMI* 24 (1926); and CCI, Ore Mining Department, *Annual Report of the General Manager,* 1934, 48, item 2005, CIMC-CCICP.

239. Elliott et al., "Mining Methods," 135.

240. *Marquette Mining Journal,* June 17, 1929.

241. Ellzey Hayden and Lucien Eaton, "Building Reinforced-Concrete Shaft Houses," *Transactions of the American Institute of Mining Engineers* 66 (1922); T. L. Condren, "Shaft House of Imposing Architecture at the Ishpeming (Michigan) Mine of the Cleveland Cliffs Iron Company," *Journal of the Western Society of Engineers* 26 (1921); and Arthur T. North, "Engineers and Architects in Artistic Collaboration," *American Architect,* Dec. 15, 1920.

242. Schroeder, *Growth,* 98; L. E. Ivens, "Market Results of Mergers, as Effected and Proposed," *Mining Congress Journal* 15 (Oct. 1929).

243. Dwight E. Woodbridge, "Lake Superior Iron-Ore Mining Improved," *Engineering and Mining Journal,* Jan. 16, 1926, 87.

244. According to Victor P. Geffine, a vice president of Cleveland Cliffs, "[F]rom 1915 to 1930 was this jockeying around for position to keep ourselves in the iron ore business. We didn't want to go in the steel business as such. Keep ourselves in the iron ore business but in a favorable position." See Geffine testimony in "Transcript of Testimony," Marquette County, Michigan, Circuit Court, "Cleveland Cliffs Iron Company vs. State of Michigan, Department of Revenue," Mar. 3, 1949, 36, folder 1, box 2, Attorney General Case Files, RG 81-27, AM.

245. See the William Gwinn Mather Family Papers, folder 378, container 47, WGMFP, for correspondence relating to steel company stock purchases and mergers. See also Mather's diary, entry for Nov. 3, 1925, in folder 14, container 2, WGMFP, for conversations with Eaton and discussions about steel mergers.

246. For example, see the analysis of CCI by Otis & Co. in 1927: "Cleveland Cliffs Iron Company," June 14, 1927 ("The management and people connected with the company are of the highest character in every way"), folder 220, container 28, WGMFP. See "New Steel Merger Plan," *New York Times,* Dec. 22, 1921, for an indication of CCI involvement in merger talks with a number of smaller steel companies. Henry Ford, attempting to vertically integrate his operations back into raw materials, may have hoped to buy out Cleveland Cliffs. See C. G. Heer to Wm. G. Mather, Feb. 16, 1924, folder 55, container 7, WGMFP; "Deny Ford Purchase of Cleveland Cliffs Co.," *Cleveland Press,* Jan. 25, 1924; and "False Rumor concerning Cleveland-Cliffs," *Skillings' Mining Review,* Feb. 2, 1924.

247. For some of the complexities of the consolidation movement in which Eaton and Mather participated, see "The Three Mathers Gravely Play Chess with Steel Men," *Business Week,* Apr. 9, 1930; and "Iron Ore—Making and Remaking the Map of Steel," *Fortune,* May 1931.

248. For a 1925 meeting, see Nov. 3, 1925, entry in Mather's diary, folder 14, container 2, WGMFP. Correspondence and documents related to the Republic Iron syndicate can also be found in folder 378, container 47, WGMFP. For discussion of the combination, see Cyrus Eaton, memorandum on a meeting with Mather, Dec. 29, 1929, folder 2094, container 89, CEP.

249. For the general background of Eaton's involvement in attempts to create a midwestern steel giant, see Marcus Gleisser, *The World of Cyrus Eaton* (New York: A. S. Barnes, 1965), 45–69; Schroeder, *Growth,* 52–53, 98–99, 133; Girdler, *Boot Straps,* 189–205; and "Catalyst in Steel [Cyrus Eaton]," *Time,* Dec. 30, 1929. The testimony by various parties in *Jensen et al. v. Republic Steel Corp.,* 426867, Common Pleas Court of Ohio, 1940 (Lexis 456), reviews the role that Eaton and others hoped Republic would play in planned steel company mergers.

250. Wm. G. Mather to Cyrus S. Eaton, May 12, 1927, folder 1863, container 78, CEP.

251. Wm. G. Mather to Cyrus S. Eaton, Apr. 6, 1929, folder 1845, container 77, CEP.

252. A copy of the agreement between Mather and the CCI Board and Eaton, dated Apr. 18, 1929, is in folder 1845, container 77, CEP.

253. Details of just how solid were the provisions made to keep CCI intact and autonomous within Cliffs Corporation are outlined in [?] to Joseph L. Weiner, Nov. 5, 1946, folder 2092, container 89, CEP. See also Cyrus S. Eaton to Wm. G. Mather, Apr. 18, 1929; and Mather to Eaton, Mar. 22, 1929, folder 1845, container 77, CEP.

254. [?] to Joseph L. Weiner, Nov. 5, 1946, attached memorandum, folder 2092, container 89, CEP, lists four different ways, including the use of a voting trust to vote CCI shares in a block, used by Mather "to continue the unquestioned control of the then management of Cleveland-Cliffs."

255. For the negotiations and elements of the agreement, see Wm. G. Mather to Cyrus S. Eaton, Mar. 22, 1929, folder 1845, container 77, CEP (with "plan discussed" attached); Eaton to Mather, Mar. 25, 1929, and Apr. 18, 1929, CEP; Wm. P. Belden to Eaton, Apr. 18, 1929, CEP; CCI, *Annual Report,* 1929, CCI-HF; Cliffs Corporation, *Annual Report,* 1929; and CCI, *Minute Book, 1920–1931,* Apr. 18 and June 4, 1929, CCI. See also "Iron Ore," 89–92, and an undated memo by Mather covering an interview with Cyrus Eaton on the purchase of Republic Iron and Steel Company stock, folder 378, container 47, WGMFP. The final plan is outlined in detail in a letter to the stockholders of Cleveland-Cliffs from Mather dated May 4, 1929, and found in folder 1845, container 77, CEP. See also Mather to Eaton, May 18, 1929, folder 1845, container 77, CEP, for details of the arrangement. Finally, see "Big Steel Merger Seen in Eaton Move," *New York Times,* May 6, 1929.

256. Geffine, a vice president of Cleveland Cliffs, testified in 1949 that Cleveland-Cliffs had purchased Corrigan, McKinney with "no idea of going into the steel business." "Our thought in buying them," he said, "was that Cleveland Cliffs would take the mines and sell to the new Mid-West Steel the furnaces and the steel plant. That was our plan." See Geffine testimony in "Transcript of Testimony," Marquette County, Michigan, Circuit Court, "Cleveland Cliffs Iron Company vs. State of Michigan, Department of Revenue," Mar. 3, 1949, 37, folder 1, box 2, Attorney General Case Files, RG 81-27, AM.

257. CCI, *Minute Book, 1920–1931,* Mar. 18, 1930, CCI. See also a Cyrus Eaton memo of June 29, 1955, regarding his recollections of the reasoning behind CCI's purchase of Corrigan, McKinney (folder 1809, container 74, CEP). Technically CCI owned all the common stock of the McKinney Steel Holding Company, which in turn owned 55.02 percent of the common stock and 20.14 percent of the preferred stock of the Corrigan, McKinney Steel Company.

CHAPTER 4

1. Numerous works provide detailed accounts of the Great Depression; among them are John Kenneth Galbraith, *The Great Crash, 1929* (Boston: Houghton Mifflin, 1955) and David M. Kennedy, *Freedom From Fear: The American People in the Great Depression* (New York: Oxford University Press, 2004).

2. CCI, Annual Report, 1932, CCI-HF.

3. CCI, Annual Report, 1931, CCI-HF; CCI, *Minute Book, 1932–36,* May 5, 1932, CCI.

4. CCI, Ore Mining Department, *Annual Report of the General Manager,* 1932, 1 (S. R. Elliott to Wm. G. Mather, Mar. 9, 1933), item 2003, CIMC-CCICP; CCI, *Minute Book,* 1932–36, May 5 and 16, 1932, CCI. See also CCIC, Ore Mining Department, *Annual Report of the General Manager,* 1933, 48, item 2004; and CCIC, Ore Mining Department, *Annual Report of the General Manager,* 1934, 91, item 2005, CIMC-CCICP.

5. CCI, Ore Mining Department, *Annual Report of the General Manager,* 1933, 37, item 2004, CIMC-CCICP.

6. CCI, *Annual Report,* 1933, CCI-HF; CCI, *Minute Book, 1932–1936,* Feb. 28, 1933, CCI.

7. CCI, Ore Mining Department, *Annual Report of the General Manager,* 1932, 265, item 2003; CCIC, Ore Mining Department, *Annual Report of the General Manager,* 1933, 278, item 2004, CIMC-CCICP.

8. CCI, Ore Mining Department, *Annual Report of the General Manager,* 1935, 43, item 2006, CIMC-CCICP.

9. CCI, Ore Mining Department, *Annual Report of the General Manager,* 1936, 36, item 2007, CIMC-CCICP

10. CCI, *Annual Report,* 1932, CCI-HF.

11. CCI, Ore Mining Department, *Annual Report of the General Manager,* 1934, 1 (S. R. Elliott to E. B. Greene, Jan. 1, 1935), item 2005, CIMC-CCICP.

12. Ibid., 96.

13. Ibid., 96, 216.

14. *Marquette Mining Journal,* Feb. 11, 12, 1931.

15. CCI, Ore Mining Department, *Annual Report of the General Manager,* 1933, 1 (S. R. Elliott to Wm. G. Mather, Mar. 7, 1934), item 2004, CIMC-CCICP.

16. CCI, Ore Mining Department, *Annual Report of the General Manager,* 1930, 1, item 2001, CIMC-CCICP.

17. CCI, Ore Mining Department, *Annual Report of the General Manager,* 1935, 269–70, item 2006, CIMC-CCICP.

18. CCI, Ore Mining Department, *Annual Report of the General Manager,* 1934, 96, item 2005; CCI, Ore Mining Department, *Annual Report of the General Manager,* 1935, 105, item 2006, CIMC-CCICP.

19. CCI, Ore Mining Department, *Annual Report of the General Manager,* 1932, 1–2 (S. R. Elliott to Wm. G. Mather, Mar. 9, 1933), item 2003, CIMC-CCICP. See also p. 33, where the Cliffs-Shaft mine superintendents also reported that the garden and wood operations were "one of the most successful and worthwhile projects ever undertaken by this Company." See also CCI, Ore Mining Department, *Annual Report of the General Manager,* 1933, 2, item 2004, CIMC-CCICP; and CCI, Land Department, *Annual Report,* 1932, 124, item 2047, CIMC-CCICP.

20. CCI, Land Department, *Annual Report,* 1932, 116, item 2047, CIMC-CCICP.

21. CCI, *Annual Report,* 1945, 4, 9, CCI-HF.

22. CCI, *Minute Book, 1932–1936,* Dec. 16, 1932, CCI.

23. CCI, Ore Mining Department, *Annual Report of the General Manager,* 1939, 572, item 2010, CIMC-CCICP.

24. CCI, *Minute Book, 1937–1945,* July 16, 1940, CCI.

25. CCI, Ore Mining Department, *Annual Report of the General Manager,* 1935, 49, 153, item 2006, CIMC-CCICP.

26. CCI, Ore Mining Department, *Annual Report of the General Manager,* 1934, 36, item 2005, CIMC-CCICP.

27. For example, CCI, Ore Mining Department, *Annual Report of the General Manager,* 1936, 47, item 2007, CIMC-CCICP.

28. Robert Joseph Goodman, "Ishpeming, Michigan: A Functional Study of a Mining Community," Ph.D. diss., Northwestern University, 1948, 113.

29. CCI, Ore Mining Department, *Annual Report of the General Manager,* 1932, 380, 379, item 2003, CIMC-CCICP.

30. CCI, *Minute Book, 1920–1931,* Sept. 2, 1930, CCI. For the company's losses in 1929, see Dec. 29, 1929.

31. CCI, *Minute Book, 1920–1931,* Oct. 2, 1931, CCI.

32. H. A. Raymond to E. B. Greene, Jan. 4, 1934, folder 2, R. Wesley Jenner Papers, MS 8, Northern Michigan University Archives, Marquette, summarizing parties, including General Chemical, that had expressed interest in purchasing the plant.

33. CCI, *Annual Report,* 1935, CCI-HF; CCI, *Minute Book, 1932–1936,* Dec. 3 and 21, 1934; Feb. 26 and May 28, 1935, CCI. See also *Marquette Mining Journal,* May 25, 1935.

34. See, for example, CCI, *Minute Book, 1932–1936,* Oct. 21 and Dec. 1, 1936, and CCI, *Minute Book, 1937–1945,* June 17, 1937, CCI.

35. CCI, *Minute Book, 1932–1936,* Dec. 18, 1936, CCI.

36. Ibid., May 28 and Aug. 5, 1935, for CCI ownership percentage. For continuing losses, see CCI, *Minute Book, 1937–1945,* Sept. 6, 1940, CCI, as well as the records of Piqua Munising Wood Products in folders 1–4, box 43, MSS 76–90, CCI-LDP.

37. CCI, *Minute Book, 1932–1936,* May 28, 1935, CCI; CCI, *Annual Report,* 1940, CCI-HF.

38. CCI, *Minute Book, 1937–1945,* June 21, 1944, CCI.

39. For discussions of dismantling the lumbering operations, see CCI, Land Department, *Annual Report,* 1938, 91–121, item 2053, CIMC-CCICP. By 1939 all the wood going into Cliffs-Dow was provided by jobbers. See CCI, Land Department, *Annual Report,* 1939, item 2054, 4.

40. CCI, *Annual Report,* 1932, CCI-HF; see also CCI, *Minute Book, 1932–1936,* Oct. 20 and Nov. 10, 1932, CCI.

41. CCI, *Annual Report,* 1933, CCI-HF; CCI, *Minute Book, 1920–1931,* July 29, 1930, CCI (on potential sale of Carpenter-Neely lease to Oliver). For other lease surrenders (Dean mine and Itasca mine on the Mesabi), see CCI, *Minute Book, 1932–1936,* Nov. 28, 1932, CCI.

42. American Iron Mining Co., *Minute Book,* Dec. 30, 1921, and June 26, 1936, item 1853, CIMC-CCICP.

43. CCI, *Minute Book, 1937–1945,* Sept. 20 and Oct. 4, 1940, CCI; CCI, Ore Mining Department, *Report of the General Manager,* 1936, 424, 457, item 2007, CIMC-CCICP.

44. CCI, *Minute Book, 1920–1931,* Sept. 30, 1926, CCI. Ford would not purchase land unless mineral rights went with it, and in this case Mather recommended making the sale anyway since he did not feel that the mineral rights on this particular tract of 55,000 acres had any real value.

45. CCI, *Minute Book, 1920–1931,* Nov. 2, 1927, CCI. There were occasional exceptions made. For instance, early in the Depression, CCI sold its abandoned Holmes mine, which still had an estimated $1 million in ore, to the Oliver Mining Company, which had an adjacent mine. Oliver had offered $1.3 million (see CCI, *Minute Book, 1920–1931,* July 29, 1930, CCI).

46. CCI, *Annual Report,* 1932, CCI-HF; CCI, *Minute Books, 1932–1936,* Nov. 10, 1932, CCI, on tie mill profits; CCI, Ore Mining Department, *Report of the General Manager,* 1936, 424, 457, item 2007, CIMC-CCICP.

47. CCI, *Annual Report,* 1932, CCI-HF.

48. CCI, *Annual Report,* 1930, CCI-HF.

49. The quote is from H. S. Harrison, "Memorandum on Retirement of Preferred Stock," Sept. 15, 1955, 1, folder 1809, container 75, CEP.

50. Cliffs Corporation, *Minutes of Board and Stockholders,* Sept. 21, 1931, CCI-HF; "Request for Ruling as to Proper Basis of 192,334 1/2 Shares of Republic Steel Corp Owned by the CCIC," Cyrus Eaton memo, Apr. 12, 1955, folder 1809, container 75, CEP.

51. CCI, *Minute Book, 1932–1936,* June 15, 1932, CCI (Banker's Committee). See also CCI, *Annual Report,* 1932, CCI-HF, and Cliffs Corporation, *Minutes of Board and Stockholder,* Sept. 21, 1931, CCI-HF.

52. In 1931 and 1932, for example, he borrowed money from the Cliffs Corporation. See *Cliffs Corporation v. Evatt, Tax Commr. et al.,* No. 28471, Supreme Court of Ohio, June 11, 1941 (1941 Ohio, LEXIS 471), 8.

53. CCI, *Minute Book, 1932–1936,* June 22, 1933 ("pressing and important"), Mar. 24, 1937 ("extremely difficult," in letter from E. B. Greene to Wm. G Mather), CCI. See also *Marquette Mining Journal,* Aug. 26, 1933.

54. William Gwinn Mather diary, Aug. 23, 1933 entry, folder 20, container 3, WGMFP.

55. "CCI Chairman," *Marquette Mining Journal,* June 29, 1950 (CCI commemorative edition).

56. CCI, *Minute Book, 1932–1936,* Aug. 23, 1933, CCI.

57. CCI, *Annual Report,* 1934, CCI-HF. See also "Request for Ruling as to Proper Basis of 192,334 1/2 Shares," 12.

58. *United States v. Republic Steel Corporation et al.,* No. 5152, District Court, N.D. Ohio, E. D. (1935 U.S. Dist., LEXIS 1539), May 2, 1935; Tom M. Girdler, *Boot Straps: The Autobiography of Tom M. Girdler* (New York: Charles Scribner, 1943), 219–22.

59. CCI, *Annual Report,* 1935, CCI-HF. See also CCI, *Minute Book, 1932–1936,* Aug. 21, 1934, CCI; and "Cleveland Cliffs Approves Funding," *Cleveland Press,* Nov. 10, 1935. The *Cleveland Press* had reported on July 11, 1934, that CCI itself might be folded into Republic Steel with Corrigan, McKinney. See also "Request for Ruling as to Proper Basis of 192,334 1/2 Shares," 18–19.

60. CCI, *Annual Report,* 1934, CCI-HF. For the phrase "especially dangerous," see E. B. Greene to Wm. G. Mather, Mar. 24, 1937, in CCI, *Minute Book, 1937–1945,* Mar. 24, 1937, CCI.

61. CCI, *Annual Report,* 1935, CCI-HF. See also CCI, *Minute Book, 1932–1936,* Feb. 26, 1935, CCI; "Eastern Group Buys Interest in Cleveland Iron," *Chicago Tribune,* Sept. 5, 1935; Otis & Co., "The Cleveland-Cliffs Iron Co. (Nov. 19, 1935)," folder 1801, container 74, CEP; "Cleveland-Cliffs Plans Bond Issue," *New York Times,* Sept. 5, 1935.

62. E. B. Greene to Wm. G. Mather, Mar. 24, 1937, in CCI, *Minute Book, 1937–1945,* Mar. 24, 1937, CCI; CCI, *Annual Report,* 1945, 9, CCI-HF.

63. Cliffs Corporation, *Annual Report,* 1931.

64. See Kitchen & Co., "Offering of Cliffs Corporation Stock, Jan. 22, 1945," folder 2091, container 88 CEP, 2.

65. CCI, *Annual Report,* 1934; CCI, *Annual Report,* 1940, CCI-HF.

66. CCI, *Annual Report,* 1936, CCI-HF.

67. The Cleveland-Cliffs Iron Co. et al., "Plan of Employees' Representation, July, 1933," LRL.

68. See, for example, S. R. Elliott to E. B. Greene, Feb. 10, 1936, 2–4, at the beginning of CCI, Ore Mining Department, *Annual Report of the General Manager,* 1935, item 2006, CIMC-CCICP.

69. *National Labor Relations Board v. Jones & Laughlin Steel Corp.,* 301 U.S. 1, 57 S. Ct. 615, 81 L. Ed. 893 (1937).

70. CCI, *Minute Book, 1937–1945,* May 14 and June 17, 1937, CCI; CCI, Ore Mining Department, *Annual Report of the General Manager,* 1938, 64, 144, 263, 366–67, item 2009, CIMC-CCICIP. While relations in the company's mining operations seemed to be going well in the mid-1930s, the same was not the case with its timber operations; see, for example, *National Labor Relations Board v. Cleveland-Cliffs Iron Co.,* no. 9162, U.S. Court of Appeals for the Sixth Circuit, Feb. 11, 1943 (Lexis 3811).

71. CCI, *Minute Book, 1937–1945,* Apr. 13, 1937, CCI.

72. Ibid., Mar. 15, 1938.

73. CCI, *Annual Report,* 1940, CCI-HF. The maximum debt figure is exclusive of the $7.25 million needed to retire the preferred stock of McKinney Steel Holding Company.

74. CCI, *Annual Report,* 1940, CCI-HF. See also CCI, *Minute Book, 1937–1945,* Jan. 16, 1940, CCI; and A. H. Hubbell, "The Mather Mine: A New Show Place for the Iron Country," *Engineering and Mining Journal* 142 (Oct. 1941). For a detailed history of the opening of this mine, see Burton H. Boyum, ed., *The Mather Mine: Negaunee and Ishpeming, Michigan* (Marquette: Marquette County Historical Society, 1979).

75. CCI, *Minute Book, 1937–1945,* Dec. 12, 1941, CCI. See also ibid., Jan. 21 and Feb. 18, 1942.

76. Ibid., Apr. 23, 1942.

77. Ibid., Jan. 20, 1943.

78. Ibid., Apr. 23, 1942. The failure to promise similar payments to employees on the iron ranges caused some disgruntlement later (ibid., Jan. 20, 1943; Apr. 21, 1943).

79. CCI, *Annual Report,* 1942, CCI-HF.

80. CCI, *Minute Book, 1937–1945,* Nov. 18, 1942, CCI.

81. Ibid., Oct. 28, 1941.

82. "Champion Mine's Old Stockpiles Yield Ore in Emergency," *Engineering and Mining Journal* 142 (Nov. 1941).

83. CCI, *Minute Book, 1937–1945,* Nov. 19, 1941, CCI; Jan. 12, 1942; and Jan. 17, 1945, CCI-HF; CCIC, *Annual Report,* 1941, CCI-HF.

84. CCI, *Minute Book, 1937–1945,* June 16 and July 29, 1943, CCI.

85. Ibid., Apr. 18, 1945.

86. *Marquette Mining Journal,* May 10, 1944. Over 80 percent of CCI's production at this time came from underground mines, and Stakel, the company's general manager, estimated that wage increases hit underground mines four times harder than open-pit mines.

87. CCI, *Annual Report,* 1943, 9, CCI-HF.

88. E. B. Greene to Wm. G. Mather, Mar. 15, 1942, folder 219, container 28, WGMFP; CCI, *Annual Report,* 1943, CCI-HF. See also CCI, *Minute Book, 1937–1945,* Apr. 21, 1943, CCI.

89. CCI, *Minute Book, 1937–1945,* Aug. 18, 1943, CCI; CCIC, *Annual Report,* 1943; and CCI, *Annual Report,* 1944, 4, CCI-HF. Greene's testimony was supported by Franklin G. Pardee of the Michigan Geological Survey: see Franklin G. Pardee to Governor Harry F. Kelly, July 2, 1943, folder 4, box 5, RG 42 (Records of the Executive Office), AM.

90. CCI, *Annual Report,* 1945, 4, CCI-HF.

91. CCI, Ore Mining Department, *Annual Report of the General Manager,* 1944, 1–2, 87, 133, 179, 183, 214, 219, 262 (repairs neglected), 296, 302, 601, item 2015, CIMC-CCICP. See also CCI, *Minute Book, 1937–1945,* Feb. 16 and Mar. 15, 1944, CCI.

92. CCI, *Minute Book, 1937–1945,* Dec. 20, 1944, CCI.

93. CCI, Ore Mining Department, *Annual Report of the General Manager,* 1942, 304, 536, item 2013, CIMC-CCICP.

94. CCI, Ore Mining Department, *Annual Report of the General Manager,* 1943, 2, item 2014, CIMC-CCICP.

95. CCI, Ore Mining Department, *Annual Report of the General Manager,* 1944, 262, 601, item 2015, CIMC-CCICP.

96. Charles Stakel to Wm. G. Mather, opening letter in CCI, Ore Mining Department, *Annual Report of the General Manager,* 1943, 4, item 2014, CIMC-CCICP. See also CCI, Ore Mining Department, *Annual Report of the General Manager,* 1945, 4, item 2015, CIMC-CCICP.

97. CCI, *Annual Report,* 1945, 4–5, 9, CCI-HF.

98. E. V. Hale to C. S. Eaton and Wm. R. Daley, Jan. 8, 1947 ("Cliffs-Hanna Comparison"), folder 1801, container 74, CEP. By 1946 Cleveland-Cliffs had dropped to fourth among the iron ore producers of the Lake Superior iron ore district. That year U.S. Steel's subsidiary Oliver was the giant, as usual, shipping almost 48 percent of the district's total ore (the Lake Superior district produced over 80 percent of American iron ores). Second place was occupied by Pickands Mather. M. A. Hanna had replaced Cleveland-Cliffs in third place. CCI shipped about 7 percent of the district's total, around 40 percent more than the fifth-highest shipper. See "1946 Lake Superior Iron Ore Shipments by Companies," *Skillings' Mining Review,* Apr. 26, 1947. In both 1929 and 1939 CCI had ranked third behind Oliver and Pickands Mather and had shipped 8 to 10 percent of total ore from the district. See "Rank of L.S. Iron Ore Producers in 1949," *Skillings' Mining Review,* Apr. 22, 1950, 2 (which provides the rankings for 1929 and 1939 as well as 1949).

99. *Marquette Mining Journal,* Feb. 25, 1943, and July 24, 1943.

100. CCI, Ore Mining Department, *Annual Report of the General Manager,* 1942, 692, 694, item 2013, CIMC-CCICP.

101. Franklin Pardee to Gov. Harry F. Kelly, May 12, 1943, folder 4, box 5, RG 42, AM.

102. J. F. Wolff, "The Economics of Underground Mining," *Skillings' Mining Review,* Apr. 6, 1946, 14.

103. H. S. Harrison, "Memorandum on Retirement of Preferred Stock," Sept. 15, 1955, folder 1809, container 75, CEP, attached to a memo from R. L. Kaiser to C. S. Eaton and Wm. R. Daley, Sept. 16, 1955.

104. For the large number of merger and recapitalization plans, see *Anderson et al. v. Cleveland-Cliffs Iron Co. et al.,* 579838, Common Pleas Court of Ohio, June 9, 1948, 10–11 (Lexis 245). For examples, see CCI, *Minute Book, 1937–1945,* Nov. 17, 1937, Dec. 14, 1937, Feb. 18, 1942, and Jan. 19, 1944, CCI; "Recapitalization Plan," July 15, 1941, folder 216, container 27, WGMFP. For a critical view of the arrangements that Mather and Greene had negotiated with Eaton in executing the formation of the Cliffs Corporation, see [?] to Joseph L. Weiner, Nov. 5, 1946 (attached memorandum), folder 2092, container 89, CEP.

105. Hartman & Craven to Cliffs Corporation, Sept. 16, 1946, folder 2092, container 89, CEP, for the claim that key members of the Cleveland-Cliffs executive team were in ill health and unable to efficiently carry out their duties.

106. *Cleveland Plain Dealer,* Feb. 20, 1947. A copy of the consolidation plan is in file 1870, container 78, CEP.

107. *Cleveland-Cliffs Iron Company v. Corporation & Securities Commission,* Docket No. 44, Supreme Court of Michigan (1958 Mich. LEXIS 550), Mar. 6, 1958, 17.

108. "Fight over Cliffs," *Business Week,* Jan. 4, 1947, 17–18; "Cliffs Fight Settled," *Business Week,* Mar. 1,

1947, 19–20; "Peace Comes to Cliffs," *Cleveland Plain Dealer,* Feb. 20, 1947. See also "Minority Group Seeks SEC Study of Plan for Cliffs Corporation," *New York Times,* Apr. 1, 1947; *Cleveland Press,* Dec. 20, 23, 24, 27, 1946; Feb. 19, Apr. 25, 1947; CCI, *Annual Report,* 1947, 3, CCI-HF; and *Anderson et al. v. Cleveland-Cliffs Iron Co. et al.* (1948 Ohio Misc. LEXIS 245). Cleveland-Cliffs' treasurer considered the process of working out the merger as the "most difficult" operation he had ever done ("Cleveland-Cliffs' Treasurer: What He Does," *Business Week,* June 7, 1952, 73). See also a variety of documents relating to the Cliffs Corporation dissolution and merger plans in CEP (folder 1801, container 74; folder 1853, container 77; folder 1870, container 78; and folders 2092 and 2094, container 89) and WGMFP (folder 244, container 31).

109. CCI, Ore Mining Department, *Annual Report of the General Manager,* 1942, 3, item 2013, CIMC-CCICP.

110. The *Annual Reports* for 1943 and 1944 suggest relations were satisfactory at the Athens, Mather, and Negaunee mines, but "not entirely satisfactory" at the Lloyd and Spies-Virgil.

111. Cyrus Eaton to Wm. G. Mather, Mar. 30, 1946, and Apr. 5, 1946, folder 1801, container 74, CEP.

112. Philip Murray (president USW) to Wm. G. Mather, Apr. 15, 1946, folder 1801, container 74, CEP.

113. A copy of the injunction with affidavits is in the clipping files of the Longyear Research Library in Marquette under "Labor Disputes."

114. For an example of a union accusation, see interview with Russell Munson by Burton Boyum, Jan. 7, 1980, transcript at Cliffs Shafts Museum Archives Room, Ishpeming; or *Miners' Voice,* special fortieth anniversary issue, May 1986 (which reprints articles from "Strike News" and local papers in 1946 regarding the strike). For a company accusation, see, for example, Charles J. Stakel to E. B. Greene, May 4, 1947, 7, CCI, Ore Mining Department, *Annual Report of the General Manager,* 1946, item 2017, CIMC-CCICP. See also CCI, Ore Mining Department, *Annual Report of the General Manager,* 110, 261, 267; interview with James Westwater by Terry Reynolds, July 17, 2007; and "Rocks Fly in Near-Riot at Mather Mine," *Marquette Mining Journal,* Apr. 8, 1946.

115. Burt Boyum, interview in *Red Dust* (1986), 43; John Lindroos, ibid., 46; Ernie Ronn, ibid., 52.

116. Ernie Ronn, interview in ibid., 52.

117. Burt Boyum, interview in ibid., 43.

118. CCI, Ore Mining Department, *Annual Report of the General Manager,* 1946, 339, 344, item 2017, CIMC-CCICP; Westwater interview.

119. CCI, Ore Mining Department, *Annual Report of the General Manager,* 1946, 1–2, item 2017, CIMC-CCICP.

120. CCI, Ore Mining Department, *Annual Report of the General Manager,* 1951, 1, item 2022; CCI, Ore Mining Department, *Annual Report of the General Manager,* 1952, 446, 470, item 2023 (for the hospital); CCI, Ore Mining Department, *Annual Report of the General Manager,* 1954, 468, item 2025 (for the nursing program), CIMC-CCICP.

121. For example, the managers at a significant number of CCI mines in 1947 reported labor relations as "very satisfactory" or the rough equivalent. CCI, Ore Mining Department, *Annual Report of the General Manager,* 1947, 107, 115, 308, 343, 425, item 2017, CIMC-CCICP.

122. Donald Richard Hakala, "An Analysis of Postwar Developments in the Upper Michigan Iron Ore Mining Industry," Ph.D. diss., Indiana University, 1966, 85.

123. CCI, *Annual Report,* 1947, CCI-HF, 5.

124. CCI, Ore Mining Department, *Annual Report of the General Manager,* 1945, 138, 140, item 2016, CIMC-CCICP.

125. CCIC, Ore Mining Department, *Annual Report of the General Manager,* 1947, 182, item 2018, CIMC-CCICP.

126. E. W. Davis, "Lake Superior Iron Ore and the War Emergency," report to the Materials Division of the War Production Board, May 20, 1942, 6 (quote), 10, folder 2, box 73, Public Relations Dept. files, Accession 1631, AISI. Warnings had been coming even earlier. See, for example, Frank J. Tolonen, *Gravity Concentration Tests on Michigan Iron Formations,* Contribution 46 (New York: American Institute of Mining Engineers, 1933), 1.

127. CCI, Ore Mining Department, *Annual Report of the General Manager,* 1942, 692, item 2013, CIMC-CCICP.

128. See A. P. Swineford, *Annual Review of the Iron Mining and Other Industries of the Upper Peninsula for the Year Ending Dec. 31, 1881* (Marquette: Mining Journal, 1882), 179–80; Michigan, Commissioner of Mineral Statistics, *Annual Report of the Commissioner of Mineral Statistics . . . for 1881* (Lansing: W. S. George, 1882), 165; Michigan, Commissioner of Mineral Statistics, *Mines and Mineral Statistics [1885],* by Charles D. Lawton (Lansing: Thorp & Godfrey, 1886), 29–30;Michigan, Commissioner of Mineral Statistics, *Mines and Mineral Statistics [1889],* by Charles D. Lawton (Lansing: Robert Smith, 1890) 150; Michigan, Commissioner of Mineral Statistics, *Mines and Mineral Statistics [1891–92],* by James P. Edwards (Lansing: Robert Smith, 1893), 64; and John Birkinbine and Thomas A. Edison, "The Concentration of Iron-Ore," *Transactions of the American Institute of Mining Engineers* 17 (1889).

129. George S. May and Victor F. Lemmer, "Thomas Edison's Experimental Work with Michigan Iron Ore," *Michigan History* 53 (1969).

130. Peter Frank Mason, "Some Changes in Domestic Iron Mining as a Result of Pelletization," *Annals of the Association of American Geographers* 59 (Sept. 1969): table 1, 537.

131. CCI, *Minute Books, 1946–1952,* Apr. 21, 1948, CCI, for reports on the possibility of joint explorations with Republic Steel in Quebec; June 16, 1948, on formation of United Dominion Mining Company. Cliffs would leave the consortium when it was clear the deposits contained no high-grade ores.

132. Boyum, "Cliffs Illustrated History," 11-8, 11-8A.

133. See table in CCI, Ore Mining Department, *Annual Report of the General Manager,* 1955, 498, item 2026, CIMC-CCICP. See also CCI, Ore Mining Department, *Annual Report of the General Manager,* 1954, tables on 513–14, item 2025, CIMC-CCICP. For the Venezuelan operations, see folder 1854, container 77, CEP.

134. CCI, Ore Mining Department, *Annual Report of the General Manager,* 1951, fig. 2 (end of volume), item 2022, CIMC-CCICP.

135. Boyum, "Cliffs Illustrated History," 11-7, 11-8.

136. CCI, Ore Mining Department, *Annual Report of the General Manager,* 1947, 304, item 2018, CIMC-CCICP.

137. CCI, Ore Mining Department, *Annual Report of the General Manager,* 1951, 4, 553, 555, item 2021, CCIC-CCICP.

138. "Cleveland-Cliffs Acquires Int'l Harvester Iron Mines," *Skillings' Mining Review,* Mar. 8, 1947; "Harvester Mines Sold to Cliffs," *Cleveland Press,* Feb. 23, 1947.

139. CCI, *Minute Book, 1938–1945,* Apr. 18, 1945; CCI, *Minute Book, 1946–1952,* May 19, 1948, CCI. Documents on the arrangement between Steep Rock and CCI can be found in folders 6800 and 6801, container 301, CEP.

140. Boyum, "Cliffs Illustrated History," 11-9, 11-13.

141. CCIC, Ore Mining Department, *Annual Report of the General Manager,* 1944, 2, 733, item 2015, CIMC-CCICP.

142. Lawrence A. Roe, *Iron Ore Beneficiation* (Lake Bluff, IL: Minerals, 1957), 104; CCIC, Ore Mining Department, *Annual Report of the General Manager,* 1944, 733, item 2015, CIMC-CCICP.

143. E. W. Davis, *Pioneering with Taconite* (St. Paul: Minnesota Historical Society, 1964), 13–64.

144. CCI, *Annual Report,* 1944, CCI-HF. Crispin Oglebay served on Cliffs' board of directors from 1931 to 1943.

145. Davis, *Pioneering,* 109. See also 224n7.

146. Ibid., 94.

147. Ibid., 92.

148. E. L. Derby, geologist, to A. C. Brown, Mar. 4, 1940, CCI file 50-73, E. W. Davis, General Office, box 125, DSC-CSU.

149. E. W. Davis, "Change in U.S. Steel Iron Ore Marketing Policy Endangers Range Communities and State," *Minnesota Municipalities,* Mar. 1940, E. W. Davis, General Office, box 125, DSC-CSU.

150. James Lewis Morrill, *Taconite! Sleeping Giant of the Mesabi* (New York: Newcomen Society of England, American Branch, 1948), 16.

151. Davis, *Pioneering,* 110.

152. Ibid., 227n1; and R. H. Ramsey, "Teamwork on Taconite: The Story of Erie Mining Co.'s Commercial Taconite Project," *Engineering and Mining Journal* 156 (Mar. 1955), 86.

153. Davis, *Pioneering,* 108. Fred D. DeVaney, "Flotation of Lake Superior District Iron Ores," *Skillings' Mining Review,* Feb. 11, 1950.

154. Davis, *Pioneering,* 121–23.

155. Ibid., 134. Reserve closed in 1986, a victim of the bankruptcy of LTV (which had merged with Republic Steel in 1984). After purchase by Cyprus Minerals, Inc., it reopened in 1989 as Northshore Mining. It was acquired by Cleveland-Cliffs in 1994. Robert V. Bartlett, *The Reserve Mining Controversy: Science, Technology, and Environmental Quality* (Bloomington: Indiana University Press, 1980) provides a brief review of the history of Reserve Mining and a detailed account of its subsequent involvement in one of the country's major environmental controversies.

156. "1943 Lake Superior Iron Ore Shipments by Companies," *Skillings' Mining Review,* Apr. 22, 1944, 1.

157. Ramsey, "Teamwork on Taconite," 86; "1954 Ore Shipping Season Opens with Movement of Taconite Concentrate Pellets," *Skillings' Mining Review,* May 8, 1954.

158. CCIC, *Minute Book, 1937–1945,* Oct. 20, 1939, and June 16, 1943, CCI.

159. CCI, *Minute Book, 1946–1952,* Sept. 17, 1947, CCI.

160. Ibid., Oct. 15, 1947.

CHAPTER 5

1. "Remarks of Alexander C. Brown, 100th Anniversary Luncheon," June 26, 1950," CCI-HF.

2. Burton H. Boyum, ed., *The Mather Mine: Negaunee and Ishpeming, Michigan* (Marquette: Marquette County Historical Society, 1979).

3. *Marquette Mining Journal,* centennial edition, June 29, 30, and July 1, 1950, contained long articles on the company's history. Editorial, *Marquette Mining Journal,* June 29, 1950.

4. CCI, *Annual Report,* 1950, 6, CCI-HF.

5. Mark L. Thompson, *Queen of the Lakes* (Detroit: Wayne State University Press, 1994), 157–62. In the winter of 1956–57 *Cliffs Victory* was lengthened again, by another ninety-three feet, making it the longest ore boat on the lakes, hence "Queen of the Lakes."

6. CCI, *Minutes of the Board of Directors,* Oct. 20, 1939, book 9, 996; CCI, *Minutes of the Board of Directors,* Oct. 15, 1947, book 1 (new series), 45; CCI, *Minutes of the Board of Directors,* Apr. 21, 1948, book 2 (new series), 101, CCI-HF.

7. "Stock in Cleveland-Cliffs Sold by Republic Steel," *New York Times,* Oct. 20, 1950.

8. CCI, *Minutes of the Board of Directors,* Nov. 16, 1949, book 3 (new series), 241–42, CCI-HF. E. W. Davis, *Pioneering with Taconite* (St. Paul: Minnesota Historical Society), 1964, 138.

9. Peter J. Kakela, "Iron Ore: From Depletion to Abundance," *Science,* n.s., Apr. 10, 1981, 133.

10. U.S. President's Materials Policy Commission, *Resources for Freedom,* vol. 2 (Washington, DC: Government Printing Office, 1952), 11–18. H. Stuart Harrison, "Where Is the Iron Ore Coming From?" *Analysts Journal* 9 (June 1953).

11. CCI, *Annual Report,* 1950, 19, CCI-HF.

12. See Daniel R. Fusfield, "Joint Subsidiaries in the Iron and Steel Industry," *American Economic Review* 48 (1958).

13. H. Stuart Harrison, *The Cleveland-Cliffs Iron Company* (New York: Newcomen Society in North America, 1974), 19.

14. See Lawrence A. Roe, *Iron Ore Beneficiation* (Lake Bluff, IL: Minerals, 1957), 1–11; and M. M. Fine, "The Beneficiation of Iron Ores," *Scientific American,* Jan. 1968.

15. Peter Frank Mason, "Some Changes in Domestic Iron Mining as a Result of Pelletization," *Annals of the Association of American Geographers* 59 (Sept. 1969).

16. C. W. Allen to VP Walter A. Sterling, Feb. 23, 1953, 1, CCI, Ore Mining Department, *Annual Report of the General Manager for the Year Ending Dec. 31, 1952,* item 2023, CIMC-CCICP.

17. William H. Mulligan Jr., "The Cliffs Shaft Mine, Ishpeming, Michigan: A Case Study in the History of American Iron Mining, 1844–1967," paper delivered at Third International Mining History Conference and Symposium on the Preservation of Historic Mining Sites, June 6–l0, l994, Golden, CO.

18. Interview with Hugh Leach by Burt Boyum, Sept. 8, 1984, CCI-HF.

19. See Dawn Bunyak, *Frothers, Bubbles and Flotation: A Survey of Flotation Milling in the Twentieth-Century Metals Industry* (Denver: National Park Service, 1998). Bunyak does not mention the use of flotation in the iron industry.

20. "Cleveland-Cliffs Iron Company to Construct Ore Laboratory in Ishpeming," *Marquette Mining Journal,* June 3, 1948; "C. C. Co. to Erect Research 'Lab' Here," *Ishpeming Iron Ore,* June 5, 1948.

21. "Further Details of CCI Research Setup Explained," *Marquette Mining Journal,* Mar. 5, 1957; "Research Lab Focuses on Ore Concentration," *Marquette Mining Journal,* June 10, 1950 (CCI commemorative edition); "Cleveland-Cliffs Laboratory Big Factor in Future of Iron Mines, *Marquette Mining Journal,* July 30, 1949; and Burton H. Boyum, "Cliffs Illustrated History," manuscript, c. 1986, 12-7–12-9, CCI-Empire.

22. Interview with Edwin (Ned) Johnson by Terry Reynolds, June 4, 2007.

23. H. Stuart Harrison, *The Cleveland-Cliffs Iron Company* (New York: Newcomen Society of North America, 1974), 23.

24. Some of the correspondence between CCI officials and between them and Michigan tax officials regarding tax matters can be found in item 2291 (State Legislate File: Ohio, Michigan & Minnesota, 1943–51), CIMC-CCICP. See also Donald Richard Hakala, , "An Analysis of Postwar Developments in the Upper Michigan Iron Ore Mining Industry," Ph.D. diss., Indiana University, 1966, 203, 207. Also *Facts about Michigan's Specific Tax on Iron Ore,* a Cleveland-Cliffs monograph, Aug. 1977, AISI; and Thomas E. McGinty, "The Pellet Revolution," talk before the Junior Membership of the New York Society of Security Analysts, Feb. 7, 1966, CCI-HF.

25. C. W. Allen to Walter A. Sterling, Feb. 23, 1953, CCI, Ore Mining Department, *Annual Report of the General Manager for the Year Ending Dec. 31, 1952,* 1–6, 83–102, item 2023, CIMC-CCICP. Also David N. Skillings Jr., "Cleveland-Cliffs' Michigan Iron Mining Operations," *Skillings' Mining Review,* Oct. 30, 1954.

26. CCI, *Minutes of the Board of Directors,* Nov. 18, 1953, book 7, 665; *Annual Report,* 1953, 7, CCI-HF.

27. *Cleveland-Cliffs Iron Company v. Corporation & Securities Commission,* Docket No. 44, Supreme Court of Michigan, June 13, 1957, Decided Mar. 6, 1958, LEXIS 550, summary, 4.

28. "Humboldt Iron Mine Opens 650,000-Ton Pellet Plant," *Daily Metal Reporter,* Sept. 23, 1960.

29. "Research Lab . . . Annual Report, 1953," CCI, Ore Mining Dept., *Annual Report of the General Manager for the Year Ending Dec. 31, 1953,* 499–515, item 2024; CIMC-CCICP.

30. Bunyak, *Frothers, Bubbles and Flotation,* 25.

31. "Iron from Jasper: Off to a Running Start," *Business Week,* June 19, 1954, 94–96, 98, 100, 103–4, quote 100. Humboldt's reserves were exhausted by 1970, but its pellet operations were kept open by transporting ore from the Republic Mine there.

32. "The Humboldt Mine," *Skillings' Mining Review,* Sept. 25, 1954. See also Robert Cochran, "New Plant Successfully Floats Michigan Jasper," *Engineering and Mining Journal* 155 (Aug. 1954).

33. Interview with R. M. DeGabriele by Burt Boyum, July 23, 1984, CCI-Empire.

34. Kenneth C. Olson, "Primary Blast Hole Drilling at the Humboldt Mine," *Mining Congress Journal* 42 (Apr. 1956).

35. E. L. Derby to Grover J. Holt, Feb. 23, 1954, CCI, Ore Mining Dept., *Annual Report of the General Manager for the Year Ending Dec. 31, 1953,* 576, item 2024, CIMC-CCICP.

36. Sterling to Eaton, Aug. 10, 1954, folder 1807, container 75, CEP.

37. CCI, *Minutes of the Board of Directors,* Oct. 20, 1954, book 8A, 743, CCI-HF.

38. Memo, Sept. 13, 1954, folder 1807, container 75, CEP.

39. "Inland-Steel Joins Cliffs and Others in Ore Project," *American Metal Market,* Feb. 15, 1956, folder: Companies—Cleveland-Cliffs Iron Co., AISI.

40. "Republic Mine and Plant of the Cleveland-Cliffs Iron Co. on Marquette Range," *Skillings' Mining Review,* Oct. 13, 1956.

41. Memo, Eagle Mills Pellet Plant, Mar. 17, 1959, folder 1810, container 75, CEP.

42. George O. Argall Jr., "Eagle Mills Pellet Plant Opens New Era for Michigan Hematite," *Mining World* 18 (Dec. 1956), 42. In 1963 the company quadrupled pellet production at Eagle Mills to 2.8 million tons per year.

43. Walter A. Sterling, "Review and Outlook for the Iron Ore Industry," *Skillings' Mining Review,* Oct. 20, 1956, 1–2, 28.

44. H. Stuart Harrison, "Iron Ore—The Big Change," *Analysts Journal* (June 1957), CCI-HF.

45. Interview with Sam Scovil by Virginia Dawson, June 11, 2007.

46. "Humboldt Pellet Plant Marks New Era for Michigan Low Grade Iron Ores," *Skillings' Mining Review,* Oct. 1, 1960. CCI, *Annual Report,* 1960, 2, CCI-HF. See also J. S. Westwater, "Pelletizing by Grate-Kiln Method at Humboldt," *Blast Furnace and Steel Plant* 49 (June 1961).

47. Folders 6799–801, container 301, CEP, contain Holt's reports.

48. "Babsons Investor's Report," Mar. 28, 1964, folder 6689, container 296, CEP.

49. "Portsmouth Steel Would Buy Stock," *New York Times,* July 19, 1953.

50. CCI, *Minutes of the Board of Directors,* Feb. 17, 1954, book 8, 690, CCI-HF. Lincoln served on Cliffs' board until his death in 1958. He was replaced by Walter J. Touhy, president of the Chesapeake and Ohio Railroad, a company controlled by Cyrus Eaton. Touhy served until his death in 1966. William R. Daley, president of Otis & Co., also served on the Cliffs board as a representative of the Eaton interests from 1955 through 1968. Both Eaton and Daley resigned in 1968 after they sold their stock in Detroit Steel. Eaton had served on the Cliffs board for thirty-eight years.

51. CCI, *Minutes of the Board of Directors,* Jan. 9, 1954, book 8, 675; Jan. 18, 1956, book 10, 835, CCI-HF. Also, miscellaneous correspondence between Cleveland-Cliffs and Cyrus Eaton, folders 6689–71, container 296, CEP.

52. Eaton to George Gund, Mar. 24, 1955, folder 1808, container 75, CEP. Gund served on the board from 1941 to 1966.

53. "U.S. Studying Bethlehem Fund's Use of 183,000 Cleveland-Cliffs Shares," *Wall Street Journal,* Mar. 31, 1955, folder 1881, container 78, CEP.

54. See Charles E. Egan, "Steel Official Faces Justice Action over Interlocking Directorships," *New York Times,* Apr. 26, 1950; "Competition Threat Seen in Steel Group," *New York Times,* Mar. 9, 1951.

55. Memo of Mr. Eaton's Meeting with R. E. McMath, Apr. 6, 1955, folder 1808, container 75, CEP.

56. Paul A. Tiffany, *The Decline of American Steel: How Management, Labor, and Government Went Wrong* (Oxford: Oxford University Press, 1988), 156.

57. The Wade family served continuously from 1867, when George Garretson Wade's great-grandfather, Jeptha Homer Wade, first became a director of the Cleveland Iron Mining Company. The Greene family's stock, acquired through the marriage of E. B. Greene to Helen Wade, great-granddaughter of Jeptha Homer Wade, also contributed to the Wade family's influence on the board. Greene continued to serve as president of the board until shortly before his death in 1957. James D. Ireland III in 2009 serves as the Mather family's sole remaining representative.

58. Letter to employees, July 23, 1959, folder 1879, container 78, CEP.

59. "Cleveland-Cliffs in Good Shape Despite Strike, Early Freeze," *American Metal Market,* Nov. 26, 1959, AISI.

60. *Iron Age,* Oct. 15, 1959; Jan 14, 1960.

61. Peter Frank Mason, "Some Aspects of Change in the Iron Mining Industry of the United States as a Result of Pelletization," Ph.D. diss., University of California, Los Angeles, 1968, 42–44.

62. Tiffany, *Decline of American Steel,* 180–81.

63. Ibid., 108.

64. CCI, *Annual Report,* 1959, 1, 21, CCI-HF.

65. F. S. Smithers & Co., *The Iron Ore Industry and the Cleveland-Cliffs Iron Company, the Hanna Mining Company, the M. A. Hanna Company* (New York: F. S. Smithers, 1960), 35–43.

66. George Melloan, "Iron Ore Upheaval," *Wall Street Journal,* July 20, 1960.

67. T. M. Rohan, "Steelmakers Plan Pellet Plants to Head Off Foreign Inroads," *Iron Age,* Sept. 15, 1960, 159–61. Also Davis, *Pioneering,* 191.

68. Rohan, "Steelmakers Plan Pellet Plants." Also Davis, *Pioneering,* 191.

69. Stanley W. Sundeen, "The Economic Role of Michigan Iron Ores," *Iron and Copper in Michigan's Economy, 6th Annual Conference Michigan Natural Resources Council,* Oct. 25, 1961.

70. Hakala, "Analysis of Postwar Developments," 203, 207. Also *Facts about Michigan's Specific Tax on Iron Ore,* AISI; and McGinty, "The Pellet Revolution," CCI-HF.

71. John S. Wilbur, "The Competitive Position of Lake Superior Iron Ores and Its Effect on Lake Shipping," *Skillings' Mining Review,* May 7, 1960, 10–11.

72. John Wilbur, "Lower Lake Railroads and the Iron Ore Industry," Jan. 10, 1961, typescript, folder 1859, container 77, CEP.

73. See Virginia P. Dawson, "Knowledge Is Power: E. G. Bailey and the Invention and Marketing of the Bailey Boiler Meter," *Technology and Culture* 37 (1996).

74. Boyum, "Cliffs Illustrated History," 8-30, 8-34. John L. Horton became manager of the company's marine division in 1976 and was nationally recognized for his contributions to marine safety. He was active in the design and construction of *Cliffs Victory* and SS *Walter A. Sterling,* and the flagship SS *Edward B. Greene.*

75. First Boston Corporation, *The Iron Ore Industry and the Cleveland-Cliffs Iron Company* (New York: First Boston, 1955), 21; "Tradition and Training Assets in Finance Post: Cleveland-Cliffs' Harrison Knows Both Producing, Fiscal Ends in Iron, Steel," *Finance,* Oct. 15, 1955, CCI-CF.

76. Stuart Harrison, "Letter to employees," Mar. 19, 1962, folder 1862, container 78, CEP.

77. CCI, *Minutes of the Board of Directors,* Feb. 21, 1962, book 14, 1162, CCI-HF.

78. "Empire Mine Could Push Marquette Iron Range to Record Production of Ore," *Marquette Mining Journal,* Apr. 23, 1963.

79. David N. Skillings Jr., "Official Opening of the Empire Iron Mine," *Skillings' Mining Review,* May 16, 1964, 1,6.

80. Philip D. Pearson, "Lake Superior Underground Iron Mines Gear for Future," *Mining Engineer* 14 (Mar. 1962): 64.

81. Hakala, "Analysis of Postwar Developments," 247.

82. "Ore Improvement Plant Eagle Mills, Michigan," *Skillings' Mining Review,* Apr. 12, 1958. Argall, "Eagle Mills Pellet Plant."

83. Interview with James Westwater by Terry Reynolds, July 17, 2007.

84. "Cleveland-Cliffs, McLouth and Bethlehem Push Mine Output for Big Pellet Project," *American Metal Market,* Dec. 28, 1963, AISI.

85. Ibid.

86. In 1976 small amounts of bentonite were added to produce a stronger pellet. See Boyum, *The Mather Mine,* 49.

87. "Cliffs' Project at Mather Mine on Marquette Range," *Skillings' Mining Review,* Dec. 28, 1963, 6–8; and David N. Skillings Jr., "Cleveland-Cliffs' Program at the Mather Mine/Pioneer Pellet Plant," *Skillings' Mining Review,* Nov. 7, 1964. Also e-mail from Richard Tuthill, Jan. 7, 2009.

88. "Cliffs' Iron Shipments at 90% Pellets," *Skillings' Mining Review,* Oct. 12, 1968, 20.

89. "Cleveland-Cliffs Hires First New Miner in Marquette Range Rebirth," *American Metal Market,* Aug. 3, 1964, AISI.

90. CCI, *Annual Report,* 1964, 13, CCI-HF. David N. Skillings Jr., "Cleveland-Cliffs/Marquette Range Iron Ore Mines," *Skillings' Mining Review,* Nov. 2, 1963.

91. H. Stuart Harrison, "The Changing Iron Ore Industry," *Skillings' Mining Review,* Oct. 16. 1965.

92. McGinty, "The Pellet Revolution," CCI-HF.

93. CCI, *Annual Report,* 1965, 2, CCI-HF.

94. *Wheeling Steel v. M. A. Hanna Co., the Hanna Mining Co, and the Cleveland-Cliffs Iron Co.,* filed Sept. 19, 1966; "3 Iron Ore Firms Here Facing Suit by Wheeling Steel," *Cleveland Plain Dealer,* Sept. 20, 1966; "Norton Simon Hits Hanna, Cliffs with Price-fixing Suit," *Cleveland Press,* Sept. 20, 1966, folders 1888, container 79, CEP.

95. See Fusfield, "Joint Subsidiaries," 586.

96. "Eaton Produces Records Showing Some of Cleveland Trust's Holdings," *Cleveland Press,* June 20, 1967, folder 1864, container 78, CEP.

97. CCI, *Minutes of the Board of Directors,* 1980, May 1, 1980, CCI-HF.

98. CCI, *Minutes of the Board of Directors,* Aug. 25, 1962, 1196–98, CCI-HF.

99. "Cliffs' Diversification with Iron Ore Backbone," *Skillings' Mining Review,* Aug. 12, 1967.

100. CCI, *Annual Report,* 1964, 11, CCI-HF.

101. CCI, *Annual Report,* 1967, 12, CCI-HF.

102. Interview with Hugh J. Leach by Burt Boyum, Sept. 9, 1984, CCI-HF.

103. Interview with R. M. DeGabriele by Burt Boyum, July 23, 1984, CCI-HF.

104. David N. Skillings Jr., "Sherman Mine Dedicated on Sept. 5 at Timagami, Ont.," *Skillings' Mining Review,* Oct. 5, 1968. In 1985 the Sherman mine began production of fluxed pellets. Note: the spelling of Timagami was later standardized to Temagami.

105. "Cliffs' Contract for Sale of Robe River Iron Ore," *Skillings' Mining Review,* May 23, 1969.

106. DeGabriele interview by Boyum.

107. "Iron Ore Pellet Production Rising: Two Firms Reveal Expansion Plans," *American Metal Market,* Dec. 17, 1969, AISI.

108. "Oliver Produces its First Taconite Concentrate," *Skillings' Mining Review,* June 20, 1953, 1. Also First Boston, *Iron Ore Industry,* 21.

109. "Great Lakes Region Iron Ore Shipments," *Skillings' Mining Review,* May 31, 1969.

110. "Rank of 1953 Iron Ore Producers—L.S. Region," *Skillings' Mining Review,* July 3, 1954, 1–4, 23.

111. Quotations and facts are from Harrison, *Cleveland-Cliffs,* 27.

112. John J. Cleary, "Cliffs Buys 78% of Detroit Steel," *Cleveland Plain Dealer,* June 13, 1970.

113. Interview with Tom McGinty by B. H. Boyum, Oct. 29, 1984, CCI-Empire.

114. "Did You Say 'The Women's Dry'?" *Cliffs News,* June–July 1973, 16, CCI-HF. Interview with Linda Bancroft LaFond by Virginia Dawson, Sept. 30, 2008.

115. A class action sexual harassment lawsuit against the owners of Eveleth, *Lois E. Jensen v. the Eveleth Taconite Company,* was settled in 1998. See Clara Bingham and Laura Leedy Gansler, *Class Action: The Story of Lois Jenson and the Landmark Case That Changed Sexual Harassment Law* (New York: Doubleday, 2002).

116. Harrison, *Cleveland-Cliffs,* 23.

117. CCI, *Annual Report* 1972, 3, CCI-HF.

CHAPTER 6

1. CCI, *Annual Report,* 1973, 4, CCI-HF.

2. CCI, *Annual Report,* 1974, 2–3, CCI-HF.

3. CCI, *Annual Report,* 1976, 2, CCI-HF.

4. William T. Hogan, *The 1970s: Critical Years for Steel* (Lexington, MA: Lexington Books, 1972), 3.

5. William Scheuerman, *The Steel Crisis: The Economics and Politics of a Declining Industry* (New York: Praeger, 1986), 131, 134.

6. Richard Thomas, "Tilden Will Double Capacity by Yearend," *Engineering and Mining Journal* 180 (Sept. 1979).

7. James W. Villar and Gilbert A. Dawe, "The Tilden Mine—A New Processing Technique for Iron Ore," *Mining Congress Journal* 61 (Oct. 1975), 41; interview with Emert W. Lindroos by B. H. Boyum, Oct. 18, 1984, CCI-HF.

8. Londroos interview; interview with James Villar by B. H. Boyum, Aug. 28, 1984, CCI-HF.

9. David N. Skillings Jr., "Cleveland-Cliffs' Tilden Iron Ore Project in Michigan," *Skillings' Mining Review,* May 26, 1973; and Villar and Dawe, "The Tilden Mine," 42. The authors are indebted to retired mining engineer Richard Tuthill for his assistance in understanding the process.

10. Interview with James W. Villar by Burt Boyum, Aug. 28, 1984.

11. Interview with James Villar by Terry Reynolds, July 10, 2008.

12. Thomas, "Tilden Will Double Capacity," 7.

13. Interview with Sam Scovil by Virginia Dawson, June 11, 2007.

14. Ibid.

15. David N. Skillings Jr., "Cleveland-Cliffs Dedicates Tilden Mine in Michigan," *Skillings' Mining Review,* Aug. 23, 1975. At this time Ling-Temco-Vought, Inc., or LTV, owned 81 percent of J&L.

16. "Ling-Temco's J&L, Cleveland-Cliffs Slate Venture to Participate in Iron Ore Facility," *Wall Street Journal,* Mar. 28, 1972.

17. See Thomas E. McGinty, *Project Organization and Finance* (Cleveland: Cleveland-Cliffs Iron Co., 1981), 43–50.

18. Ibid., 49.

19. According to ibid., 14–15, percentage depletion for tax purposes was questioned by the IRS in 1956. After this resulted in a favorable ruling, iron ore pellet projects were financed using accelerated depreciation, as well as percentage depletion once the projects were profitable.

20. Ibid., 6–7.

21. Christopher G. L. Hall, *Steel Phoenix: The Fall and Rise of the U.S. Steel Industry* (New York: St. Martin's, 1997), 97–98.

22. Duane A. Smith, *Mining America: The Industry and the Environment, 1800–1980* (Lawrence: University Press of Kansas, 1987), 46–47, 136–38.

23. Robert V. Bartlett, *The Reserve Mining Controversy: Science, Technology, and Environmental Quality* (Bloomington: Indiana University Press, 1980), 25–27.

24. Ibid., 146–49. See also Thomas R. Huffman, "Exploring the Legacy of Reserve Mining: What Does the Longest Environmental Trial in History Tell Us about the Meaning of American Environmentalism?" *Journal of Policy History* 12 (2000).

25. John Wilbur, "A Statement of Cleveland-Cliffs Environmental Concerns in the Upper Peninsula," [1972], CCI-HF.

26. "Open-pit Mining: Vise for Upper Michigan," *New York Times,* Mar. 20, 1972.

27. Robert J. Labelle, "Cleveland-Cliffs Experience in Monitoring, Analysis and Control of Effluent Water Quality," *Skillings' Mining Review,* Oct. 19, 1974.

28. "Visit to Marquette Iron Range Michigan Department of Natural Resources," Sept. 6, 1973, information booklet, box 10, Cleveland-Cliffs unprocessed collection, DSC-CSU.

29. Transcript of public hearing May 15, 1972, folder 1, box 1, RG 89-343, Dept. of Natural Resources, Fisheries Division, 1971–84, AM.

30. "For Generations to Come" [Cleveland-Cliffs, 1975], LRL.

31. Julia K. Tibbitts, *"Let's Go around the Island"* (Marquette: Lake Superior Press, 1992).

32. Ibid., 46–48; "Ecology Enlivens Cliffs Meeting," *Cleveland Press,* Apr. 25, 1974, Cleveland Press files, Special Collections, Cleveland State University; interview with Don Ryan by Terry Reynolds, June 25, 2008.

33. William K. Stevens, "Profit vs. the Land in Upper Michigan," *New York Times,* Dec. 22, 1975.

34. Interview with Tom Petersen by Terry Reynolds, July 10, 2008.

35. Interview with Richard Tuthill by Virginia Dawson, Apr. 1, 2008.

36. Jenna Alderton, "A Slow Start with a Promising End: Tsu-Ming Han," *Red Dust* (1994); "Remembering Two of Our Own," *Cliffs Connections,* Summer 2005, 6.

37. Interview with Edwin Johnson by Terry Reynolds, June 4, 2007.

38. CCI, *Minutes of the Board of Directors,* book 25, Sept. 1, 1971, and Nov. 3, 1971, CCI-HF.

39. "Your Memo of March 20—Keeping Vessel Rate Down A/C Fleet," J. S. Wilbur to H. S. Harrision, Apr. 6, 1973; and "Big Boats, Technological Breakthrough for Republic Float," J. S. Wilbur to S. K. Scovil, July 17, 1973, box 72, DSC-CSU.

40. "Republic Float Contract," H. Thomas Moore to H. S. Harrison, July 9, 1973; and Harrison to Scovil, July 11, 1973, box 72, DSC-CSU.

41. "Reasons for Recommending Phase out of Marine Operations," and "Marine Department," J. S. Wilbur to H.S. Harrison, Jan. 16, 1976 (same date for both memos), box 72, DSC-CSU.

42. Mark L. Thompson, *Queen of the Lakes* (Detroit: Wayne State University Press, 1994), 161.

43. "Republic Float Contract," H. S. Harrison to S. K. Scovil, July 11, 1973, box 72, and unpublished manuscript, John Horton, "The Cleveland Cliffs Marine History," 14-3, no box number, unprocessed John L. Horton Collection, DSC-CSU.

44. Interview with Elton Hoyt III by Virginia Dawson, June 14, 2007.

45. Thompson, *Queen of the Lakes,* 201.

46. "President's Letter to Shareholders," CCI, *Annual Report,* 1976, 2, CCI-HF.

47. Jerry Flint, "First Big Strike in 18 Years Idles Some Steel Locals," *New York Times,* Aug. 2, 1977.

48. 1977 Employee Annual Report, 16, CCI, *Annual Report,* 1977, 8, CCI-HF.

49. CCI, *Annual Report* 1977, 9, CCI-HF.

50. Ibid., 3.

51. CCI, *Annual Report* 1978, 9, CCI-HF; Thomas, "Tilden Will Double Capacity."

52. Burton H. Boyum, ed., *The Mather Mine: Negaunee and Ishpeming, Michigan* (Marquette: Marquette County Historical Society, 1979), 8–9.

53. Peter J. Kakela, Allen K. Montgomery, and William C. Patric, "Factors Influencing Mine Location: An Iron Ore Example," *Land Economics* 58 (Nov. 1982): 533.

54. Interview with Gilbert Dawe by Terry Reynolds, June 26, 2008.

55. Kakela, Montgomery, and Patric, "Factors," 534. For a comparison of costs between Minnesota and Michigan, see Terry D. Monson, *The Effects of Domestic and International Competition upon Michigan's Iron Ore and Steel Industries* (Houghton: Bureau of Industrial Development, School of Business and Engineering Administration, Michigan Technological University, 1980).

56. Peter Brown, "Pioneer Plant Dust Rapped," *Marquette Mining Journal,* July 21, 1976; and "DNR Rates Shutdown Date Crucial Part of CCI Order," *Marquette Mining Journal,* Jan. 3., 1977.

57. CCI, *Annual Report,* 1979, 12, CCI-HF.

58. Ibid., 13.

59. CCI, *Annual Report,* 1978, 3, CCI-HF.

60. CCI, *Annual Report,* 1981, 2, CCI-HF.

61. Tom Moore, Address to Harvard Business School Club of Cleveland, Mar. 30, 1989, section 8, bound volume of M. Thomas Moore's speeches, 1987–97, box 58, DSC-CSU.

62. Interview with John Brinzo by Virginia Dawson, Oct. 19, 2006; and CCI, *Annual Report,* 1981, 4, CCI-HF.

63. Paul A. Tiffany, *The Decline of American Steel: How Management, Labor, and Government Went Wrong* (Oxford: Oxford University Press, 1988), 133.

64. Ibid., 173–74.

65. Robert P. Rogers, *An Economic History of the American Steel Industry* (London: Routledge, 2009), 108–9.

66. See Donald F. Barnett and Robert W. Crandall, *Up from the Ashes: The Rise of the Steel Minimill in the United States* (Washington, DC: Brookings Institution, 1986).

67. Rogers, *Economic History,* 110, 132.

68. Tiffany, *Decline of American Steel,* 131; Rogers, *Economic History,* 132–33.

69. Rogers, *Economic History,* 155.

70. Scheuerman, *Steel Crisis,* 67, 94 (quote).

71. Hall, *Steel Phoenix,* 121.

72. Ibid., 124.

73. William T. Hogan, *World Steel in the 1980s: A Case for Survival* (Lexington, MA: Lexington Books, 1983), 7.

74. Hall, *Steel Phoenix,* 121.

75. CCI, *Annual Report,* 1978, 10, CCI-HF.

76. CCI, *Annual Report,* 1982, 2–3, CCI-HF.

77. Interview with William Calfee by Virginia Dawson, Nov. 8, 2007.

78. Brinzo interview.

79. "Presentation by William Calfee to New York Society of Security Analysts," Oct. 28, 1988, bound volume of M. Thomas Moore's speeches, 1987–97, box 58, DSC-CSU.

80. Harvard Business School, "Cleveland-Cliffs, Inc," case 9-293-051, *Harvard Business School Review,* Sept. 21, 1993.

81. "Presentation by William Calfee, John Brinzo, and Thomas Moore to New York Society of Security Analysts," Oct. 28, 1988, bound volume of M. Thomas Moore's speeches, 1987–97, box 58, DSC-CSU; Calfee interview.

82. Scovil interview.

83. Thompson, *Queen of the Lakes,* 203.

84. Terry D. Monson, *The New Challenge of Foreign Imports to the Steel, Manufacturing, and Iron Ore Industries of Michigan and the Great Lakes States* (Houghton: Bureau of Industrial Development, School of Business and Engineering Administration, Michigan Technological University, 1980), 7.

85. Peter F. Marcus, Karlis M. Kirsis, and Peter J. Kakela, *Cleveland Cliffs [sic] Viability in a Restructured North American Iron Ore Industry,* World Steel Dynamics, Report to the State of Michigan (New York: Paine Webber, 1987). The report has no pagination.

86. CCI, *Annual Report,* 1996, 9, CCI-HF.

87. "Profiles: M. Thomas Moore," *Cliffs News,* Oct. 1989, 7.

88. Marcus, Kirsis, and Kakela, *Cleveland Cliff's Viability,* 2, 13.

89. CCI, *Minutes of the Board of Directors,* Sept. 8, 1987, CCI-HF.

90. Petersen interview.

91. "Self-fluxing Iron Pellet Developed," *Blast Furnace and Steel Plant* 50 (Aug. 1962); "Cleveland-Cliffs Sees Potential for Self-fluxed Iron Pellets," *American Metal Market,* Feb. 28, 1963; Tom Hayes, "Some Steelmakers Say, 'Make Mine Fluxed,'" *Cliffs News,* Fall 1987, 6.

92. Interview with Don Gallagher by Virginia Dawson, Jan. 19, 2009.

93. Ibid.

94. Great Lakes Commission, *Steel and the Great Lakes States: A Policy Statement for the Region* (Ann Arbor, MI: Great Lakes Commission, 1989), 10.

95. Harvard Business School, "Cleveland-Cliffs, Inc," 3.

96. M. Thomas Moore, *Cleveland-Cliffs, Inc.: Pioneers and Partners* (New York: Newcomen Society of the United States, 1997), 16.

97. Ibid., 17.

98. CCI, *Annual Report,* 1987, 2, CCI-HF.

99. CCI, *Annual Report,* 1988, 2, CCI-HF.

100. M. Thomas Moore, Address to Minnesota Section of AIME, reprinted in *Cliffs News,* Mar. 1988, 4.

101. Harvard Business School, "Cleveland-Cliffs, Inc," 5. Also Tom Moore, "Industry Turnaround—Real or Aberration?" speech given Jan. 13, 1988, Duluth, MN, AIME Minnesota Section, bound volume of Moore's speeches, 1987–97, box 58, DSC-CSU.

102. Hall, *Steel Phoenix,* 136, quote, 143.

103. "Bankruptcy Cases Inch toward 1990 Recovery," *Cliffs News,* June 1989, 1, 3.

104. Memo from J. S. Brinzo to M. T. Moore, "'Grand Slam' North American Iron Ore Strategy," Oct. 15, 1990, CCI-HF.

105. Ibid.

106. CCI, *Minutes of the Board of Directors,* Jan. 8, 1991; CCI, *Minutes of the Board of Directors,* Apr. 9, 1991, CCI-HF.

107. Moore, *Cleveland-Cliffs, Inc.,* 17.

108. Quoted in Harvard Business School, "Cleveland-Cliffs, Inc," 7.

109. Brinzo interview.

110. CCI, *Minutes of the Board of Directors,* Nov. 10, 1992, CCI-HF.

111. Jeanne Weitman, "Tilden Partners Change Mine Plan," *Cliffs News,* Apr. 1994, 1.

112. Dale Hemmila, "Tilden Mine Tries Once Again for Elusive Production Record," *Cliffs News,* Feb. 1999, 1.

113. Interview with A. Stanley West by Virginia Dawson, Aug. 16, 2007.

114. "The Options: Republic or Tilden," *Cliffs News,* June 1989, 1.

115. CCI, *Minutes of the Board of Directors,* July 10, 1990, CCI-HF.

116. CCI, *Annual Report,* 1997, 9, CCI-HF; interview with John Brinzo and Dana Byrne by Virginia Dawson and Terry Reynolds, Feb. 9, 2007.

117. Brinzo and Byrne interview.

118. CCI, *Minutes of the Board of Directors,* Sept. 13, 1994, CCI-HF.

119. A. Stanley West, "1996, the Year of Contradiction: Steel Sees High Volume, More Competition," *Cliffs News,* Oct. 1995, 2.

120. CCI, *Annual Report,* 1998, 6, CCI-HF.

121. Ibid., 8.

CHAPTER 7

1. Clare Ansberry, "Seizing the Moment: Steelmakers' Troubles Create an Opening for an Iron-Ore Giant," *Wall Street Journal,* Oct. 17, 2001.

2. Interview with William Calfee by Virginia Dawson, Nov. 8, 2007.

3. Ibid.

4. Interview with John Brinzo by Virginia Dawson, Oct. 19, 2006.

5. Ibid.

6. CCI, *Annual Report,* 1997, 3–4, CCI-HF.

7. John Brinzo, "Face Up to Realities," *Cliffs News,* Summer 1999, 1.

8. Address by John Brinzo, Shareholders' Meeting, May 9, 2000, CCI-HF.

9. John Brinzo, "What's Wrong with This Picture?" *Cliffs News,* Dec. 1998, 1.

10. Robert P. Rogers, *An Economic History of the American Steel Industry* (London: Routledge, 2009), 182–84.

11. Ibid., 184.

12. "Steel Vice," *Wall Street Journal,* Oct. 1, 1998. See also "Ailing Steel Industry Launches a Battle against Imports" in the same issue.

13. Leslie Wayne, "American Steel at the Barricades," *New York Times,* Dec. 10, 1998.

14. "We Stand Up for Steel," *Marquette Mining Journal,* special edition, Mar. 9, 1999.

15. Ibid., 14.

16. Rogers, *Economic History,* 185.

17. CCI, *Minutes of the Board of Directors,* Jan. 27, 2000, CCI-HF.

18. CCI, *Minutes of the Board of Directors,* Sept. 12, 2000, CCI-HF. "Testimony of Michael P. Mlinar," Public Hearing on Section 232, July 15, 2001, 232 Investigation files, Files of Dana Byrne, CCI-HF.

19. CCI, *Annual Report,* 2000, 5, CCI-HF.

20. "Brinzo Reports on State of the Company at Cleveland-Cliffs Annual Shareholders Meeting," May 8, 2001, CCI-HF.

21. *Cliffs News,* Spring 2001, 1.

22. CCI, *Minutes of the Board of Directors,* Nov. 13, 2001, CCI-HF.

23. CCI, *Annual Report,* 2001, 2, CCI-HF.

24. Peter Krouse, "Crisis in Steel Industry Creates Mine Ownership Opportunities," *Cleveland Plain Dealer,* July 29, 2001.

25. Ansberry, "Seizing the Moment."

26. Letter to Shareholders, CCI, *Annual Report,* 2001, 3, CCI-HF. See also Robert Casey, "Blast Furnace," in *The Iron and Steel Industry in the Twentieth Century: Encyclopedia of American Business History and Biography,* ed. Bruce E. Seely (New York: Facts on File, 1994).

27. "Brinzo: Industry Can Be Saved, but Some Jobs Will Be Lost," *Marquette Mining Journal,* Aug. 29, 2001, 1A.

28. James Oberstar and Bart Stupak to Norman Mineta, Jan. 16, 2001, 232 Investigation files, Files of Dana Byrne, CCI-HF. See *Fed. Register,* Feb. 6, 2001, 9067.

29. Louis Cappo, "An Open Letter to the President of the United States," *Roll Call,* Apr. 26, 2001.

30. Carl L. Valdiserri to Brad Botwin, Apr. 6, 2001; Bart Stupak to Brad Botwin, Apr. 9, 2001, 232 Investigation, Files of Dana Byrne, CCI-HF.

31. Keith Busse, CEO Steel Dynamics, Inc., to Brad Botwin, U.S. Dept. Commerce, Apr. 9, 2001, 232 Investigation, Files of Dana Byrne, CCI-HF.

32. "Another Earful for the DOC: Public Hearing in Marquette," *Skillings' Mining Review,* July 21, 2001, 4–5, 11 (quotation, 4). Statement of Senator Carl Levin, U.S. Dept. of Commerce Section 232 Field Hearing, July 15, 2001, 232 Investigation, Files of Dana Byrne, CCI-HF.

33. See Peter J. Kakela, "National Security and US Imports of Semi-finished Steel," part 1, "Federal Policies That Insured a Domestic Iron Ore Supply," and part 2, "US Iron Ore as a Building Block in the American Economy," *Skillings' Mining Review,* Oct. 27, 2001, and Nov. 3, 2001.

34. CCI, *Annual Report,* 2001, CCI-HF.

35. Carol Poh Miller, "Iron and Steel Industry," in *Encyclopedia of Cleveland History* (Bloomington: Indiana University Press, 1996), 580.

36. Interview with John Brinzo and Dana Byrne by Virginia Dawson and Terry Reynolds, Feb. 9, 2007.

37. Nicholas Stein, "Wilbur Ross Is a Man of Steel . . .," *Fortune,* May 28, 2003, http://money.cnn.com/magazines/fortune/fortune_archive. See also Daniel Gross, "The Bottom-feeder King," *New York Maga-*

zine, Nov. 1, 2004; Claudia H. Deutsch, "Got an Ailing Business? He Wants to Make It Right," *New York Times,* Oct. 26, 2005.

38. Stein, "Wilbur Ross."

39. Calfee interview.

40. Brinzo interview.

41. "Web" Webster, "Cliffs CEO Upbeat regarding 2002," *Skillings' Mining Review,* Mar. 23, 2002, 12.

42. Andrea Holecek, "WL Ross Wins LTV Backing; U.S. Steel Drops Out," *Times,* Munster, IN, Knight Ridder / Tribune Business News, Feb. 28, 2002.

43. CCI, *Annual Report,* 2002, CCI-HF.

44. Todd Shyrock, "Man of Steel: How Rodney Mott and ISG Reinvented the Domestic Steel Industry," *SmartBusiness,* Jan. 2005, www.sbnonline.com; "Contract Accord at International Steel," *New York Times,* Dec. 24, 2002.

45. Bud Sargent, "Miners Leave Message for Bush," *Marquette Mining Journal,* Mar. 2, 2002.

46. James Toedtman, "Bush to Lift Sanctions on Steel," *Newsday,* Dec. 3, 2003, www.chicgotribune.com.

47. "Iron-willed competitors," *Cliffs News,* Dec. 1998, 5.

48. John Brinzo, "We Are Investing in Our Future," *Cliffs News,* June 1998, 7.

49. Don Ryan, "Mines Could Witness 50% turnover," *Cliffs News,* July 1998, 2.

50. *Cliffs News,* Spring 2000.

51. James Lake, "Groups Continue to Fight Mine Expansion in Court," *Marquette Mining Journal,* Nov. 22, 2000.

52. John G. Meier and Charles L. Wolverton, "Wetland Mitigation: Understanding This Dynamic Issue," *Skillings' Mining Review,* Apr. 20, 2002. See also the website of the National Mining Congress, www. nma.org/issues/environment/reclamation/reclamation_mod.asp.

53. "Letter to Shareholders," CCI, *Annual Report,* 2001, 2, CCI-HF.

54. Brinzo and Byrne interview; CCI, *Minutes of the Board of Directors,* July 9, 2002, CCI-HF. EVTAC declared bankruptcy in May 2003.

55. CCI, *Annual Report,* 2002, 2, CCI-HF. "Cliffs Increases Ownership of Empire Mine," *Skillings' Mining Review,* Feb. 2003.

56. CCI, *Annual Report,* 2002, 4, CCI-HF.

57. "Empire and Tilden Mines Operate as A Combined Unit," news release, Cleveland-Cliffs Iron Co., June 11, 2003, CCI-HF; John E. Sacco, "Cleveland-Cliffs Cuts Management by 22%," *American Metal Market,* Aug. 1, 2003; "Cliffs Moves to Improve Bottom Line," *Skillings' Mining Review,* Sept. 2003.

58. "LS&I Railroad to Become Part of Cliffs Michigan Mining Company," news release, Sept. 11, 2003, CCI-HF.

59. Larry Fortner, "Don Prahl, Northshore's New VP & MM Is Bullish on Iron Ore," *Skillings' Mining Review,* Dec. 2003.

60. William S. Kirk, "China's Emergence as the World's Leading Iron Ore–Consuming Country," parts 1 and 2, *Skillings' Mining Review,* July 2003, and Aug. 2003.

61. Peter J. Kakela, "China's Impact on North American Iron Mines," *Skillings' Mining Review,* July 2005.

62. Gail Rosenquist, "Cleveland-Cliffs, Laiwu Steel Bid for Evtac," and "Evtac to China; Possible Shipping Routes," *Skillings' Mining Review,* Nov. 2003, 17; Calfee interview.

63. John E. Sacco, "Cliffs, Laiwu Get Court OK on Evtac Joint Acquisition," *American Metal Market,* Dec. 1, 2003, General OneFile. Gale, Michigan Tech. University. 7 Dec. 2007.

64. CCI, *Annual Report,* 2003, 2, CCI-HF; Ivan Hohnstadt, "United Taconite—East Meets West," *Skillings' Mining Review,* Jan. 2004, 3; Kakela, "China's Impact," 22–23; and interview with Dana Byrne by Virginia Dawson, Oct. 30, 2007.

65. CCI, *Annual Report,* 2003, 11, CCI-HF.

66. Thomas W. Gerdel, "ISG Buying Caribbean Iron-Making Plant Operation Owned by Cleveland Cliffs, German Firm," *Cleveland Plain Dealer,* May 13, 2004. Cleveland-Cliffs had purchased LTV's stake in the HBI venture in 2000 for $2 million, allowing LTV to write off most of its $84 million investment. ISG purchased the idled HBI plant from Cleveland-Cliffs in June 2004.

67. Ivan Hohnstadt, "When It Rains, It Pours! Cliffs Updates 2003 Outlook," *Skillings' Mining Review,* June 2003.

68. Interview with Don Gallagher by Virginia Dawson, Jan. 19, 2009.

69. "A Letter to Employees of Cleveland-Cliffs and Its Managed Mines," *Marquette Mining Journal,* July 7, 2004; "USWA Blasts Cleveland Cliffs' Decision to Operate Mines in the Event of a Labor Dispute, Urges Company to Live Up to Retiree Promises," press release, July 7, 2004, CCI-HF; "CCI Training Replacement Workers," *Marquette Mining Journal,* July 16, 2004.

70. "CCI Training Replacement Workers."

71. Jacqueline Perry, "State Senator Weighs in on Iron Mining Labor Negotiations," *Marquette Mining Journal,* July 24, 2004.

72. Stephen Kinzer, "In Minnesota's Iron Range, a Rare Victory for Labor," *New York Times,* Aug. 6, 2004; "Iron Producer Settles with Steelworkers," *New York Times,* Oct 12, 2004.

73. Phyllis Berman, "Iron Monger," *Forbes,* Apr. 11, 2005.

74. Larry Fortner, "Cliffs/Union Settle New Labor Contracts," *Skillings' Mining Review,* Sept. 2004.

75. CCI, *Annual Report,* 2004, CCI-HF.

76. Ibid., 3; Brinzo interview.

77. Rogers, *Economic History,* 192.

78. CCI, *Annual Report,* 2004, 3–4, CCI-HF.

79. "Opportunity 'Down Under': Portman Offers Attractive Growth Opportunity," *Cliffs Connections,* Summer 2005, 1.

80. "Australian Iron Miner Has Long List of Suitors," *American Metal Market,* Jan. 12, 2005; "CSFB Ups Portman Stake as Cliffs Cuts Threshold," *American Metal Market,* Feb. 23, 2005, www.amm.com.

81. "Portman White Knight Said Waiting in Wings," *American Metal Market,* Mar. 11, 2005. www.amm.com.

82. "Portman's Eldridge Resigns Post as Cleveland-Cliffs Tightens Grip," *American Metal Market,* Apr. 14, 2005, www.amm.com.

83. John Brinzo, "Chairman's Letter to Shareholders," CCI, *Annual Report,* 2005, 2, CCI-HF.

84. CCI, *Annual Report,* 2005, 9, CCI-HF.

85. Samuel P. Orth, *A History of Cleveland, Ohio* (Chicago: S. J. Clarke, 1910), 2:31–32.

BIBLIOGRAPHY

PUBLISHED BOOKS AND ARTICLES

Adams, Walter, and Joel B. Dirlam. "Steel Imports and Vertical Oligopoly Power." *American Economic Review* 54 (Sept. 1964): 626–55.

——. "Steel Imports and Vertical Oligopoly Power: Reply." *American Economic Review* 56 (Mar.–May 1966): 160–68.

"Ailing Steel Industry Launches a Battle against Imports." *Wall Street Journal,* Oct. 1, 1998.

Alanen, Arnold R. "Companies as Caretakers: Paternalism, Welfare Capitalism, and Immigrants in the Lake Superior Mining Region." In *A Century of European Migrations, 1830–1930,* edited by Rudolph J. Vecoli and Suzanne M. Sinke, 364–91. Urbana: University of Illinois Press, 1991.

——. "The Settlements of Marquette County, Michigan." In *Historic Resources of the Iron Range in Marquette County, Michigan, 1844–1941,* edited by William Mulligan, sec. 9, 46pp. Marquette: Economic Development Corporation of the County of Marquette, 1991.

Alderton, Jenna. "A Slow Start with a Promising End: Tsu-Ming Han." *Red Dust* (1994): 1–3.

Aldrich, Mark. *Safety First: Technology, Labor, and Business in the Building of American Work Safety, 1870–1939.* Baltimore: Johns Hopkins University Press, 1997.

American Iron and Steel Association. *Annual Statistical Report,* 1940.

American Iron Mining & Manufacturing Co. *Prospectus Preparatory to the Organization of the American Iron Mining & Manufacturing Co. . . .* New York: Francis Hart, 1864.

Anderson et al. v. Cleveland-Cliffs Iron Co. et al. 579838, Common Pleas Court of Ohio, June 9, 1948. Lexis 245.

Ansberry, Clare. "Seizing the Moment: Steelmakers' Troubles Create an Opening For an Iron-Ore Giant." *Wall Street Journal,* Oct. 17, 2001.

Argall, George O., Jr. "Eagle Mills Pellet Plant Opens New Era for Michigan Hematite." *Mining World* 18 (Dec. 1956): 42–46.

"Austin Farrell." *Proceedings of the Lake Superior Mining Institute* 25 (1926): 259.

"Australian Iron Miner Has Long List of Suitors." *American Metal Market,* Jan. 12, 2005.

Bacon, Don H. "The Development of Lake Superior Iron Ore." *Transactions of the American Institute of Mining Engineers* 27 (1897): 341–44.

Balch, William Ralston, comp. *The Mines, Miners and Mining Interests of the United States in 1882.* Philadelphia: Mining Industrial Publishing Bureau, 1882.

Barger, Harold, and Sam H. Schurr. *The Mining Industries, 1899–1939: A Study of Output, Employment and Productivity.* 1944. Reprint, New York: Arno, 1972.

Barloon, Marvin. "The Question of Steel Capacity." *Harvard Business Review* 27, no. 2 (1949): 209–36.

Barnett, Donald F., and Robert W. Crandall. *Up from the Ashes: The Rise of the Steel Minimill in the United States.* Washington, DC: Brookings Institution, 1986.

Barnett, Donald F., and Louis Schorsch. *Steel: Upheaval in a Basic Industry.* Cambridge, MA: Ballinger, 1983.

Bartlett, Robert V. *The Reserve Mining Controversy: Science, Technology, and Environmental Quality.* Bloomington: Indiana University Press, 1980.

Baxter, William, Jr. "The Electric Light in Mining Operations." *Engineering and Mining Journal,* July 4, 1896, 6.

Beard's Directory and History of Marquette County, with Sketches of the Early History of Lake Superior, Its Mines, Furnaces, etc. Detroit: Hadger & Bryce, 1873.

Beck, John P. "Homegrown." *Michigan History* 79 (1995): 19–25.

Bell, Lowthian. "On the American Iron Trade and Its Progress during Sixteen Years." In Iron and Steel Institute (London), *The Iron and Steel Institute in America in 1890,* 1–208. London: E. & F. N. Spon, 1890.

Berman, Phyllis. "Iron Monger." *Forbes,* Apr. 11, 2005, 108–9.

Berteling, John F. "Notes on Mining the North Palms Orebody." *Proceedings of the Lake Superior Mining Institute* 25 (1926): 52–62.

Bingham, Clara, and Laura Leedy Gansler. *Class Action: The Story of Lois Jenson and the Landmark Case That Changed Sexual Harassment Law.* New York: Doubleday, 2002.

Birkinbine, John. "From Mine to Furnace." *Cassier's Magazine* 4 (1893): 163–74, 243–56, 345–54, 425–35.

———. "The Iron Ore Supply." *Transactions of the American Institute of Mining Engineers* 27 (1897): 519–28.

———. "Notes upon American Iron Deposits." In Iron and Steel Institute (London), *The Iron and Steel Institute in America in 1890,* 361–602. London: E. & F. N. Spon, 1890.

Birkinbine, John, and Thomas A. Edison. "The Concentration of Iron-Ore." *Transactions of the American Institute of Mining Engineers* 17 (1889): 728–44.

Black, John D., and L. C. Gray. "Land Settlement and Colonization in the Great Lakes States." USDA Bulletin 1295, Mar. 23, 1925.

Blake, William P. "Provision for the Health and Comfort of Miners—Miners' Homes." *Transactions of the American Institute of Mining Engineers* 3 (1874–75): 218–28.

"A Bond of Interest." *Harlow's Wooden Man* 13, no. 5 (1978).

Boyd, Ford E. "Slushing Practice in the Mines of the Oliver Iron Mining Company." *Proceedings of the Lake Superior Mining Institute* 23 (1923): 68–77.

Boyum, Burton H. *The Marquette District of Michigan.* Ishpeming: Cleveland-Cliffs, 1975.

———, ed. *The Mather Mine: Negaunee and Ishpeming, Michigan.* Marquette: Marquette County Historical Society, 1979.

Brandes, Stuart D. *American Welfare Capitalism, 1880–1940.* Chicago: University of Chicago Press, 1976.

Bridge, James Howard. *The Inside History of the Carnegie Steel Company.* New York: Aldine, 1903.

Brinks, Herbert J. "Marquette Iron Range Strike, 1895." *Michigan History* 50 (1966): 293–305.

Brinzo, John. "A Letter to Employees of Cleveland-Cliffs and Its Managed Mines." *Marquette Mining Journal,* July 7, 2004.

Brooks, Thomas B., et al. *Geological Survey of Michigan: Upper Peninsula, 1869–1873.* 2 vols. and atlas. New York: Julius Bien, 1873.

Brown, Peter. "Pioneer Plant Dust Rapped." *Marquette Mining Journal,* July 21, 1976.

Bunyak, Dawn. *Frothers, Bubbles and Flotation: A Survey of Flotation Milling in the Twentieth-Century Metals Industry.* Denver: National Park Service, 1998.

Burt, John S. *They Left Their Mark: William Austin Burt and His Sons, Surveyors of the Public Domain.* Rancho Cordova, CA: Landmark Enterprises, 1985.

Calomiris, Charles W., *and Larry Schweikart.* "The Panic of 1857: Origins, Transmission, and Containment." *Journal of Economic History* 51 (1991): 807–34.

Campbell, Alexander. "The Upper Peninsula: An Address on the Climate, Soil, Resources, Development, Commerce and Future of the Upper Peninsula of Michigan" [delivered Feb. 6, 1861]. *Michigan Pioneer and Historical Collections* 3 (1881): 247–65.

Campbell, Tom. "Long-Term Ore Projects Stirring." *Iron Age,* Mar. 6, 1958, 81–82.

———. "Why American Steel Companies Search World-wide for Ore." *Iron Age,* Mar. 31, 1960, 112–13.

Cappo, Louis. "An Open Letter to the President of the United States." *Roll Call,* Apr. 26, 2001.

Cascade Iron Co. *Prospectus of the Cascade Iron Company of Lake Superior.* Philadelphia: W. P. Kildare, 1865.

Casey, Robert. "Blast Furnace." In *The Iron and Steel Industry in the Twentieth Century: Encyclopedia of American Business History and Biography,* edited by Bruce E. Seely, 47–48. New York: Facts on File, 1994.

"CCI Training Replacement Workers." *Marquette Mining Journal,* July 16, 2004.

"Champion Mine's Old Stockpiles Yield Ore in Emergency." *Engineering and Mining Journal* 142 (Nov. 1941): 41.

Chandler, Alfred D., Jr. *Scale and Scope: The Dynamics of Industrial Capitalism.* Cambridge, MA: Harvard University Press, Belknap Press, 1990.

———. *The Visible Hand: The Managerial Revolution in American Business.* Cambridge, MA: Harvard University Press, Belknap Press, 1977.

Channing, J. Park. "Mine Accidents: Address of the Retiring President." *Proceedings of the Lake Superior Mining Institute* 3 (1895): 34–48.

Cleary, John J. "Cliffs Buys 78% of Detroit Steel." *Cleveland Plain Dealer,* June 13, 1970.

"Cleveland-Cliffs Acquires Int'l Harvester Iron Mines." *Skillings' Mining Review,* Mar. 8, 1947, 1.

Cleveland-Cliffs Iron Co. *The Cleveland-Cliffs Iron Company and Its Extensive Operations in the Lake Superior District.* Cleveland: Cleveland-Cliffs, 1929. Also published as an issue of *Mining Congress Journal,* Oct. 1929.

———. *The Cleveland-Cliffs Iron Company: An Historical Review of This Company's Development and Resources Issued in Commemoration of Its Seventieth Anniversary, 1850–1920.* Cleveland: Cleveland-Cliffs, 1920.

———. *Description of the Exhibit of the Cleveland-Cliffs Iron Co. and Pioneer Iron Co. at the Louisiana Purchase Exposition . . . St. Louis, 1904.* N.p., [1904?] Copy in Cleveland Public Library.

Cleveland-Cliffs Iron Co. et al. *Plan of Employees' Representation, July, 1933.* [Ishpeming?]: Cleveland-Cliffs Iron et al., 1933. Copy in Longyear Research Library, Marquette County Historical Society, Marquette.

"Cleveland-Cliffs Iron Co. Marks Centennial in Iron Ore Mining." *Skillings' Mining Review,* June 24, 1950, 1–2.

Cleveland-Cliffs Iron Company v. Department of Revenue. Docket no. 31, Supreme Court of Michigan, Dec. 6, 1950. Lexis 303.

"Cleveland-Cliffs, McLouth and Bethlehem Push Mine Output for Big Pellet Project." *American Metal Market,* Dec. 28, 1963.

"Cleveland-Cliffs Sees Potential for Self-fluxed Iron Pellets." *American Metal Market,* Feb. 28, 1963.

"Cleveland-Cliffs' Spies-Virgil Iron Mine Has Important Future." *Skillings' Mining Review,* July 28, 1923, 1.

"Cleveland-Cliffs' Treasurer: What He Does." *Business Week,* June 7, 1952, 68–75.

Cleveland Iron Mining Co. *Act of Incorporation with the Articles of Association and By Laws of the Cleveland Iron Mining Company.* Cleveland: Sanford & Hayward, 1853.

———. *Annual Report of the Directors to the Stockholders . . . for the Year Ending May 18, 1864.* Cleveland: Sanford & Hayward, 1864.

———. *Annual Report of the Directors to the Stockholders . . . for the Year Ending May 19, 1869.* Cleveland: Sanford & Hayward, 1869.

———. *Annual Report of the Directors to the Stockholders . . . for the Year Ending May 17, 1871.* Cleveland: Sanford & Hayward, 1871.

———. *Annual Report of the Directors to the Stockholders . . . for the Year Ending May 21, 1873.* Cleveland: Sanford & Hayward, 1873.

———. *Annual Report of the Directors to the Stockholders . . . for the Year Ending May 17, 1876.* Cleveland: Sanford & Hayward, 1876.

———. *Charter and By-laws of the Cleveland Iron Mining Company of Michigan.* Cleveland: Sanford & Hayward, 1851.

"Cliffs' Contract for Sale of Robe River Iron Ore." *Skillings' Mining Review,* May 23, 1969, 4.

Cliffs Corporation v. Evatt, Tax Commr. et al. No. 28471, Supreme Court of Ohio, June 11, 1941. Lexis 471.

"Cliffs' Diversification with Iron Ore Backbone." *Skillings' Mining Review,* Aug. 12, 1967, 8.

"Cliffs Fight Settled." *Business Week,* Mar. 1, 1947, 19–20.

"Cliffs Increases Ownership of Empire Mine." *Skillings' Mining Review,* Feb. 2003, 22.

"Cliffs Moves to Improve Bottom Line." *Skillings' Mining Review,* Sept. 2003, 26.

Cochran, Robert. "New Plant Successfully Floats Michigan Jasper." *Engineering and Mining Journal* 155 (Aug. 1954): 100–104, 114.

Condren, T. L. "Shaft House of Imposing Architecture at the Ishpeming (Michigan) Mine of the Cleveland Cliffs Iron Company." *Journal of the Western Society of Engineers* 26 (1921): 22–32.

Conibear, William. "Development of Safety in the Iron Mines of Michigan." *Proceedings of the Lake Superior Mining Institute* 29 (1936): 180–84.

———. "System of Safety Inspection of the Cleveland-Cliffs Iron Company." *Proceedings of the Lake Superior Mining Institute* 17 (1912): 94–111.

Counties of White and Pulaski, Indiana, Historical and Biographical. Chicago: F. A. Battey, 1883.

"The Course of the Iron Markets in 1896." *Engineering and Mining Journal,* Jan. 2, 1897, 12–16.

Cowpland, C. C. "A Resume of the Early History of the Explosives Industry in Marquette County." *Proceedings of the Lake Superior Mining Institute* 29 (1935): 21–28.

Crowell & Murray, Inc. *The Iron Ores of Lake Superior.* 7th ed. Cleveland: Penton, 1930.

"CSFB Ups Portman Stake as Cliffs Cuts Threshold." *American Metal Market,* Feb. 23, 2005.

Cummings, George P. "Reminiscences of the Early Days on the Marquette Range." *Proceedings of the Lake Superior Mining Institute* 14 (1909): 214–15.

Daddow, Samuel H., and Benjamin Bannan. *Coal, Iron, and Oil; or, The Practical American Miner. . . .* Pottsville, PA: Benjamin Bannan, 1866.

Davis, E. W. "Change in U.S. Steel Iron Ore Marketing Policy Endangers Range Communities and State." *Minnesota Municipalities,* Mar. 1940, 78–82.

———. *Pioneering with Taconite.* St. Paul: Minnesota Historical Society, 1964.

Dawson, Christopher J. *Steel Remembered: Photographs from the LTV Steel Collection.* Kent, OH: Kent State University Press, in cooperation with Western Reserve Historical Society, 2008.

Dawson, Virginia P. "Knowledge Is Power: E. G. Bailey and the Invention and Marketing of the Bailey Boiler Meter." *Technology and Culture* 37 (1996): 493–526.

DeMark, Judith Boyce. "Iron Mining and Immigrants, Negaunee and Ishpeming, 1870–1910." In *A Sense of Place: Michigan's Upper Peninsula,* edited by Russell M. Magnaghi and Michael T. Marsden, 35–42. Marquette: Northern Michigan University Press, 1997.

Dennison, S. R. "Vertical Integration and the Iron and Steel Industry." *Economic Journal* 49 (June 1939): 224–58.

Deutsch, Claudia H. "Got an Ailing Business? He Wants to Make It Right." *New York Times,* Oct. 26, 2005.

DeVaney, Fred D. "Flotation of Lake Superior District Iron Ores." *Skillings' Mining Review,* Feb. 11, 1950, 1, 4, 6, 15.

Dickinson, John N. *To Build a Canal: Sault Ste. Marie, 1853–54 and After.* Columbus: Published for Miami University by the Ohio State University Press, 1981.

Disturnell, John. *Trip through the Lakes of North America.* New York: J. Disturnell, 1857.

"DNR Rates Shutdown Date Crucial Part of CCI Order." *Marquette Mining Journal,* Jan. 3, 1977.

Drinker, Henry S. *A Treatise on Explosive Compounds, Machine Rock Drills and Blasting.* New York: John Wiley, 1883.

Dunlap, Thomas, ed. *Wiley's American Iron Trade Manual. . . .* New York: John Wiley, 1874.

Earney, Fillmore C. F. "New Ores for Old Furnaces: Pelletized Iron." *Annals of the Association of American Geographers* 59 (Sept. 1969): 512–34.

Eaton, Lucien. "Present Scraping Practices at the Cliffs Shaft Mine." *Proceedings of the Lake Superior Mining Institute* 24 (1926): 121–25.

"Eaton Produces Records Showing Some of Cleveland Trust's Holdings." *Cleveland Press,* June 20, 1967.

Eckel, Edwin C. "Iron Ore, Pig Iron, and Steel." In *Mineral Resources of the United States, 1906,* 67–102. Washington, DC: Government Printing Office, 1907.

"Ecology Enlivens Cliffs Meeting." *Cleveland Press,* Apr. 25, 1974.

Elliott, S. R. "Athens System of Mining." *Transactions of the American Institute of Mining Engineers* 66 (1922): 220–24.

Elliott, S. R., J. E. Jopling, R. J. Cenneour, and E. L. Derby. "Mining Methods of Marquette District, Michigan." *Transactions of the American Institute of Mining Engineers* 72 (1925): 122–38.

Ellis, William Donohue. *The Cuyahoga.* New York: Holt, Rinehart & Winston, 1967.

"Empire Builders—Business Biographies, No. 8: The Cleveland Cliffs Iron Co." *Clevelander,* Apr. 1939, 10, 58, 60.

"Empire Mine Could Push Marquette Iron Range to Record Production of Ore." *Marquette Mining Journal,* Apr. 23, 1963.

Engineering and Mining Journal. *Handbook of Mining Details.* New York: McGraw-Hill, 1912.

Evans, Henry Oliver. *Iron Pioneer: Henry W. Oliver, 1840–1904.* New York: E. P. Dutton, 1942.

Everett, Philo M. "Recollections of the Early Explorations and Discovery of Iron Ore on Lake Superior." *Michigan Pioneer Historical Collections* 11 (1887): 161–74.

Facts about Michigan's Specific Tax on Iron Ore. Cleveland-Cliffs monograph, Aug. 1977. Copy in American Iron and Steel Institute Vertical File, Hagley Museum and Library, Wilmington, DE.

"False Rumor concerning Cleveland-Cliffs." *Skillings' Mining Review,* Feb. 2, 1924, 9.

"Fight over Cliffs." *Business Week,* Jan. 4, 1947, 17–18.

Fine, M. M. "The Beneficiation of Iron Ores." *Scientific American,* Jan. 1968, 28–35.

First Boston Corporation. *The Iron Ore Industry and the Cleveland-Cliffs Iron Company.* New York: First Boston, 1955.

Fortner, Larry. "Cliffs/Union Settle New Labor Contracts." *Skillings' Mining Review,* Sept. 2004, 14.

———. "Don Prahl, Northshore's New VP and MM Is Bullish on Iron Ore." *Skillings' Mining Review,* Dec. 2003, 4–5.

Foster, J. W. *The Geology and Metallurgy of the Iron Ores of Lake Superior, Being a Report Addressed to the Board of Directors of the Iron Cliffs Company.* New York, 1865.

"Frank P. Mills, Sr." *Proceedings of the Lake Superior Mining Institute* 15 (1910): 212–13.

French, Benjamin F. *History of the Rise and Progress of the Iron Trade of the United States from 1621 to 1857.* New York: Wiley & Halsted, 1858.

Friggens, Thomas G. *No Tears in Heaven: The 1926 Barnes-Hecker Mine Disaster.* 2nd ed. Lansing: Michigan History Magazine, 2002.

Fritz, John. *Autobiography of John Fritz.* New York: John Wiley, 1912.

F. S. Smithers & Co. *The Iron Ore Industry and the Cleveland-Cliffs Iron Company, the Hanna Mining Company, the M. A. Hanna Company.* New York: F. S. Smithers, 1960.

Fusfield, Daniel R. "Joint Subsidiaries in the Iron and Steel Industry." *American Economic Review* 48 (1958): 586.

Gaertner, John T. *The Duluth, South Shore and Atlantic Railway: A History of the Lake Superior District's Pioneer Iron Ore Hauler.* Bloomington: Indiana University Press, 2009.

Gandre, Donald A. "Recent Changes in the Flow Pattern of Iron Ore on the Great Lakes." *Inland Seas* 27 (1971): 247–59.

Gerdel, Thomas W. "ISG Buying Caribbean Iron-Making Plant Operation Owned by Cleveland Cliffs, German Firm." *Cleveland Plain Dealer,* May 13, 2004.

Girdler, Tom M. *Boot Straps: The Autobiography of Tom M. Girdler.* New York: Charles Scribner, 1943.

Gleisser, Marcus. *The World of Cyrus Eaton.* New York: A. S. Barnes, 1965.

Gold, Bela, William S. Pierce, Gerhard Rosegger, and Mark Perlman. *Technological Progress and Industrial Leadership: The Growth of the U.S. Steel Industry, 1900–1970.* Lexington, MA: Lexington Books, 1984.

Goodman, Robert J. "The Future of Iron Mining on the Marquette Range, Marquette County, Michigan." *Papers of the Michigan Academy of Science, Arts, and Letters* 35 (1949): 189–95.

Gordon, Robert B. *American Iron, 1607–1900.* Baltimore: Johns Hopkins University Press, 1996.

———. *A Landscape Transformed: The Ironmaking District of Salisbury, Connecticut.* Oxford: Oxford University Press, 2001.

Great Lakes Commission. *Steel and the Great Lakes States: A Policy Statement for the Region.* Ann Arbor, MI: Great Lakes Commission, 1989.

"Great Lakes Region Iron Ore Shipments." *Skillings' Mining Review,* May 31, 1969, 1, 4, 19.

Greenwald, William I. "Supply Shifts and Iron Ore Pricing." *Journal of Industrial Economics* 9 (Apr. 1961): 170–80.

Greenwood, John O. *The Fleets of Cleveland-Cliffs, Detroit and Cleveland Navigation. . . .* Fleet Histories Series 6. Cleveland: Freshwater, 1998.

Hall, Christopher G. L. *Steel Phoenix: The Fall and Rise of the U.S. Steel Industry.* New York: St. Martin's, 1997.

Harrigan, Kathryn Rudie. *Joint Ventures, Alliances, and Corporate Strategy.* Washington, DC: Beard, 2003.

Harrison, H. Stuart. "The Changing Iron Ore Industry." *Skillings' Mining Review,* Oct. 16, 1965, 1, 6–7.

———. *The Cleveland-Cliffs Iron Company.* New York: Newcomen Society in North America, 1974.

———. "Iron Ore—The Big Change." *Analysts Journal* 13 (June 1957): 67–68.

———. "Where Is the Iron Ore Coming From?" *Analysts Journal* 9 (June 1953): 98–101.

Harrison, W. Douglas, Peter F. Mason, and Fillmore C. F. Earney. "Iron Ore Pellets." *Annals of the Association of American Geographers* 61 (June 1971): 422–26.

Harte, Charles Rufus. "Connecticut's Iron and Copper." *60th Annual Report of the Connecticut Society of Civil Engineers* (1944): 143–45.

Harvard Business School. "Cleveland Cliffs Inc." Case 9-293-051. *Harvard Business School Review,* Sept. 21, 1993.

Hatcher, Harlan. *A Century of Iron and Men.* Indianapolis: Bobbs-Merrill, 1950.

Havighurst, Walter. *Vein of Iron: The Pickands Mather Story.* Cleveland: World, 1958.

Hayden, Ellzey, and Lucien Eaton. "Building Reinforced-Concrete Shaft Houses." *Transactions of the American Institute of Mining Engineers* 66 (1922): 225–34.

Head, Jeremiah, and Archibald P. Head. "The Lake Superior Iron Mines: Their Influence upon the Production of Iron and Steel." *Cassier's Magazine* 16 (1899): 623–46.

Hearding, John. "Early Methods of Transporting Iron Ore in the Lake Superior Region." *Proceedings of the Lake Superior Mining Institute* 29 (1935): 174–79.

"Hematite Is Mined in Michigan to Make Pellets for Steel Mills." *New York Times,* Feb. 24, 1957, 1, 4.

Hoerr, John P. *And the Wolf Finally Came: The Decline of the American Steel Industry.* Pittsburgh: University of Pittsburgh Press, 1988.

Hogan, William T. *Economic History of the Iron and Steel Industry in the United States.* 5 vols. Lexington, MA: Lexington Books, 1971.

———. *Minimills and Integrated Mills: A Comparison of Steelmaking in the United States.* Lexington, MA: Lexington Books, 1987.

———. *The 1970s: Critical Years for Steel.* Lexington, MA: Lexington Books, 1972.

———. *Steel in the 21st Century: Competition Forges a New World Order.* Lexington, MA: Lexington Books, 1994.

———. *World Steel in the 1980s: A Case of Survival.* Lexington, MA: Lexington Books, 1983.

Hohnstadt, Ivan. "When It Rains, It Pours! Cliffs Updates 2003 Outlook." *Skillings' Mining Review,* June 2003, 20.

Holbrook, Stewart H. *Iron Brew: A Century of American Ore and Steel.* New York: Macmillan, 1939.

Holecek, Andrea. "WL Ross Wins LTV Backing: U.S. Steel Drops Out." *Times,* Munster, IN, Knight Ridder / Tribune Business News, Feb. 28, 2002.

Hone, G. A., and D. S. Schoenbrod. "Steel Imports and Vertical Oligopoly Power: Comment." *American Economic Review* 56 (Mar.–May 1966): 156–60.

Hubbell, A. H. "The Mather Mine: A New Show Place for the Iron Country." *Engineering and Mining Journal* 142 (Oct. 1941): 43–46.

Huffman, Thomas R. "Exploring the Legacy of Reserve Mining: What Does the Longest Environmental

Trial in History Tell Us about the Meaning of American Environmentalism?" *Journal of Policy History* 12 (2000): 339–68.

"Humboldt Iron Mine Opens 650,000-Ton Pellet Plant." *Daily Metal Reporter,* Sept. 23, 1960.

"The Humboldt Mine." *Skillings' Mining Review,* Sept. 25, 1954, 1–2, 4.

"Humboldt Pellet Plant Marks New Era for Michigan Low Grade Iron Ores." *Skillings' Mining Review,* Oct. 1, 1960, 1, 4–5.

Hurd, Rukard. *Hurd's Iron Ore Manual.* St. Paul, MN: Rukard Hurd, 1911.

Huston, James L. *The Panic of 1857 and the Coming of the Civil War.* Baton Rouge: Louisiana State University Press, 1987.

Hybels, Robert James. "The Lake Superior Copper Fever, 1841–47." *Michigan History* 34 (1950): 97–119, 224–44, 309–26.

Ingham, John N. *Biographical Dictionary of American Business Leaders.* Westport, CT: Greenwood, 1983.

———. *The Iron Barons: A Social Analysis of an American Urban Elite, 1874–1965.* Westport, CT: Greenwood, 1978.

Iron Mining and Agriculture. Souvenir edition of the *Marquette Mining Journal,* Mar. 1, 1913.

"Iron Ore—Making and Remaking the Map of Steel." *Fortune,* May 1931, 85–92, 133–38.

"Iron Ore Pellet Production Rising: Two Firms Reveal Expansion Plans." *American Metal Market,* Dec. 17, 1969.

"Iron Ore Search Planned." *New York Times,* July 18, 1957.

"Iron Region of Lake Superior." *Mining Magazine,* July 1854, 103–4.

The Iron Resources of Michigan and General Statistics of Iron. . . . Detroit: H. Barns, 1856.

Ivens, L. E. "Market Results of Mergers, as Effected and Proposed." *Mining Congress Journal* 15 (Oct. 1929): 793, 797.

Jackson Iron Co. *16th Annual Report, June 1869.* New York: C. O. Jones, 1869.

Jeffreys-Jones, Rhodri. "Profits over Class: A Study in American Industrial Espionage." *Journal of American Studies* 6 (1972): 233–48.

Jenner, R. W. *Hardwood Distillation in Michigan: Background for Cliffs-Dow Chemical Company.* Marquette: Cliffs-Dow, n.d.

Jensen et al. v. Republic Steel Corp. 426867, Common Pleas Court of Ohio, 1940. Lexis 456.

Joblin, Maurice, pub. *Cleveland, Past and Present: Its Representative Men.* Cleveland: Fairbanks, Benedict, 1869.

Jopling, James E. "A Brief History of the Cleveland-Cliffs Iron Company." *Michigan History* 5 (1921): 150–72.

———. "Charcoal Furnaces of Michigan." *Proceedings of the Lake Superior Mining Institute* 25 (1926): 249–51.

———. "The Marquette Range—Its Discovery, Development and Resources." *Transactions of the American Institute of Mining Engineers* 27 (1897): 541–55.

Jorgenson, J. D. *Challenges Facing the North American Iron Ore Industry.* Open-File Report 2006-1061. Reston, VA: U.S. Department of the Interior, U.S. Geological Survey, 2006.

Kakela, Peter J. "China's Impact on North American Iron Mines." *Skillings' Mining Review,* July 2005, 22–24.

———. "Iron Ore: Energy, Labor, and Capital Changes with Technology." *Science,* Dec. 15, 1978, 1151–57.

———. "Iron Ore: From Depletion to Abundance." *Science,* n.s., Apr. 10, 1981, 132–36.

———. "National Security and US Imports of Semi-finished Steel," part 1, "Federal Policies That Insured a Domestic Iron Ore Supply," and part 2, "US Iron Ore as a Building Block in the American Economy." *Skillings' Mining Review,* Oct. 27, 2001, 4–6, and Nov. 3, 2001, 4–5.

Kakela, Peter J., Allen K. Montgomery, and William C. Patric. "Factors Influencing Mine Location: An Iron Ore Example." *Land Economics* 58 (Nov. 1982): 524–36.

Kelly, Robert. "A Trip to Lake Superior." *Proceedings of the Lake Superior Mining Institute* 19 (1914): 309–23.

Kelly, William. "Address of President William Kelly." *Proceedings of the Lake Superior Mining Institute* 6 (1900): 13–22.

Kennedy, James Harrison. *A History of the City of Cleveland.* Cleveland: Imperial, 1896.

———. *A History of the City of Cleveland: Biographical Volume.* Cleveland: Imperial, 1897.

———. "The Opening of the Lake Superior Iron Region." *Magazine of Western History* 2 (Aug. 1885): 346–69.

Kimball, J. P. *Notes on the Iron Ores of Marquette, Michigan: Extracts from a Report to the Iron Cliffs Mining Co., June 1, 1864.* Marquette: n.p., 1864. Copy in Bentley Historical Library, Ann Arbor, MI.

Kinzer, Stephen. "In Minnesota's Iron Range, a Rare Victory for Labor." *New York Times,* Aug. 6, 2004, 14A.

Kirby, Ralph C., and Andrew S. Prokopovitsh. "Technological Insurance against Shortages in Minerals and Metals." *Science,* Feb. 20, 1976, 713–19.

Kirk, William S. "China's Emergence as the World's Leading Iron Ore–Consuming Country." *Skillings' Mining Review,* July 2003, 6–8, 13, and Aug. 2003, 5–7, 26.

Kohn, Clyde F., and Raymond E. Specht. "The Mining of Taconite, Lake Superior Iron Mining District." *Geographical Review* 48 (Oct. 1958): 528–39.

Korman, Richard. "An Iron Merchant Presses Fast Forward." *New York Times,* Apr. 26, 1998.

Krause, David J. *The Making of a Mining District: Keweenaw Native Copper, 1500–1870.* Detroit: Wayne State University Press, 1992.

Labelle, Robert J. "Cleveland-Cliffs Experience in Monitoring, Analysis and Control of Effluent Water Quality." *Skillings' Mining Review,* Oct. 19, 1974, 8–11.

LaFayette, Kenneth D. *Flaming Brands: Fifty Years of Iron Making in the Upper Peninsula of Michigan, 1848–1898.* Marquette: Northern Michigan University Press, 1990.

Lake, James. "Groups Continue to Fight Mine Expansion in Court." *Marquette Mining Journal,* Nov. 22, 2000, 3A.

Lake Superior Iron Co. *Annual Report, December 31, 1874.* Boston: Rand, Avery, 1875.

———. *Annual Report, December 31, 1875.* Boston: Rand, Avery, 1876.

"Lake Superior Iron Mines: Many Changes in Ownership under the Consolidations." *Iron Trade Review,* Nov. 29, 1900, 14–15.

Lake Superior Iron Ore Association. *Lake Superior Iron Ores, 1938.* Cleveland: Lake Superior Iron Ore Association, 1938.

"Lake Superior Iron Ore Notes." *Mines and Minerals* 31 (July 1911): 714.

"Lake Superior Iron Ores." *Mines and Minerals* 31 (Dec. 1910): 297.

Lamoreaux, Naomi. *The Great Merger Movement in American Business, 1895–1904.* Cambridge: Cambridge University Press, 1985.

"Landscaping on an Iron Range: The Cleveland-Cliffs Iron Co. Believes in the Utility. . . ." *Explosives Engineer* 2 (Aug. 1924): 271–76.

Lane, Frederic C. "Family Partnerships and Joint Ventures in the Venetian Republic." *Journal of Economic History* 4 (Nov. 1944): 178–96.

Lankton, Larry D. *Cradle to Grave: Life, Work, and Death at the Lake Superior Copper Mines.* Oxford: Oxford University Press, 1991.

———. "The Machine under the Garden: Rock Drills Arrive at the Lake Superior Copper Mines, 1868–1883." *Technology and Culture* 24 (1983): 10–22.

Lawton, Charles D. "The Cleveland Cliffs Iron Company and Its Mine." *Michigan Miner,* Oct. 1, 1900, 9–13.

Lewis, Robert S. *Elements of Mining.* New York: John Wiley, 1933.

Libby, E. M. "The Company Surgeon." *Proceedings of the Lake Superior Mining Institute* 15 (1910): 195–200.

"Ling-Temco's J&L, Cleveland-Cliffs Slate Venture to Participate in Iron Ore Facility." *Wall Street Journal,* Mar. 28, 1972.

Loya, Timothy J. "William Gwinn Mather, the Man." *Inland Seas* 46 (1990): 117–33.

Lubetkin, M. John. *Jay Cooke's Gamble: The Northern Pacific Railroad, the Sioux, and the Panic of 1873.* Norman: University of Oklahoma Press, 2006.

Magnaghi, Russell M. *The Way It Happened: Settling Michigan's Upper Peninsula.* Iron Mountain, MI: Mid-peninsula Library Cooperative, 1982.

Mancke, Richard B. "Iron Ore and Steel: A Case Study of the Economic Causes and Consequences of Vertical Integration." *Journal of Industrial Economics* 20 (1972): 220–29.

Manners, Gerald. *The Changing World Market for Iron Ore, 1950–1980.* Baltimore: Johns Hopkins University Press, 1971.

Manning, Warren H. "Villages and Homes for Working Men." *Western Architect* 16 (Aug. 1910): 84–88.

Mansfield, John B. *History of the Great Lakes.* Chicago: J. H. Beers, 1899.

Marcus, Peter F., Karlis M. Kirsis, and Peter J. Kakela. *Cleveland Cliffs [sic] Viability in a Restructured North American Iron Ore Industry.* World Steel Dynamics, Report to the State of Michigan. New York: Paine Webber, 1987.

Marquette Iron Co. *Prospectus for the Organization of the Marquette Iron Co. of Lake Superior, Michigan.* New York: L. H. Biglow, 1864.

Mason, Peter Frank. "Some Changes in Domestic Iron Mining as a Result of Pelletization." *Annals of the Association of American Geographers* 59 (Sept. 1969): 535–51.

Mather, William G. "Charcoal Iron Industry of the Upper Peninsula of Michigan." *Proceedings of the Lake Superior Mining Institute* 9 (1903): 63–85.

———. "Some Observations on the Principle of Benefit Funds and Their Place in the Lake Superior Iron Mining Industry." *Proceedings of the Lake Superior Mining Institute* 5 (1898): 10–20.

May, George S., and Victor F. Lemmer. "Thomas Edison's Experimental Work with Michigan Iron Ore." *Michigan History* 53 (1969): 108–30.

McGinty, Thomas E. *Project Organization and Finance.* Cleveland: Cleveland-Cliffs, 1981.

McHugh, Jeanne. *Alexander Holley and the Makers of Steel.* Baltimore: Johns Hopkins University Press, 1980.

Meier, John G., and Charles L. Wolverton. "Wetland Mitigation: Understand This Dynamic Issue." *Skillings' Mining Review,* Apr. 20, 2002, 4–6.

Melloan, George. "Iron Ore Upheaval." *Wall Street Journal,* July 20, 1960.

Mennie, John S. "A New Change House at the Cliffs Shaft Mine." *Proceedings of the Lake Superior Mining Institute* 9 (1903): 121–26.

Merritt, D. H. "History of Marquette Ore Docks." *Proceedings of the Lake Superior Mining Institute* 19 (1914): 305–8.

Miller, Carol Poh. "Iron and Steel Industry." In *Encyclopedia of Cleveland History,* 578–81. Bloomington: Indiana University Press, 1996.

Milliken, William M. *A Time Remembered: A Cleveland Memoir.* Cleveland: Western Reserve Historical Society, 1975.

Mills, Frank P. "Cost of Mine Lighting." *Transactions of the American Institute of Electrical Engineers* 2 (1885), paper 3.

Miners' Voice. Special fortieth anniversary issue, May 1986. Copy in Longyear Research Library, Marquette County Historical Society, Marquette.

Misa, Thomas J. *A Nation of Steel: The Making of Modern America, 1865–1925.* Baltimore: Johns Hopkins University Press, 1995.

Monson, Terry D. *The Effects of Domestic and International Competition upon Michigan's Iron Ore and Steel Industries.* Houghton: Bureau of Industrial Development, School of Business and Engineering Administration, Michigan Technological University, 1980.

———. *The Effects of New Steelmaking Technologies upon Michigan's Iron Ore and Steel Industries: Final Report, Feb. 1983.* Houghton: Bureau of Industrial Development, School of Business and Engineering Administration, Michigan Technological University, 1983.

———. *The New Challenge of Foreign Imports to the Steel, Manufacturing, and Iron Ore Industries of Michigan and the Great Lakes States.* Houghton: Bureau of Industrial Development, School of Business and Engineering Administration, Michigan Technological University, 1980.

Moore, M. Thomas. *Cleveland Cliffs, Inc.: Pioneers and Partners.* New York: Newcomen Society of the United States, 1997.

More, L. C. "Spherical-Bottom Skip Self-cleaning." *Proceedings of the Lake Superior Mining Institute* 30 (1939): 81–89.

Morrill, James Lewis. *Taconite! Sleeping Giant of the Mesabi.* New York: Newcomen Society of England, American Branch, 1948.

Moulton, W. H. "Sanitation for the Mine Locations." *Proceedings of the Lake Superior Mining Institute* 18 (1913): 38–42.

———. "The Sociological Side of the Mining Industry." *Proceedings of the Lake Superior Mining Institute* 14 (1909): 82–98.

Mulligan, William H., Jr. "The Cliffs Shaft Mine, Ishpeming, Michigan: A Case Study in the History of American Iron Mining, 1844–1967." Paper delivered at Third International Mining History Conference and Symposium on the Preservation of Historic Mining Sites, June 6–l0, l994, Golden, CO.

———, ed. *Historic Resources of the Iron Range in Marquette County, Michigan, 1844–1941.* Marquette: Economic Development Corporation of the County of Marquette, 1991.

Mussey, Henry Raymond. *Combination in the Mining Industry: A Study of Concentration in Lake Superior Iron Ore Production.* New York: Columbia University Press, 1905.

National Labor Relations Board v. Cleveland-Cliffs Iron Co. No. 9162, U.S. Court of Appeals for the Sixth Circuit, Feb. 11, 1943. Lexis 3811.

Neil, Robert M. "The Barnes-Hecker, 1926: Michigan's Worst Mine Disaster." *Mining History Journal* 15 (2008): 83–90.

Newberry, J. S. *The Iron Resources of the United States.* New York: A. S. Barnes, 1874.

Newett, George A. "The Early History of the Marquette Iron Ore Range." *Proceedings of the Lake Superior Mining Institute* 19 (1914): 297–304.

"New Life in the Iron Country." *Business Week,* Oct. 12, 1957, 132–36, 139–40.

"New 600-Ton Blast Furnace Plant." *Blast Furnace and Steel Plant* 9 (Oct. 1921): 588–97.

Nichols, Henry W. "Joint Ventures." *Virginia Law Review* 36 (May 1950): 425–59.

"1954 Ore Shipping Season Opens with Movement of Taconite Concentrate Pellets." *Skillings' Mining Review,* May 8, 1954, 4.

"1946 Lake Superior Iron Ore Shipments by Companies." *Skillings' Mining Review,* Apr. 26, 1947, 1–2, 4, 9, 13.

"1943 Lake Superior Iron Ore Shipments by Companies." *Skillings' Mining Review,* Apr. 22, 1944, 1–2, 9, 11.

Noble, George. "Cleveland Cliffs: Not Your Typical Cyclical." *Hedge Fund Confidential,* May 9, 2005, 1–11.

North, Arthur T. "Engineers and Architects in Artistic Collaboration." *American Architect,* Dec. 15, 1920, 783–87, 791.

"Norton Simon Hits Hanna, Cliffs with Price-fixing Suit." *Cleveland Press,* Sept. 20, 1966.

Olson, Kenneth C. "Primary Blast Hole Drilling at the Humboldt Mine." *Mining Congress Journal* 42 (Apr. 1956): 70–73, 119.

"Open-pit Mining: Vise for Upper Michigan." *New York Times,* Mar. 20, 1972.

"Ore Improvement Plant Eagle Mills, Michigan." *Skillings' Mining Review,* Apr. 12, 1958, 1, 4–5.

Orth, Samuel P. *A History of Cleveland, Ohio.* 3 vols. Chicago: S. J. Clarke, 1910.

Osborne, Richard J. *Celebrating 150 Years: Cleveland-Cliffs Inc., 1847–1997.* Cleveland: Cleveland-Cliffs, 1997.

Palmer, Frank L. *Spies in Steel: An Expose of Industrial War.* Denver: Labor, 1928.

Parks, Roland D. "Recent Developments in Methods of Mining in the Marquette Iron Range." *Proceedings of the Lake Superior Mining Institute* 26 (1928): 115–52.

Pearse, John B. *A Concise History of the Iron Manufacture of the American Colonies up to the Revolution and of Pennsylvania until the Present Time.* Philadelphia: Allen, Lane & Scott, 1876.

Pearson, Philip D. "Lake Superior Underground Iron Mines Gear for Future." *Mining Engineer* 14 (Mar. 1962): 63–66.

"A Pellet Gives Iron Ore Industry Shot in the Arm." *Business Week,* Dec. 4, 1965, 106–8, 110, 112, 114.

"Pelletizing Is Encouraged by Tax Law." *New York Times,* Oct. 16, 1966.

Perry, Jacqueline. "State Senator Weighs in on Iron Mining Labor Negotiations." *Marquette Mining Journal,* July 24, 2004.

Peterson, Florence. *Strikes in the United States, 1880–1936.* Department of Labor Bulletin 651. Washington, DC: Government Printing Office, 1938.

"Pioneer Furnace No. 2." *Proceedings of the Lake Superior Mining Institute* 9 (1903): 89–93.

"Portman's Eldridge Resigns Post as Cleveland-Cliffs Tightens Grip." *American Metal Market,* Apr. 14, 2005.

"Portman White Knight Said Waiting in Wings." *American Metal Market,* Mar. 11, 2005.

"Portsmouth Steel Would Buy Stock." *New York Times,* July 19, 1953.

Pumpelly, Raphael. *Report on the Mining Industries of the United States (Exclusive of the Precious Metals), with Special Investigations into the Iron Resources of the Republic. . . .* Washington, DC: Government Printing Office, 1886.

Ramsey, Robert H. *Men and Mines of Newmont: A Fifty-Year History.* New York: Octagon, 1973.
———. "Teamwork on Taconite: The Story of Erie Mining Co.'s Commercial Taconite Project." *Engineering and Mining Journal* 156 (Mar. 1955): 71–93.
Rankin, Ernest H. "Marquette's Iron Ore Docks." *Inland Seas* 23 (1967): 231–37.
"Republic Mine and Plant of the Cleveland-Cliffs Iron Co. on Marquette Range." *Skillings' Mining Review,* Oct. 13, 1956, 6–7.
Reynolds, Terry S. "Flirting with Vertical Integration: The Cleveland Iron Mining Company and Great Lakes Shipping, 1855–1880." *Inland Seas* 63 (2007): 22–35.
———. "'Quite an Experiment': A Mining Company Attempts to Promote Agriculture on Michigan's Upper Peninsula, 1895–1915." *Agricultural History* 80 (2006): 64–98.
———. "'We Were Satisfied with It': Corporate Paternalism on the Michigan Iron Ranges." *Michigan History* 78 (1994): 24–32.
Robertson, Scott. "Steel's Financial Plight Springboards Asset Sales." *American Metal Market,* May 14, 2001.
Rodgers, Bradley A. *Guardian of the Great Lakes: The U.S. Paddle Frigate Michigan.* Ann Arbor: University of Michigan Press, 1996.
Roe, F. A. "The United States Steamer *Michigan* and the Lake Frontier during the War of the Rebellion." *United Service: A Monthly Review of Military and Naval Affairs* 6 (Dec. 1891): 544–51.
Roe, Lawrence A. *Iron Ore Beneficiation.* Lake Bluff, IL: Minerals, 1957.
Rogers, Robert P. *An Economic History of the American Steel Industry.* London: Routledge, 2009.
Rohan, T. M. "Steelmakers Plan Pellet Plants to Head Off Foreign Inroads." *Iron Age,* Sept. 15, 1960, 159–61.
Rosegger, Gerhard. "Diffusion and Technological Specificity: The Case of Continuous Casting." *Journal of Industrial Economics* 28 (Sept. 1979): 39–53.
Rosenbaum, Joe B. "Minerals Extraction and Processing: New Developments." *Science,* Feb. 20, 1976, 720–23.
Sacco, John E. "Cleveland-Cliffs Cuts Management by 22%." *American Metal Market,* Aug. 1, 2003.
Sargent, Bud. "Miners Leave Message for Bush." *Marquette Mining Journal,* Mar. 2, 2002.
Schallenberg, Richard H. "Evolution, Adaptation and Survival: The Very Slow Death of the American Charcoal Iron Industry." *Annals of Science* 32 (1975): 341–58.
Scheuerman, William. "Joint Ventures in the U.S. Steel Industry: Steel's Restructuring Includes Efforts to Achieve Tighter Control over Raw Materials and Markets." *American Journal of Economics and Sociology* 49 (Oct. 1990): 413–29.
———. *The Steel Crisis: The Economics and Politics of a Declining Industry.* New York: Praeger, 1986.
Schneider, Nancy A. "Breaking the Bottleneck." *Inland Seas* 55 (1999): 124–31.
———. "The Struggle to Establish Civilization on the Michigan Iron Range." *Inland Seas* 54 (1998): 24–31.
Schroeder, Gertrude G. *The Growth of Major Steel Companies, 1900–1950.* Baltimore: Johns Hopkins University Press, 1953.
Seely, Bruce E., ed. *The Iron and Steel Industry in the Twentieth Century: Encyclopedia of American Business History and Biography.* New York: Bruccoli Clark Layman, 1994.
"Self-fluxing Iron Pellet Developed." *Blast Furnace and Steel Plant* 50 (Aug. 1962): 804.
"Semi-centennial: The Cleveland Iron Mining Company." *Iron Trade Review,* July 26, 1900, 16–18.
Sharon Iron Co. *The Sharon Iron Company, Sharon, Mercer Co., PA, Connected with the Jackson Iron Mountain, Lake Superior.* New York: Baker, Godwin, 1852.
Shyrock, Todd. "Man of Steel: How Rodney Mott and ISG Reinvented the Domestic Steel Industry." *Smart Business,* Jan. 2005. www.sbnonline.com.
Skillings, David N., Jr. "Cleveland-Cliffs Dedicates Tilden Mine in Michigan." *Skillings' Mining Review,* Aug. 23, 1975, 10–14.
———. "Cleveland-Cliffs/Marquette Range Iron Ore Mines." *Skillings' Mining Review,* Nov. 2, 1963, 1, 4.
———. "Cleveland-Cliffs' Michigan Iron Mining Operations." *Skillings' Mining Review,* Oct. 30, 1954, 2.
———. "Cleveland-Cliffs' Program at the Mather Mine/Pioneer Pellet Plant." *Skillings' Mining Review,* Nov. 7, 1964, 6–7.
———. "Cleveland-Cliffs Starts Up Tilden II Processing Facilities." *Skillings' Mining Review,* Nov. 10, 1979, 8–9, 12–13, 16.

———. "Cleveland-Cliffs' Tilden Iron Ore Project in Michigan." *Skillings' Mining Review,* May 26, 1973, 4–10.

———. "Cliffs' Pioneer Pellet Plant, Marquette Iron Range." *Skillings' Mining Review,* Sept. 4, 1965, 1, 4–5.

———. "Empire Mine Project." *Skillings' Mining Review,* Mar. 23, 1963, 4–5.

———. "Official Opening of the Empire Iron Mine." *Skillings' Mining Review,* May 16, 1954, 1, 6.

———. "Sherman Mine Dedicated on Sept. 5 at Timagami, Ont." *Skillings' Mining Review,* Oct. 5, 1968, 1, 4–7.

———. "Ventures into Low Grade Iron Jasper Active on Marquette Range." *Skillings' Mining Review,* Dec. 26, 1953, 1–2.

Smith, Duane A. *Mining America: The Industry and the Environment, 1800–1980.* Lawrence: University Press of Kansas, 1987.

Stakel, Charles J. "Detachable Bits at the Cleveland-Cliffs Iron Company's Cliffs Shaft Mine on the Marquette Range." *Proceedings of the Lake Superior Mining Institute* 29 (1936): 197–213.

———. *Memoirs of Charles J. Stakel: Cleveland-Cliffs Mining Engineer, Mine Superintendent, and Mining Manager.* Marquette: Marquette County Historical Society, 1994.

Stanford, F. C. "The Electrification of the Mines of the Cleveland-Cliffs Iron Company." *Proceedings of the Lake Superior Mining Institute* 19 (1914): 189–222.

———. "Training of Workmen for Positions of Higher Responsibility." *Transactions of the American Institute of Mining Engineers* 59 (1918): 612–18.

"Steel Vice." *Wall Street Journal,* Oct. 1, 1998.

Steeples, Douglas, and David Whitten. *Democracy in Desperation: The Depression of 1893.* Westport, CT: Greenwood, 1998.

Stein, Nicholas. "Wilbur Ross Is a Man of Steel . . ." *Fortune,* May 28, 2003. http://money.cnn.com/magazines/fortune/fortune_archive.

Sterling, Walter. "Review and Outlook for the Iron Ore Industry." *Skillings' Mining Review,* Oct. 20, 1956, 1–2, 28–29.

Stevens, William K. "Profit vs. the Land in Upper Michigan." *New York Times,* Dec. 22, 1975.

Stoek, H. H. "Marquette-Range Caving Methods." *Mines and Minerals* 30 (Nov. 1909): 193–200.

Stone, Frank B. *Philo Marshall Everett: Father of Michigan's Iron Industry and Founder of the City of Marquette.* Baltimore: Gateway, 1997.

Sundeen, Stanley W. "The Economic Role of Michigan Iron Ores." *Iron and Copper in Michigan's Economy, 6th Annual Conference Michigan Natural Resources Council,* Oct. 25, 1961, 19–25.

Swank, James M. *History of the Manufacture of Iron in All Ages. . . .* Philadelphia: James M. Swank, 1884.

Swineford, A. P. *Annual Review of the Iron Mining and Other Industries of the Upper Peninsula for the Year Ending Dec. 1880.* Marquette: Mining Journal, 1881.

———. *Annual Review of the Iron Mining and Other Industries of the Upper Peninsula for the Year Ending Dec. 31, 1881.* Marquette: Mining Journal, 1882.

———. *Annual Review of the Iron Mining and Other Industries of the Upper Peninsula for the Year Ending Dec. 1882.* Marquette: Mining Journal, 1883.

———. *Appendix to Swineford's History of the Lake Superior Iron District: Being a Review of Its Mines and Furnaces for 1873.* Marquette: Mining Journal, 1873.

———. *History and Review of the Copper, Iron, Silver, Slate and Other Mineral Interests of the South Shore of Lake Superior.* Marquette: Mining Journal, 1876.

———. *Swineford's History of the Lake Superior Iron District.* 2nd ed. Marquette: Mining Journal, 1871.

Symon, Charles A., ed. *Alger County: A Centennial History, 1885–1985.* Munising, MI: Bayshore, 1986.

Temin, Peter. *Iron and Steel in Nineteenth-Century America: An Economic Inquiry.* Cambridge, MA: MIT Press, 1964.

Thomas, Richard. "Tilden Will Double Capacity by Yearend." *Engineering and Mining Journal* 180 (Sept. 1979): 2.

Thompson, Mark L. *Queen of the Lakes.* Detroit: Wayne State University Press, 1994.

"3 Iron Ore Firms Here Facing Suit by Wheeling Steel." *Cleveland Plain Dealer,* Sept. 20, 1966.

"The Three Mathers Gravely Play Chess with Steel Men." *Business Week,* Apr. 9, 1930, 25–26.

Tibbitts, Julia K. *"Let's Go around the Island."* Marquette: Lake Superior Press, 1992.

Tiffany, Paul A. *The Decline of American Steel: How Management, Labor, and Government Went Wrong.* Oxford: Oxford University Press, 1988.

Timberlake, Richard H. "Private Production of Scrip-Money in the Isolated Community." *Journal of Money, Credit and Banking* 19 (Nov. 1987): 437–47.

Toedtman, James. "Bush to Lift Sanctions on Steel." *Newsday,* Dec. 3, 2003. www.chicgotribune.com.

Tolonen, Frank J. *Gravity Concentration Tests on Michigan Iron Formations.* Contribution 46. New York: American Institute of Mining Engineers, 1933.

Tone, Andrea. *The Business of Benevolence: Industrial Paternalism in Progressive America.* Ithaca, NY: Cornell University Press, 1997.

"Tradition and Training Assets in Finance Post: Cleveland-Cliffs' Harrison Knows Both Producing, Fiscal Ends in Iron, Steel." *Finance,* Oct. 15, 1955, 57–59.

Twitty, Eric. *Blown to Bits in the Mine: A History of Mining and Explosives in the United States.* Ouay, CO: Western Reflections, 2001.

United States v. Republic Steel Corporation et al. No. 5152, District Court, N.D. Ohio, May 2, 1935. Lexis 1539.

"U.S. Studying Bethlehem Fund's Use of 183,000 Cleveland-Cliffs Shares." *Wall Street Journal,* Mar. 31, 1955.

Van Gelder, Arthur Pine, and Hugo Schlatter. *History of the Explosives Industry in America.* 1927. Reprint, New York: Arno, 1972.

Van Tassel, David D., and John J. Grabowski, eds. *Encyclopedia of Cleveland History.* Bloomington: Indiana University Press, 1996.

Villar, James W., and Gilbert A. Dawe. "The Tilden Mine—A New Processing Technique for Iron Ore." *Mining Congress Journal* 61 (Oct. 1975): 40–48.

Wall, Joseph. *Andrew Carnegie.* New York: Oxford University Press, 1970.

Warren, Kenneth. *The American Steel Industry, 1850–1970: A Geographical Interpretation.* Oxford: Clarendon, 1973.

———. *Bethlehem Steel: Builder and Arsenal of America.* Pittsburgh: University of Pittsburgh Press, 2008.

Wayne, Leslie. "American Steel at the Barricades." *New York Times,* Dec. 10, 1998.

"We Stand Up for Steel." *Marquette Mining Journal,* special edition, Mar. 9, 1999.

Westwater, J. S. "Pelletizing by Grate-Kiln Method at Humboldt." *Blast Furnace and Steel Plant* 49 (June 1961): 513–18, 528.

Whitaker, Joe Russell. *Negaunee, Michigan: An Urban Center Dominated by Iron Mining.* Chicago: University of Chicago Libraries, 1931.

White, Peter. "A Brief Attempt at the History of the Mining Industry of Northern Michigan." *Publications of the Michigan Political Science Association* 3 (Jan. 1899): 143–61.

———. "The First Fifty Years: The Cleveland Cliffs Iron Company Celebrates Its Semi-centennial at Ishpeming." *Michigan Miner,* Aug. 1, 1900, 12–18.

———. "A Mere Sketch of Iron Money in the Upper Peninsula." *Michigan Pioneer Historical Collections* 10 (1905): 283–95.

"Why Money Keeps Flowing from Iron Ore." *Business Week,* Aug. 15, 1977 (reprint).

Wilbur, John S. "The Competitive Position of Lake Superior Iron Ores and Its Effect on Lake Shipping." *Skillings' Mining Review,* May 7, 1960, 10–11, 23.

Williams, Ralph D. *The Honorable Peter White.* Cleveland: Penton, 1907.

Winchell, Alexander N. *Handbook of Mining in the Lake Superior Region.* Minneapolis: Byron & Learned, 1920.

Wolff, J. F. "The Economics of Underground Mining." *Skillings' Mining Review,* Apr. 6, 1946, 13–16.

Woodbridge, Dwight E. "Lake Superior Iron-Ore Mining Improved." *Engineering and Mining Journal,* Jan. 16, 1926, 86–87.

Yaworski, N., O. E. Kiessling, C. H. Baxter, Lucien Eaton, and E. W. Davis. *Technology, Employment, and Output Per Man in Iron Mining.* Philadelphia: Works Projects Administration, 1940.

THESES AND DISSERTATIONS

Benison, Saul. "Railroads, Land and Iron: A Phase in the Career of Lewis Henry Morgan." Ph.D. diss., Columbia University, 1953.

Bowlus, Bruce. "'Changes of Vast Magnitude': The Development of an Iron Ore Delivery System on the Great Lakes during the Nineteenth Century." Ph.D. diss., Bowling Green State University, 1992.

Goodman, Robert Joseph. "Ishpeming, Michigan: A Functional Study of a Mining Community." Ph.D. diss., Northwestern University, 1948.

Hakala, Donald Richard. "An Analysis of Postwar Developments in the Upper Michigan Iron Ore Mining Industry." Ph.D. diss., Indiana University, 1966.

Mason, Peter Frank. "Some Aspects of Change in the Iron Mining Industry of the United States as a Result of Pelletization." Ph.D. diss., University of California, Los Angeles, 1968.

Waters, Albert L. "A Thesis on the Modern Methods of Prospecting for Iron Ores in the Lake Superior Region." Thesis, Michigan College of Mines, 1893.

Wilson, Anne Elizabeth. "Michigan Iron Ore Mining Safety: Policies and Fatalities, 1880–1979." Ph.D. diss., Michigan State University, 2001.

GOVERNMENT REPORTS AND DOCUMENTS

Davis, E. W. "Lake Superior Iron Ore and the War Emergency." Report to the Materials Division of the War Production Board, May 20, 1942. Copy in American Iron and Steel Institute Vertical File, Hagley Museum and Library, Wilmington, DE.

Michigan Adjutant General. *Report of the Adjutant General of the State of Michigan for the Years 1895–1896.* Lansing: Robert Smith, 1896.

Michigan Bureau of Labor and Industrial Statistics. *Annual Report,* 1894, 1896, 1904.

Michigan Commissioner of Mineral Statistics. *Annual Report of the Commissioner of Mineral Statistics of the State of Michigan,* 1877–85.

———. *Mines and Mineral Statistics,* 1885–1909.

Michigan Corporation Securities Division. Annual Mining Reports, 1853, "Cleveland Iron Mining Company." Michigan Technological University Archives and Copper Country Historical Collections, Houghton.

Michigan Department of Natural Resources. Fisheries Division, 1971–84. Transcript of Public Hearing, May 15, 1972. RG 89-343, box 1, folder 1, Archives of Michigan, Lansing.

Michigan Supreme Court. *Jeremy Compo v. Jackson Iron Company, State of Michigan, Supreme Court.* Copy in Longyear Research Library, Marquette County Historical Society, Marquette.

U.S. Department of Commerce, Bureau of Corporations. *The Lumber Industry.* Vol. 1, pt. 3, "Landholdings of Large Timber Owners." Washington, DC: Government Printing Office, 1914.

U.S. Federal Trade Commission. *Report of the Federal Trade Commission on the Control of Iron Ore for the Antitrust Subcommittee of the Committee on the Judiciary, House of Representatives,* esp. "Background of Domestic Ore Holdings," 6–30. Washington, DC: Government Printing Office, 1952. Copy in American Iron and Steel Institute, Public Relations Department, folder 1, box 71, ass. no. 1631, Hagley Museum and Library, Wilmington, DE.

U.S. House of Representatives. *Mineral Lands of Lake Superior.* 29th Cong., 1st sess., 1846. H.R. 211.

U.S. President's Materials Policy Commission. *Resources for Freedom.* Vol. 2. Washington, DC: Government Printing Office, 1952. Also known as the Paley Report.

U.S. Senate. *Report of General Walter Cunningham.* 28th Cong., 2nd sess., Jan. 8, 1845. S. 98, vol. 7.

Young, H. O. *Supplementary Report on the Facts, by H. O. Young, a Member of the Special Committee to Investigate the United States Steel Corporation.* 61st Cong., 2nd sess., 1912. H.R. 1127, pt. 3.

MANUSCRIPT, ARCHIVAL, AND UNPUBLISHED CORPORATE SOURCES

Archives of Michigan, Lansing, MI
RG 81-27, Attorney General Case Files
RG 70-41, Attorney General's Office
Brooks, Thomas B. "Report on the Iron Interests of Lake Superior in Relation to the Lands of the Lake Superior Ship Canal Iron & RR Co." Dec. 15, 1877. Folder 2, box 1, Report on Menominee Range, 1877.
RG 89-343, Department of Natural Resources, Fisheries Division
RG 81-29, Department of State
RG 76-91, Jackson Iron Company Papers
RG 42, Records of the Executive Office

Archives of Michigan, Northern Michigan University Repository, Marquette, MI
MS 76-90, Cleveland-Cliffs Iron Company, Land Department Papers
MS 86-100, Cleveland Iron Mining Company and Cleveland-Cliffs Iron Company Papers, containing papers from some companies acquired, including the Marquette Iron Company, the Michigamme Mining Company, the McComber Iron Company, and the Upper Peninsula Land Company
RG 65-37, Iron Cliffs Company Papers
RG 66-36, Iron Cliffs Company Papers
RG 68-102, Iron Cliffs Company Papers

Bentley Historical Library, Ann Arbor, MI
Harlow Family Papers, 1836–50
Peter White Papers

Cleveland-Cliffs Iron Company, Cleveland, OH
Annual Reports, Cleveland-Cliffs, Inc., 1947–2006.
Brinzo, John. "Brinzo Reports on State of the Company at Cleveland-Cliffs Annual Shareholders Meeting." Cleveland-Cliffs, news release, May 8, 2001.
Harrison, H. Stuart. "The Changing Story of Cliffs." Speech before the New York Society of Security Analysts, June 1, 1967.
———. "Iron Ore Revisited." Presented at the Mining Show of the American Mining Congress, Oct. 11–14, 1971.
———. "No Room for Dinosaurs." Remarks before the New York Society of Security Analysts, June 18, 1964.
Harrison, H. Stuart, and T. E. McGinty. "The Cleveland-Cliffs Iron Company." Remarks before the Cleveland Society of Security Analysts, Mar. 14, 1973.
———. "The Cleveland-Cliffs Iron Company." Presentation to the New York Society of Security Analysts, Dec. 19, 1975.
Kakela, Peter J. "The Shift to Taconite Pellets; Necessary Evil or Lucky Break?" Paper presented at "Forged from Iron" conference, Sept. 23, 1994, Marquette, with critique by John L. Kelley.
Leach, Hugh J., John E. Stukel, Walter Nummela, Emert W. Lindroos, and Roy A. Koski. "Developments and Realization in the Flotation of Iron Ores in North America." Paper presented at the Twelfth International Mineral Processing Congress, Aug. 1977, Sao Paulo, Brazil.
McGinty, Thomas E. "The Pellet Revolution." Talk before the Junior Membership of the New York Society of Security Analysts, Feb. 7, 1966.
Memo from J. S. Brinzo to M. T. Moore. "'Grand Slam' North American Iron Ore Strategy," Oct. 15, 1990.
Minute Books, Cleveland-Cliffs Iron Company, 1890–.
Minute Books, Cliffs Corporation, 1929–47.
Minute Books, Iron Cliffs Company, 1864–90.
Minutes of Meetings of the Board of Directors and Annual Shareholders' Meetings, Cleveland-Cliffs, Inc., 1947–2002.

Moore, M. Thomas. Interview by Edward Walsh and Richard Osborne, 1996.

Cleveland-Cliffs Iron Company, Empire mine, Palmer, MI
    Boyum Collection
        Boyum, Burton H. "Cliffs Illustrated History," manuscript, c. 1986. Additional copies at Michigan Iron Industry Museum, Negaunee, MI; Cliffs Shaft Museum, Ishpeming, MI; Cleveland-Cliffs Iron Co., Historical Files, Cleveland, OH.

Cleveland State University, Michael Schwartz Library, Department of Special Collections, Cleveland, OH
    Cleveland Press Collection: clippings files
    John L. Horton Collection: pamphlets, correspondence, photographs, and memorabilia relating to Cleveland-Cliffs, including:
        Horton, John. "The Cleveland Cliffs Marine History," n.d.
        Moore, Tom. Bound volume of speeches, 1987–97
        Photo Collection: Cleveland Cliffs 10, series D: photographs relating to company mines and vessel management

Longyear Research Library, Marquette County Historical Society, Marquette, MI
    Armstrong, William F. "Historical Sketch of the Marquette Iron Range," 1932.
    Burt, John, "Autobiography," c. 1883.
    William Austin Burt Papers
    J. G. Chamberlain Letters, 1854
    Everett, Philo, to brother, Nov. 10, 1845.
    Gordon, L. I., to "Friend Pharo," Apr. 2, 1857.
    Olive Harlow Diary, early 1850s
    Morgan L. Hewitt Letters
    "Iron Mountain Railroad Report of Charles T. Harvey, Referee, Marquette, Michigan, October 2, 1855."
    Iron Mountain Railway Company Articles of Incorporation, Mar. 14, 1855.
    Jeremy Compo v. Jackson Iron Company, transcript, State of Michigan Supreme Court, 1881.
    Jopling, James E., "Dr. Morgan L. Hewitt," 1935.
    Mather, William G., miscellaneous newspaper articles, memorabilia
    Joseph Henry Sawbridge Diary, 1866

Michigan Technological University Archives and Copper Country Historical Collections, Houghton, MI
    "Final Report, First Conference: Upper Peninsula County Mine Inspectors, Upper Peninsula Company Safety Engineers, Lake Superior Mining Division Staff of the United States Bureau of Mines, Houghton, Michigan, April 23–25, 1919."
    Michigan Corporation Securities Division, Annual Mining Reports

Northern Michigan University Archives, Marquette, MI
    MS 8, R. Wesley Jenner Papers
    Thoren, Clarence J. "Shade of Old Negaunee: A 19th Century History of Negaunee, Michigan." Unpublished manuscript, 1969.

Peter White Library, Marquette, MI
    Peter White Papers, microfilm

Western Reserve Historical Society, Cleveland, OH
    MS 3136, Cleveland Iron Mining Company Correspondence
    MS 3913, Cyrus Eaton Papers
    MS 3735, Samuel Livingston Mather Family Papers
    MS 4578, William Gwinn Mather Family Papers

NEWSPAPERS, TRADE JOURNALS, AND COMPANY NEWSLETTERS

American Journal of Mining
American Metal Market
American Railroad Journal
American Railway Times
Bulletin of the American Iron and Steel Association
Chicago Tribune
Cleveland Herald
Cleveland Leader
Cleveland Press
Cliffs Connection
Cliffs News
Congressional Globe
Daily Metal Reporter
Detroit Free Press
Engineering and Mining Journal
Iron Trade Review
Ishpeming Iron Agitator
Ishpeming Iron Home
Ishpeming Iron Ore
Lake Superior Journal (Sault Ste. Marie; Marquette, MI)
Marquette Mining Journal
Mining Congress Journal
Mining Magazine
Negaunee Mining News
New York Times
New York Weekly Tribune
Portage Lake Mining Gazette (Houghton, MI)
Red Dust [oral histories conducted by students at the National Mine School, MI]
Roll Call
Sharpsville (OH) Advertiser
Skillings' Mining Review (Duluth, MN)

INTERVIEWS

Transcripts of Interviews by Burt Boyum, Cleveland-Cliffs Iron Company, Empire mine, Palmer, MI
Steve Crowley, June 21, 1984
R. M. DeGabriele, July 23, 1984
Hugh J. Leach, Oct. 9, 1984
Tom McGinty, Oct. 29, 1984
Kenneth Ramsey, Oct. 7, 1984
James W. Villar, Aug. 28, 1984

Interviews with Current and Former CCI Employees by Virginia Dawson and Terry Reynolds
John Brinzo, Oct. 19, 2006, by Virginia Dawson, Cleveland HQ
Dana Byrne, Oct. 30, 2007, by Virginia Dawson, Cleveland HQ
William Calfee, Nov. 8, 2007, by Virginia Dawson, Cleveland HQ
Joe Carrabba, Aug. 23, 2008, by Virginia Dawson, Cleveland HQ
Gilbert Dawe, June 26, 2008, by Terry Reynolds, Cliffs Shaft Mine, Ishpeming
Don Gallagher, Jan. 19, 2009, by Virginia Dawson, Cleveland HQ

Dale Hemmila, Sept. 22, 2008, by Virginia Dawson, Empire mine, Palmer
Elton (Pete) Hoyt III, June 14, 2007, by Virginia Dawson, Cleveland HQ
Edwin (Ned) Johnson, June 4, 2007, by Terry Reynolds, Ishpeming
Linda Bancroft LaFond, Sept. 30, 2008, by Virginia Dawson, by telephone
Tom Petersen, July 10, 2008, by Terry Reynolds, Empire mine, Palmer
Don Ryan, June 25, 2008, by Terry Reynolds, Marquette
Sam Scovil, June 11, 2007, by Virginia Dawson, Cleveland HQ
Richard Tuthill, Apr. 1, 2008, by Virginia Dawson, Cleveland HQ
James Villar, July 10, 2008, by Terry Reynolds, Republic, MI
A. Stanley West, Aug. 16, 2007, by Virginia Dawson, Bay Village, OH
James Westwater, July 17, 2007, by Terry Reynolds, Michigamme, MI

INDEX

Great Lakes Books

*A complete listing of the books in this series can be
found online at wsupress.wayne.edu*